Annals of Mathematics Studies

Number 212

Supersingular p-adic L-functions, Maass-Shimura Operators and Waldspurger Formulas

Daniel J. Kriz

PRINCETON UNIVERSITY PRESS

PRINCETON AND OXFORD

2021

Requests for permission to reproduce material from this work should be sent to permissions@press.princeton.edu

Published by Princeton University Press
41 William Street, Princeton, New Jersey 08540
6 Oxford Street, Woodstock, Oxfordshire OX20 1TR

press.princeton.edu

Library of Congress Control Number: 2021944468

ISBN 9780691216478
ISBN (pbk.) 9780691216461
ISBN (e-book) 9780691225739

British Library Cataloging-in-Publication Data is available

Editorial: Susannah Shoemaker and Kristen Hop
Production Editorial: Nathan Carr
Production: Jacquie Poirier

This book has been composed in LaTeX

10   9   8   7   6   5   4   3   2   1

To my family.

Contents

Preface

The theory of p-adic L-functions is an important refinement of the study of special values of L-functions and their arithmetic. It is a crucial tool in the study of Bloch-Kato and Beilinson-Bloch type conjectures such as the Birch and Swinnerton-Dyer conjecture. Arising in the analytic side of Iwasawa theory, p-adic L-functions give a way to transfer arithmetic information along rigid analytic families $\mathbb{V} \to X$ of Galois representations through means of interpolation: For fibers $\mathbb{V}_x$ where x belongs to a certain "critical locus," the associated Bloch-Kato Selmer groups are controlled by critical values $L(\mathbb{V}_x)$ of the L-function attached to $\mathbb{V}_x$. One expects these critical values, after slight modification such as removal of Euler factors at p, to be interpolated by a p-adic L-function $\mathcal{L}_p(\mathbb{V})$, whose special values at *any* point $x \in X$ governs, through a main conjecture, the arithmetic of the corresponding $\mathbb{V}_x$. In certain cases, formulas for special values can be made explicit and give fundamental arithmetic information, for example, through being related to a p-adic regulator of a cohomology class forming the first step of an Euler system. Hence the nonvanishing of these p-adic L-values, through the machinery of Euler systems, gives bounds on the Selmer group and other arithmetic invariants such as Tamagawa numbers, and even the finiteness of the Shafarevich-Tate group in certain settings.

A setting in which p-adic L-functions have found exceedingly fruitful recent arithmetic applications is for the "Rankin-Selberg setting," with $\mathbb{V} \to X$ parametrizing a family of anticyclotomic twists of a fixed $\mathrm{Gal}(\overline{K}/K)$-representation V attached to a modular form f and an imaginary quadratic field K, and where X is a p-adic Lie group parametrizing p-adic anticyclotomic characters. The special value of the trivial character $\mathcal{L}_p(\mathbb{V}\ )$ then is related to the formal logarithm along f of a certain "Heegner point," and hence the nonvanishing of this value gives, by the work of Gross-Zagier ([27]) and Kolyvagin ([40]), the precise Mordell-Weil rank and finiteness of Shafarevich-Tate over K of the GL_2-type abelian variety attached to f. This "p-adic Waldspurger formula" has found applications in proving "converse theorems" in the style of Skinner, Zhang and others, which produce bounds on analytic ranks given bounds on algebraic ranks, which have applications to arithmetic statistics. For example, these converse theorems were used in an essential way in recent work of Bhargava-Skinner-Zhang ([6]) in showing that 66.48% of elliptic curves over $\mathbb{Q}$ satisfy the rank part of the Birch and Swinnerton-Dyer conjecture.

The first examples of Rankin-Selberg type p-adic L-functions appeared in fundamental work of Katz in the '70s ([33], [34], [35]), in which Katz addressed the case where f is an Eisenstein series, established a p-adic Kronecker limit formula involving p-adic logarithms of elliptic units, and subsequently generalized his construction to K general CM fields. The Katz p-adic L-function was shown by Rubin ([53]) to govern the arithmetic of elliptic curves of complex multiplication, in his proof of the main conjectures for GL_1/K. Later, in the late 2000s, Bertolini-Darmon Prasanna ([5]) made another breakthrough in constructing a Rankin-Selberg type p-adic L-function in the case where f is a newform, as well as establishing the p-adic Waldspurger formula alluded to in the previous paragraph. Finally, in the early 2010s Liu-Zhang-Zhang ([46]) generalized Bertolini-Darmon-Prasanna's p-adic L-function to CM fields, as well as establishing a p-adic Waldspurger formula in this setting.

One key restriction of all the Rankin-Selberg type p-adic L-functions constructed thus far is that they all make an "ordinariness" assumption that p is split in K, which imposes a certain condition on Hodge-Tate weights of the Galois representation and forces the associated automorphic representations in the critical locus to be ordinary. Geometrically, this ordinariness assumption is used to ensure the construction of the p-adic L-functions takes place on the ordinary locus of the underlying Shimura curve, where the p-adic analysis of modular forms is well-developed and many classical results such as Serre-Tate coordinates, the unit root splitting of the Hodge filtration and the Atkin-Serre weight-raising, or "θ", operator are available. Accordingly, many major results in number theory, including the formulation and demonstration of anticyclotomic main conjectures for Rankin-Selberg families, are currently restricted to this ordinary setting. However, for many arithmetic applications such as the study of Birch and Swinnerton-Dyer for elliptic curves with complex multiplication by an order of K, or converse theorems for such curves, the arithmetic at supersingular primes is often quite subtle and encodes the most interesting information.

The main goal of this book is to introduce a form of Rankin-Selberg type p-adic L-functions which do not necessarily satisfy the above ordinariness assumption, i.e., allow p to be inert or ramified in K. In order to do this, we develop a new theory of "generalized p-adic modular forms" which encompasses both the ordinary and supersingular locus, by realizing (and generalizing the notion of) p-adic modular forms as sections of certain de Rham period sheaves on the infinite-level modular curve $\mathcal{Y}$ which transform by certain weight characters under the natural $GL_2(\mathbb{Q}_p)$ action on $\mathcal{Y}$. This theory is a basic foundational tool toward extending the full range of results currently known for ordinary locus to the supersingular locus, including extensions of the notions of Serre-Tate coordinates and splittings of the Hodge filtration. An important feature of the theory presented in this book is that we define an analogue (and extension) of the θ operator in our setting, which we call the "p-adic Maass-Shimura operator" because it raises the weights of generalized p-adic modular forms. We also present a detailed p-adic analysis of the operator, and how it gives rise to the supersingular Rankin-Selberg type p-adic L-function. As a further application of our p-adic analysis on infinite-level, we also prove a p-adic Waldspurger formula for the supersingular p-adic L-functions, the arithmetic applications of which have been described above.

The construction presented here rests on understanding Katz's original approach to constructing p-adic L-functions in the context of the geometry of Scholze's infinite-level modular curve, as well as understanding how extending sheaves of coefficients from the (rigid) structure sheaf to larger period sheaves and considering various periods belonging to these sheaves, leads to an extension of these ideas to the supersingular locus. Our aim is to present enough background on this material to make this book reasonably self-contained. Our presentation is organized as follows: We begin with an introduction and overview discussing Katz's original ideas and how they run parallel and can be extended in the infinite-level setting which lead to the main constructions and results covered in this book. In Chapter 2, we give some background on the category-theoretic abstract language we use throughout the book, as well as an overview of Huber's theory of adic spaces, the proétale site introduced by Scholze and various period sheaves on this site, which can be seen as relative versions of Fontaine's period rings. In Chapter 3, we present key properties of the geometry of the infinite level modular curve which are used in our construction. In Chapter 4, we define the fundamental de Rham periods which are essential to establishing a theory of p-adic analysis on the supersingular locus and the construction of the p-adic Maass-Shimura. In Chapter 5, we give the construction of the p-adic Maass-Shimura operator and examine some of its properties, as well as introduce generalized p-adic modular forms, while comparing these notions with their classical parts on the ordinary locus (in particular showing how the ordinary theory can be recovered from ours). In Chapter 6, we examine p-adic analytic properties of the p-adic Maass-Shimura operator, in particular through defining a certain "q_{dR}-expansion map" on generalized p-adic modular forms, which serves as an analogue (and extension) of Serre-Tate coordinates. The integrality of the coefficients of these q_{dR}-expansions is crucial to controlling the p-adic behavior of Maass-Shimura derivatives, and for this a variation of a classical lemma of Dieudonné-Dwork forms the technical heart of this chapter. In Chapter 7, we bound the p-adic valuations of the specializations of various periods in our consideration to supersingular CM points, which is also necessary for the construction of p-adic L-functions. In Chapter 8, we construct the supersingular Rankin-Selberg type p-adic L-functions. Finally, in Chapter 9 we show how the p-adic analytic properties of our p-adic Maass-Shimura operator, along with Coleman's theory of locally analytic integration, can be used to establish a p-adic Waldspurger formula for our p-adic L-function.

Acknowledgments

I thank Shouwu Zhang and Christopher Skinner for their overwhelming support and many helpful discussions throughout this work, as well as for their encouragement and advice on my Ph.D. thesis [43] on this subject. I also thank the anonymous referees for their helpful comments and suggestions. This work was partially supported by the National Science Foundation under grant DGE 1148900.

Supersingular *p*-adic *L*-functions, Maass-Shimura Operators and Waldspurger Formulas

Chapter One

Introduction

1.1 PREVIOUS CONSTRUCTIONS AND KATZ'S THEORY OF p-ADIC MODULAR FORMS ON THE ORDINARY LOCUS

Let us start by giving a brief account of Katz ([34], [33] and [35]) and Bertolini-Darmon-Prasanna's ([5]) construction of p-adic L-functions over imaginary quadratic fields K in which p splits in K. The splitting assumption of Katz allows one to make use of his theory of p-adic modular forms in order to construct his and Bertolini-Darmon-Prasanna/Liu-Zhang-Zhang's p-adic L-functions, now colloquially known as the *Katz* and *BDP p-adic L-functions*, respectively. The former is also constructed for CM extensions K/L (i.e., where $L/\mathbb{Q}$ is totally real and K/L is imaginary quadratic) for which all primes of L above p split in K, and the latter was generalized by Liu-Zhang-Zhang ([46]) to the case of CM fields and weight 2 newforms. Namely, the p-adic L-functions over K which Katz, Bertolini-Darmon-Prasanna and Liu-Zhang-Zhang construct are linear functionals on the space of (p-adic) modular forms, which are obtained by evaluating p-adic differential operators applied to modular forms at ordinary CM points associated with K. This means the CM points belong to the *ordinary locus*

$$Y^{\mathrm{ord}} \subset Y,$$

which is the affinoid subdomain of (the rigid analytification of) Y obtained by removing all points which reduce to supersingular points on the special fiber (this latter locus being isomorphic to a finite union of rigid analytic open unit discs). Here, the ordinariness assumption is crucial in order to establish nice analytic properties of the p-adic L-function, namely that (p-adic) modular forms have local coordinates in neighborhoods of CM points with respect to which the differential operators alluded to above have a nice, clearly analytic expression. In Katz's setting, one views p-adic modular forms as functions on a proétale cover called the *Igusa tower*

$$Y^{\mathrm{Ig}} \to Y^{\mathrm{ord}}$$

using an explicit trivialization of the Hodge bundle (by a so-called *canonical differential*). Then on Y^{Ig}, he defines a differential operator θ called the *Atkin-Serre operator*, which sends p-adic modular forms of weight k to forms of weight $k+2$, and the nice coordinates are provided by Serre-Tate coordinates. One can express

$$\theta_{\mathrm{AS}} = (1+T)\frac{d}{dT}$$

in terms of the Serre-Tate coordinate T, and using this expression one can show easily that the family

$$\{\theta_{\mathrm{AS}}^j f\}_{j\in\mathbb{Z}_{\geq 0}}$$

for a given p-adic modular form f gives rise to a "nearly-analytic" function of j: after applying a certain Hecke operator known as p-*stabilization* to f (which corresponds to removing an Euler factor in the p-adic L-function), one can show that

$$\theta_{\mathrm{AS}}^j f^{(p)},$$

where $f^{(p)}$ denotes the p-stabilization, is an analytic function (valued in the space of p-adic modular forms) of $j \in \mathbb{Z}_p^\times$.

One could also use coordinates provided by q-expansions, if one compactifies all modular curves under our consideration; we stick to the open modular curve in this article in order to avoid boundary issues occurring at cusps, which present bigger technical issues when defining the proétale topology later.

The key property of $Y^{\mathrm{Ig}} \to Y^{\mathrm{ord}}$ which allows one to construct the differential operator θ_{AS} is the existence of the *unit root splitting* of the Hodge filtration on Y^{Ig}. Namely, one can find sections of the relative de Rham cohomology

$$\mathcal{H}_{\mathrm{dR}}^1(\mathcal{A})|_{Y^{\mathrm{Ig}}}$$

which are *horizontal* with respect to the algebraic Gauss-Manin connection

$$\nabla : \mathcal{H}_{\mathrm{dR}}^1(\mathcal{A}) \to \mathcal{H}_{\mathrm{dR}}^1(\mathcal{A}) \otimes_{\mathcal{O}_Y} \Omega_Y^1$$

(here a section α being horizontal means that $\nabla(\alpha)=0$), and which are also eigenvectors for the canonical (Frobenius-linear) Frobenius endomorphism

$$F : Y^{\mathrm{Ig}} \to Y^{\mathrm{Ig}}$$

over W. (The reason for the terminology "unit root" is because one of the eigenvalues for F is a p-adic unit, i.e., a section of $\mathcal{O}_{Y^{\mathrm{Ig}}}^\times$, since we restrict to a covering of the ordinary locus Y^{ord}.) The unit root splitting is a functorial, F-equivariant splitting of the Hodge filtration, which allows one to then define the differential operator θ_{AS}. This uses a standard formalism of Katz which produces such a differential (weight-raising) operator, whenever a splitting of the Hodge filtration with nice properties (e.g., $\mathrm{Gal}(Y^{\mathrm{Ig}}/Y^{\mathrm{ord}})$-equivariance) exists.

Another key property of the unit root splitting is that for CM elliptic curves A, which by the theory of complex multiplication always have models over $\overline{\mathbb{Q}}$, it is induced by the splitting of $H^1_{dR}(A)$ defined over $\overline{\mathbb{Q}}$ given by the eigen-decomposition under the CM action. This CM splitting over $\overline{\mathbb{Q}}$ also gives rise to the real analytic Hodge decomposition over $\mathbb{C}$ from classical Hodge theory, which in that setting gives rise to the real analytic Maass-Shimura operator $\eth$ sending nearly holomorphic modular forms of weight k to nearly holomorphic forms of weight $k+2$. The consequence is that after normalizing by appropriate "canonical" periods

$$\Omega_p \text{ and } \Omega_\infty,$$

one can show that given an algebraic modular form w of weight k, the values

$$\theta^j_{AS} w(y)/\Omega_p^{k+2j} \text{ and } \eth^j w(y)/\Omega_\infty^{k+2j}$$

at ordinary CM points $y \in Y^{ord}$ belong to $\overline{\mathbb{Q}}$ and *coincide*. This observation of Katz is essential to establishing interpolation properties of the Katz and BDP/LZZ p-adic L-functions, i.e., to relate them to critical values of complex L-functions in the interpolation (Panchishkin) range. This is because such critical L-values can be expressed as period integrals over the CM torus (or finite sums over orbits of CM points) of $\eth^j w$, and hence by the above discussion these can be related to such p-adic period sums of $\theta^j_{AS} w$ over CM points, which themselves give rise to the Katz and BDP/LZZ p-adic L-functions.

Let us elaborate on Serre-Tate coordinates and Katz's notion of p-adic modular forms, and expound on the above discussion. To fix ideas, suppose that a modular curve Y represents a fine moduli space (for example, if its topological fundamental group as an analytic space over $\mathbb{C}$ is *neat* in the sense that it has no torsion), and so it admits a universal object

$$\pi: \mathcal{A} \to Y.$$

The *Hodge bundle* is then defined as

$$\omega := \pi_* \Omega^1_{\mathcal{A}/Y}$$

and weight k modular forms can be identified with sections of $\omega^{\otimes k}$. Katz's theory of modular forms arises by constructing a nonvanishing section known as the *canonical differential*

$$\omega_{can}^{Katz} \in \omega(Y^{Ig}),$$

and using the induced trivialization

$$\omega|_{Y^{Ig}} \cong \mathcal{O}_{Y^{Ig}}$$

to view modular forms on Y as functions on Y^{Ig} transforming by some weight character under the action of

$$\mathrm{Gal}(Y^{\mathrm{Ig}}/Y^{\mathrm{ord}}) \cong \mathbb{Z}_p^{\times}.$$

To obtain the trivialization of ω, Katz uses the simple structure of the p-divisible groups of ordinary elliptic curves, namely that they are isomorphic to

$$\mu_{p^{\infty}} \times \mathbb{Q}_p/\mathbb{Z}_p.$$

By the Weil pairing (or Cartier duality), such a trivialization for a given p-divisible group $A[p^{\infty}]$ of an ordinary elliptic curve A is determined by fixing an isomorphism

$$A[p^{\infty}]^{\mathrm{\acute{e}t}} \cong \mathbb{Q}_p/\mathbb{Z}_p.$$

In fact, Y^{Ig} is exactly the cover of Y^{ord} defined over $W = W(\overline{\mathbb{F}}_p)$ parametrizing such trivializations

$$\alpha : \mathbb{Q}_p/\mathbb{Z}_p \xrightarrow{\sim} A[p^{\infty}]^{\mathrm{\acute{e}t}},$$

or equivalently (by the previous discussion), trivializations

$$\alpha : \mu_{p^{\infty}} \times \mathbb{Q}_p/\mathbb{Z}_p \xrightarrow{\sim} A[p^{\infty}]$$

of the entire p-divisible group.

Let $A_0/\overline{\mathbb{F}}_p$ be an elliptic curve corresponding to a closed geometric point y_0 on the special fiber

$$Y_0^{\mathrm{ord}} = Y^{\mathrm{ord}} \otimes_W \overline{\mathbb{F}}_p,$$

and let A/W denote any lift of A_0, i.e., with

$$A \otimes_W \overline{\mathbb{F}}_p \cong A_0,$$

corresponding to a point y on Y^{ord}. Formally completing Y^{Ig} along y_0 hence gives the formal moduli space $\hat{D}(y_0)$ of deformations of A_0 (with some level structure, which we will suppress for brevity). Since there is a canonical isomorphism

$$A[p^{\infty}]^{\mathrm{\acute{e}t}} \cong A_0[p^{\infty}](\overline{\mathbb{F}}_p),$$

then a choice of trivialization

$$\alpha_0 : \mathbb{Q}_p/\mathbb{Z}_p \xrightarrow{\sim} A_0[p^{\infty}](\overline{\mathbb{F}}_p)$$

fixes $A[p^\infty]^{\text{ét}}$ in the formal neighborhood $\tilde{D}(\tilde{y}_0)$ of $\tilde{y}_0 = (A_0, \alpha_0)$ in Y^{Ig}. Hence $\tilde{D}(\tilde{y}_0)$ is parametrized exactly by the connected component $A[p^\infty]^0$ of $A[p^\infty]$, and so there is an (in fact, canonical) isomorphism

$$\tilde{D}(\tilde{y}_0) \cong \hat{\mathbb{G}}_m.$$

The canonical coordinate T on the torus gives rise to the *Serre-Tate coordinate*, also denoted by T, on

$$\tilde{D}(\tilde{y}_0),$$

and on the associated residue disc

$$\tilde{D}(\tilde{y}_0) \otimes_W W[1/p]$$

(viewed as the rigid analytic generic fiber of $\tilde{D}(\tilde{Y}_0)$).

Katz uses the above description of formal neighborhoods on Y^{Ig} around closed points of the special fiber as being canonically isomorphic to $\hat{\mathbb{G}}_m$ in order to construct the canonical differential ω_{can} mentioned before; in terms of the Serre-Tate coordinate on a residue disc D, the canonical differential is just given by

$$\omega_{\text{can}}^{\text{Katz}}|_D = dT/T.$$

Using tensorial powers of the canonical differential, modular forms, viewed as sections of powers $\omega^{\otimes k}$ of the Hodge bundle ω restricted to Y^{ord}, can be identified as functions on Y^{Ig}. Since the canonical differential transforms by

$$d^* \omega_{\text{can}}^{\text{Katz}} = d\omega_{\text{can}}^{\text{Katz}}$$

for

$$d \in \mathbb{Z}_p^\times \cong \text{Gal}(Y^{\text{Ig}}/Y^{\text{ord}}),$$

then we can even identify a modular form of weight k, i.e., a section of $w \in \omega^{\otimes k}(Y^{\text{ord}})$, as a function f of *weight k* on Y^{Ig}, via

$$w|_{Y^{\text{Ig}}} = f \cdot \omega_{\text{can}}^{\text{Katz}, \otimes k},$$

where weight $k \in \mathbb{Z}$ means that f transforms as

$$d^* f = d^{-k} f \tag{1.1}$$

for

$$d \in \mathbb{Z}_p^\times = \text{Gal}(Y^{\text{Ig}}/Y^{\text{ord}}).$$

Katz also uses this viewpoint to generalize modular forms to *p-adic modular forms of weight* $k \in \mathbb{Z}_p^\times$, which are functions on Y^{Ig} which have weight $k \in \mathbb{Z}_p^\times$ in the same way as defined above.

1.2 OUTLINE OF OUR THEORY OF p-ADIC ANALYSIS ON THE SUPERSINGULAR LOCUS AND CONSTRUCTION OF p-ADIC L-FUNCTIONS

The key question addressed by this article is that of developing a satisfactory theory of p-adic analysis of modular forms on the supersingular locus of modular curves, and subsequently to construct "supersingular" p-adic L-functions for Rankin-Selberg families $\mathbb{V}$ of twist families of automorphic representations

$$(\pi_w)_K \times \chi^{-1}$$

for anticyclotomic characters χ over an imaginary quadratic field $K/\mathbb{Q}$, where

$$\pi_w$$

is the automorphic representation of $GL_2(\mathbb{A}_\mathbb{Q})$ attached to a normalized new eigenform w (i.e., a newform or Eisenstein series),

$$(\pi_w)_K$$

denotes its base change to an automorphic representation of $GL_2(\mathbb{A}_K)$, and χ varies through a family of anticyclotomic Hecke characters over K. Here, "supersingular" means that we assume that p is inert or ramified in K. This is analogous, outside the splitting assumption on p, to the "ordinary" setting in which Katz and Bertolini-Darmon-Prasanna/Liu-Zhang-Zhang construct their one-variable p-adic L-functions. In fact our theory addresses the ordinary and supersingular settings uniformly by working on an affinoid subdomain

$$\mathcal{Y}_x \subset \mathcal{Y}$$

of the p-adic universal cover

$$\mathcal{Y} \to Y$$

(defined below); in fact, $\mathcal{Y}_x$ contains a natural cover

$$\mathcal{Y}^{\mathrm{Ig}} \to Y^{\mathrm{Ig}},$$

and restricting our theory to $\mathcal{Y}^{\mathrm{Ig}}$ allows one to recover the one-variable p-adic L-functions in the ordinary case, as well as Katz's theory of p-adic modular forms on Y^{Ig}.

One motivation for the construction of supersingular Rankin-Selberg p-adic L-functions is to develop special value formulas in the same style as those of Katz and Bertolini-Darmon-Prasanna, where in the former case a special value of the Katz p-adic L-function is related to the p-adic logarithm of elliptic units attached to K, and in the latter case the special value formula is a "p-adic Waldspurger formula" (following the terminology of [46]) involving the p-adic formal logarithm of a Heegner point attached to K (when a Heegner hypothesis holds for K and level N of w). Indeed, we succeed in proving such a formula in the case $p \nmid N$ in Section 9, though in future work we expect to remove both $p \nmid N$ as well as relax the Heegner hypothesis on N, which would simply necessitate considering more general quaternionic Shimura curves than modular curves.

We seek to develop a satisfactory theory of p-adic analysis on the supersingular locus, namely a notion of p-adic modular forms on the supersingular locus

$$Y^{\mathrm{ss}} = Y \setminus Y^{\mathrm{ord}}$$

which "behaves well" with respect to some differential operator d; more precisely, this means there is some notion of "weight" which is raised by 2 under the action of d, and given a p-adic modular form f,

$$d^j f$$

or some stabilization

$$(d^j f)^\flat$$

gives rise to some p-adic analytically well-behaved family. To do this, there are several technical difficulties which must be overcome. One of which is that there is no obvious canonical differential with which to trivialize ω over a cover in order to view modular forms as functions on the cover (in the same way as $\omega_{\mathrm{can}}^{\mathrm{Katz}}$ does so for ω on $Y^{\mathrm{Ig}} \to Y^{\mathrm{ord}}$). It is also a difficulty that there is no "canonical line" in the p-divisible group of a supersingular curve as there is for

$$\mu_{p^\infty} \subset A[p^\infty]$$

when A is ordinary. Hence there is no natural splitting of the Hodge filtration with which to define a differential operator d analogous to the Atkin-Serre operator in the ordinary setting, and even if one were to construct such an operator, there is no obvious analogue of the Serre-Tate coordinate under which to locally express p-adic modular forms f and study the analytic properties of $d^j f$.

Another difficulty with defining a satisfactory p-adic Maass-Shimura operator on Y^{ss} comes from the lack of unit root splitting, whose construction comes from a horizontal basis for the Gauss-Manin connection defined as sections of the relative étale cohomology $\mathcal{H}_{\text{ét}}^1(\mathcal{A})$ over Y^{Ig} which are eigenvectors of the canonical Frobenius. This unit root splitting in the ordinary case gives a splitting of the Hodge filtration

$$0 \to \omega|_{Y^{\mathrm{ord}}} \to \mathcal{H}^1_{\mathrm{dR}}(\mathcal{A})|_{Y^{\mathrm{ord}}} \to \omega^{-1}|_{Y^{\mathrm{ord}}} \to 0$$

as an exact sequence of $\mathcal{O}_{Y^{\mathrm{ord}}}$-modules, where $\mathcal{O}_Y$ denote the rigid analytic structure sheaf on Y. It is this functorial splitting, which is algebraically defined and coincides with the real analytic Hodge splitting at CM points, which gives rise to the ordinary p-adic Maass-Shimura operator θ_{AS} with the desired algebraicity properties. Note that unlike in the complex analytic setting, we do not have to extend the sheaf of rigid functions (the analogue of holomorphic functions) to a large sheaf (of "real analytic functions") in order to obtain the Hodge decomposition, as long as we restrict to $Y^{\mathrm{ord}} \subset Y$.

To overcome these difficulties, we generalize the strategy of Katz and in some sense emulate the construction of the complex analytic Maass-Shimura operator by working on the full p-adic universal cover

$$\mathcal{Y} \to Y$$

and by extending our coefficients from the structure sheaf to some larger sheaf of periods containing it (viewing this as analogous with extending holomorphic functions to real analytic functions).

Let us elaborate a little on the motivation of this strategy and how it works. As no unit root basis of the de Rham cohomology exists outside of Y^{ord}, we instead consider the moduli space of all horizontal bases of étale cohomology. This moduli space is representable by the *p-adic universal cover* $\mathcal{Y} \to Y$ (which we define more explicitly in the next paragraph), and with universal object being given by

$$(\mathcal{A}, \alpha_\infty) \to \mathcal{Y}$$

where α_∞ is the universal full p^∞-level structure. We then use a relative p-adic de Rham comparison theorem to view α_∞ as a universal horizontal basis for relative de Rham cohomology; unlike in the ordinary case, this comparison involves extending the structure sheaf to a certain period sheaf $\mathcal{O}\mathbb{B}^+_{\mathrm{dR},Y}$ (where this is really a sheaf on the proétale site $Y_{\mathrm{proét}}$) first constructed by Scholze in [56]. From this horizontal "framing" of the relative de Rham cohomology $\mathcal{H}^1_{\mathrm{dR}}(\mathcal{A})$, we get a "Hodge–de Rham period" measuring the position of the Hodge filtration and the "Hodge-Tate period" measuring the position of the Hodge-Tate filtration, as considered by Scholze in loc. cit., and use these periods to construct a relative Hodge decomposition which we use as a substitute for the unit root splitting. This splitting is in fact defined over an "intermediate period sheaf"

$$\mathcal{O}_\Delta := \mathcal{O}\mathbb{B}^+_{\mathrm{dR},\mathcal{Y}}/(t),$$

equipped with natural connection

$$\nabla : \mathcal{O}_\Delta \to \mathcal{O}_\Delta \otimes_{\mathcal{O}_\mathcal{Y}} \Omega^1_{\mathcal{Y}}$$

which is $\mathbb{B}^+_{dR,\mathcal{Y}}/(t)$-linear, induced by the natural connection

$$\nabla : \mathcal{OB}^+_{dR,\mathcal{Y}} \to \mathcal{OB}^+_{dR,\mathcal{Y}} \otimes_{\mathcal{O}_{\mathcal{Y}}} \Omega^1_{\mathcal{Y}}$$

which is $\mathbb{B}^+_{dR,\mathcal{Y}}$-linear. Moreover, there is a natural map

$$\mathcal{O}_{\mathcal{Y}} \subset \mathcal{OB}^+_{dR,\mathcal{Y}} \overset{\text{mod } t}{\twoheadrightarrow} \mathcal{O}_{\Delta}$$

which is in fact an inclusion compatible with connections, and such that its composition with the natural map

$$\theta : \mathcal{O}_{\Delta} \twoheadrightarrow \hat{\mathcal{O}}_{\mathcal{Y}}$$

where $\hat{\mathcal{O}}_{\mathcal{Y}}$ is the p-adically completed structure sheaf on $\mathcal{Y}$ is the natural map

$$\mathcal{O}_{\mathcal{Y}} \to \hat{\mathcal{O}}_{\mathcal{Y}}.$$

Here θ is induced by the natural relative analogue

$$\theta : \mathcal{OB}^+_{dR,\mathcal{Y}} \twoheadrightarrow \hat{\mathcal{O}}_{\mathcal{Y}}$$

of Fontaine's map $\theta : B^+_{dR} \twoheadrightarrow \mathbb{C}_p$. Here,

$$t \subset \mathbb{B}^+_{dR,Y}(\mathcal{Y})$$

is a global analogue of Fontaine's "$2\pi i$" and is a global section of a period sheaf $\mathbb{B}^+_{dR,\mathcal{Y}}$ on $\mathcal{Y}$, which is itself a relative version of Fontaine's ring of periods B^+_{dR}. We call $\mathcal{O}_{\Delta}$ "intermediate" because it, in the sense above, lies in between $\mathcal{O}_{\mathcal{Y}}$ and $\mathcal{OB}^+_{dR,\mathcal{Y}}$. In analogy with having to extend from holomorphic to real analytic functions on the complex universal cover $\mathcal{H}$ in order to define the complex analytic Hodge decomposition, we view $\mathcal{O}_{\Delta}$ as a sheaf of "p-adic nearly holomorphic (or rigid) functions on the p-adic universal cover $\mathcal{Y}$."

Let us go into more detail on the construction of $\mathcal{Y}$. On geometric points, it has a moduli-theoretic interpretation moduli space parametrizing elliptic curves with full p^∞-level structure represented by a $GL_2(\mathbb{Z}_p)$-profinite-étale cover $\mathcal{Y}$ of Y (viewing the latter as an adic space over $\mathrm{Spa}(\mathbb{Q}_p, \mathbb{Z}_p)$) called the *p-adic universal cover* (or *infinite-level modular curve*)

$$\mathcal{Y} = \varprojlim_i Y(p^i),$$

as considered by Scholze in [56] and Scholze-Weinstein in [62]. Here $Y(p^i)$ is the modular curve obtained by adding full p^i-level structure to the moduli space represented by Y, and $\mathcal{Y}$ is an adic space over $\mathrm{Spa}(\mathbb{Q}_p, \mathbb{Z}_p)$ which is an object in the proétale site $Y_{\text{proét}}$. Here, the full universal p^∞-level structure α_∞ is just a trivialization of the Tate module of $\mathcal{A}$

$$\alpha_\infty : \hat{\mathbb{Z}}_{p,\mathcal{Y}}^{\oplus 2} \xrightarrow{\sim} T_p\mathcal{A}|_{\mathcal{Y}},$$

here $\hat{\mathbb{Z}}_{p,\mathcal{Y}}$ is the "constant sheaf" on $\mathcal{Y}$ associated with $\mathbb{Z}_p$, except that sections are continuous functions into $\mathbb{Z}_p$ where the latter has the p-adic (and not discrete) topology. Now let $\mathcal{O}_Y$ denote the proétale structure sheaf on $Y_{\text{proét}}$. Using the Hodge–de Rham comparison theorem of Scholze ([57]), we then have a natural inclusion

$$\mathcal{H}_{\text{dR}}^1(\mathcal{A}) \otimes_{\mathcal{O}_Y} \mathcal{O}\mathbb{B}_{\text{dR},Y}^+ \overset{\iota_{\text{dR}}}{\subset} \mathcal{H}_{\text{ét}}^1(\mathcal{A}) \otimes_{\hat{\mathbb{Z}}_{p,Y}} \mathcal{O}\mathbb{B}_{\text{dR},Y}^+$$

on $Y_{\text{proét}}$ compatible with filtrations (on the left, the convolution of the Hodge filtration on $\mathcal{H}_{\text{dR}}^1(\mathcal{A})$ and the natural filtration on $\mathcal{O}\mathbb{B}_{\text{dR},Y}^+$, and on the right is just the filtration on $\mathcal{O}\mathbb{B}_{\text{dR},Y}^+$) and connections (on the left, the convolution of the Gauss-Manin connection on $\mathcal{H}_{\text{dR}}^1(\mathcal{A})$ and the natural connection on $\mathcal{O}\mathbb{B}_{\text{dR},Y}^+$ via the Leibniz rule, and on the right is just the connection on $\mathcal{O}\mathbb{B}_{\text{dR},Y}^+$). Pulling back to $\mathcal{Y}$, we then have

$$\mathcal{H}_{\text{dR}}^1(\mathcal{A}) \otimes_{\mathcal{O}_Y} \mathcal{O}\mathbb{B}_{\text{dR},Y}^+|_{\mathcal{Y}} \overset{\iota_{\text{dR}}}{\hookrightarrow} \mathcal{H}_{\text{ét}}^1(\mathcal{A}) \otimes_{\hat{\mathbb{Z}}_{p,Y}} \mathcal{O}\mathbb{B}_{\text{dR},Y}^+|_{\mathcal{Y}} \overset{\alpha_\infty^{-1}}{\underset{\sim}{\longrightarrow}} (\mathcal{O}\mathbb{B}_{\text{dR},\mathcal{Y}}^+ \cdot t^{-1})^{\oplus 2}$$

$$(1.2)$$

where the last isomorphism uses the universal p^∞-level structure α_∞ and the isomorphism

$$\mathcal{H}_{\text{ét}}^1(\mathcal{A}) \cong T_p A(-1)$$

given by the Weil pairing. We also use the fact that there is a natural isomorphism

$$\hat{\mathbb{Z}}_{p,\mathcal{Y}}(-1) = \hat{\mathbb{Z}}_{p,\mathcal{Y}} \cdot t^{-1},$$

as t is a period for the cyclotomic character.

We note that there is a natural sublocus

$$\mathcal{Y}^{\text{Ig}} = \{\hat{\mathbf{z}} = \infty\} \subset \mathcal{Y}$$

which parametrizes ordinary elliptic curves A together with a trivialization $\alpha : \mathbb{Z}_p^{\oplus 2} \xrightarrow{\sim} T_p A$ of their Tate modules $T_p A$ and with

$$\alpha_{\mathbb{Z}_p \oplus \{0\}} : \mathbb{Z}_p \cong T_p A^0 \subset T_p A$$

trivializing the canonical line in $T_p A$, viewed as arithmetic p^∞-level structures

$$\alpha : \mathbb{Z}_p(1) \oplus \mathbb{Z}_p \xrightarrow{\sim} T_p A,$$

together with a trivialization

$$\mathbb{Z}_p(1) \cong \mathbb{Z}_p,$$

and it is clear that these two data are equivalent to a full p^∞-level structure

$$\alpha : \mathbb{Z}_p \oplus \mathbb{Z}_p \xrightarrow{\sim} T_p A.$$

Thus,

$$\mathcal{Y}^{\mathrm{Ig}} \to Y^{\mathrm{Ig}}$$

is a natural $\mathbb{Z}_p \rtimes \mathbb{Z}_p^\times$-cover of the Igusa tower Y^{Ig}, and a B-cover of Y^{ord}, where $B \subset GL_2(\mathbb{Z}_p)$ denotes the subgroup of upper triangular matrices.

Using (1.2), one sees that

$$\alpha_{\infty,1} := \alpha_\infty|_{\hat{\mathbb{Z}}_{p,\mathcal{Y}} \oplus \{0\}} \quad \text{and} \quad \alpha_{\infty,2} := \alpha_\infty|_{\{0\} \oplus \hat{\mathbb{Z}}_{p,\mathcal{Y}}}$$

form a horizontal basis for the connection ∇. Moreover, upon making the identification (via the Weil pairing)

$$T_p \mathcal{A} \cong \mathrm{Hom}(\mathcal{A}[p^\infty], \mu_{p^\infty}),$$

we get a natural map

$$HT_\mathcal{A} : T_p \mathcal{A} \otimes_{\hat{\mathbb{Z}}_{p,Y}} \mathcal{O}_Y \to \omega, \qquad \alpha \mapsto \alpha^* \frac{dT}{T}$$

where dT/T is the canonical invariant differential on μ_{p^∞}. (It is also sometimes customary to denote $HT_\mathcal{A} = d\log$, as $d\log T = dT/T$.) We then define the *fake Hasse invariant* as

$$\mathfrak{s} := HT_\mathcal{A}(\alpha_{\infty,2}),$$

and in fact we have that on the restriction to $\mathcal{Y}^{\mathrm{Ig}}$,

$$\mathfrak{s}|_{\mathcal{Y}^{\mathrm{Ig}}} = \omega_{\mathrm{can}}^{\mathrm{Katz}}|_{\mathcal{Y}^{\mathrm{Ig}}}. \tag{1.3}$$

Consider the affinoid subdomain

$$\mathcal{Y}_x = \{\mathfrak{s} \neq 0\} \subset \mathcal{Y}.$$

We note that $\mathfrak{s} \in \omega(\mathcal{Y}_x)$ is a nonvanishing global section, i.e., a generator. Then let $\mathfrak{s}^{-1} \in \omega^{-1}(\mathcal{Y}_x)$ the generator corresponding to $\mathfrak{s}$ under Poincaré duality. The trivialization

$$\omega|_{\mathcal{Y}^{\mathrm{Ig}}} \cong \mathcal{O}_{\mathcal{Y}^{\mathrm{Ig}}}$$

induced by (1.3), along with the universal level structure on $\mathcal{Y}^{\mathrm{Ig}}$ given by $\alpha_\infty|_{\mathcal{Y}^{\mathrm{Ig}}}$, gives rise to a p-adic differential operator (the Atkin-Serre operator)

$$\theta_{\mathrm{AS}} : \mathcal{O}_{\mathcal{Y}^{\mathrm{Ig}}} \to \mathcal{O}_{\mathcal{Y}^{\mathrm{Ig}}}$$

with nice p-adic analytic properties, as seen using Serre-Tate coordinates. The key to these nice p-adic properties is the identity (see [34, Main Theorem 3.7.2])

$$\sigma(\omega_{\mathrm{can}}^{\mathrm{Katz}, \otimes 2}) = d \log T$$

where T is the Serre-Tate coordinate, and

$$\sigma : \omega^{\otimes 2} \xrightarrow{\sim} \Omega_Y^1 \tag{1.4}$$

is the Kodaira-Spencer isomorphism.

By the above discussion, $\mathfrak{s}$ seems like a natural candidate to extend Katz's idea of viewing p-adic modular forms (sections of ω) as functions to the (non-Galois) covering $\mathcal{Y}_x \to Y$. However, unlike in Katz's situation on $\mathcal{Y}^{\mathrm{Ig}}$, in our situation the splitting of (a lift of) the Hodge filtration which we define and use will require extending coefficients from $\mathcal{O}_{\mathcal{Y}_x}$ to a larger sheaf $\mathcal{O}_{\Delta, \mathcal{Y}_x}$ (which can be viewed as "the sheaf of nearly rigid functions," in analogy to extending coefficients from the sheaf of real analytic functions to the sheaf of nearly holomorphic functions in order to define the Hodge decomposition in the complex analytic situation), and with respect to this splitting $\mathfrak{s}$ will *not* be the most convenient choice for trivializing $\omega \otimes_{\mathcal{O}_Y} \mathcal{O}_{\Delta, \mathcal{Y}_x}$. Instead, we will trivialize using the generator

$$\omega_{\mathrm{can}} := \frac{\mathfrak{s}}{y_{\mathrm{dR}}} \in (\omega \otimes_{\mathcal{O}_Y} \mathcal{O}_{\Delta, \mathcal{Y}_x})(\mathcal{Y}_x)$$

where $y_{\mathrm{dR}} \in \mathcal{O}_{\Delta, \mathcal{Y}_x}(\mathcal{Y}_x)^\times$ is a certain p-adic period associated with $\mathfrak{s}$. Hence this induces a trivialization

$$\omega \otimes_{\mathcal{O}_Y} \mathcal{O}_{\Delta, \mathcal{Y}_x} \cong \mathcal{O}_{\Delta, \mathcal{Y}_x}.$$

One can show that $y_{\mathrm{dR}} = 1$ on the sublocus $\mathcal{Y}^{\mathrm{Ig}} \subset \mathcal{Y}$, and so we have

$$\omega_{\mathrm{can}}|_{\mathcal{Y}^{\mathrm{Ig}}} = \omega_{\mathrm{can}}^{\mathrm{Katz}}|_{\mathcal{Y}^{\mathrm{Ig}}}. \tag{1.5}$$

In analogy with (1.4), we also have

$$\sigma(\omega_{\mathrm{can}}^{\otimes, 2}) = dz_{\mathrm{dR}} \tag{1.6}$$

where $z_{\mathrm{dR}} = \mathbf{z}_{\mathrm{dR}} \pmod{t}$ for the fundamental de Rham period $\mathbf{z}_{\mathrm{dR}}$, which we describe in more detail below. We note that the analogy between (1.4) and (1.6), along with (1.5), suggests that z_{dR} provides the correct analogue of $\log T$. It is this observation which later leads to our notion of the $q_{\mathrm{dR}} = \exp(z_{\mathrm{dR}} - \bar{z}_{\mathrm{dR}})$-coordinate as an analogue (and extension) for the Serre-Tate coordinate T, and q_{dR}-expansions as analogues (and extensions) of Serre-Tate T-expansions.

We can also use ω_{can} to generalize Katz's notion of p-adic modular forms. Let $\mathcal{U} \subset \mathcal{Y}_x$ be a subadic space, let $\lambda : \mathcal{Y} \to Y$ denote the natural projection, and let $\lambda(\mathcal{U}) = U$. Then letting

$$\Gamma = \mathrm{Gal}(\mathcal{U}/U) \subset \mathrm{Gal}(\mathcal{Y}/Y) = GL_2(\mathbb{Z}_p),$$

we have a natural map

$$\omega^{\otimes k}|_U(U) \xrightarrow{\lambda^*} \omega^{\otimes k}|_{\mathcal{U}}(\mathcal{U}) \hookrightarrow (\omega \otimes_{\mathcal{O}_Y} \mathcal{O}_{\Delta, \mathcal{Y}_x}|_{\mathcal{U}})^{\otimes k}(\mathcal{U}) \xrightarrow[\sim]{\omega_{\mathrm{can}}^{\otimes k}} \mathcal{O}_{\Delta, \mathcal{Y}_x}|_{\mathcal{U}}(\mathcal{U}).$$

In fact, the image under this map consists of sections $f \subset \mathcal{O}_{\Delta, \mathcal{Y}_x}|_{\mathcal{U}}(\mathcal{U})$ such that

$$\begin{pmatrix} a & b \\ c & d \end{pmatrix}^* f = (ad - bc)^{-k}(cz_{\mathrm{dR}} + a)^k f \qquad (1.7)$$

for any $\begin{pmatrix} a & b \\ c & d \end{pmatrix} \in \Gamma$. In this situation, we say that f has *weight k for Γ on $\mathcal{U}$*. We note that when $\mathcal{U} = \mathcal{Y}^{\mathrm{Ig}}$ and so $U = Y^{\mathrm{ord}}$, and

$$\Gamma = B \subset GL_2(\mathbb{Z}_p)$$

the subgroup of upper triangular matrices, and then (1.7) becomes

$$\begin{pmatrix} a & b \\ 0 & d \end{pmatrix}^* f = d^{-k} f. \qquad (1.8)$$

In particular, f descends to a section in $\mathcal{O}_Y(Y^{\mathrm{Ig}})$ and we recover Katz's notion (1.1) of a p-adic modular form of weight k. Our main interest, which is defining a satisfactory notion of p-adic modular form on the supersingular locus, will involve the case $\mathcal{U} = \mathcal{Y}^{\mathrm{ss}}, U = Y^{\mathrm{ss}}$ and $\Gamma = GL_2(\mathbb{Z}_p)$.

Let us now elaborate on the construction of our splitting of (a lift of) the Hodge filtration alluded to above, which is crucial to the construction of the p-adic Maass-Shimura operator and its algebraicity properties. Unlike in Katz's theory, outside of $\mathcal{Y}^{\mathrm{Ig}}$, $\alpha_{\infty,1}$ and $\alpha_{\infty,2}$ do not generate either the Hodge or Hodge-Tate filtrations, and instead we are led to consider certain relative periods

$$\mathbf{z}_{\mathrm{dR}} \in \mathcal{O}\mathbb{B}_{\mathrm{dR},Y}^+(\mathcal{Y}_x), \hat{\mathbf{z}} \in \hat{\mathcal{O}}_Y(\mathcal{Y}_x),$$

where $\hat{\mathcal{O}}_Y = \hat{\mathcal{O}}_Y^+[1/p]$ and where

$$\hat{\mathcal{O}}_Y^+ = \varprojlim_n \mathcal{O}_Y^+/p^n \subset \hat{\mathcal{O}}_Y = (\varprojlim_n \mathcal{O}_Y^+/p^n)[1/p]$$

denotes the p-adic completion integral structure sheaf

$$\mathcal{O}_Y^+ \subset \mathcal{O}_Y = \mathcal{O}_Y^+[1/p].$$

Both of the above periods can be viewed as sections of an ambient period presheaf

$$\hat{\mathcal{O}}\mathbb{B}_{\mathrm{dR},Y}^+ := \mathcal{O}\mathbb{B}_{\mathrm{dR},Y}^+ \otimes_{\mathcal{O}_Y} \hat{\mathcal{O}}_Y,$$

where $\mathcal{O}\mathbb{B}_{\mathrm{dR},Y}^+$ is the usual de Rham period sheaf as considered in [57]. The *Hodge–de Rham period* $\mathbf{z}_{\mathrm{dR}} \in \mathcal{O}\mathbb{B}_{\mathrm{dR},Y}^+(\mathcal{Y}_x)$ measures the position of the Hodge filtration

$$\omega|_{\mathcal{Y}_x} = \mathfrak{s} \cdot \hat{\mathcal{O}}_{\mathcal{Y}_x} \subset \mathcal{H}_{\mathrm{dR}}^1(\mathcal{A}) \otimes_{\mathcal{O}_Y} \hat{\mathcal{O}}\mathbb{B}_{\mathrm{dR},\mathcal{Y}_x}^+ \xrightarrow{\iota_{\mathrm{dR}}} \mathcal{H}_{\mathrm{\acute{e}t}}^1(\mathcal{A}) \otimes_{\hat{\mathbb{Z}}_{p,Y}} \hat{\mathcal{O}}\mathbb{B}_{\mathrm{dR},\mathcal{Y}_x}^+$$
$$\xrightarrow[\sim]{\alpha_\infty^{-1}} (\hat{\mathcal{O}}\mathbb{B}_{\mathrm{dR},\mathcal{Y}_x}^+ \cdot t^{-1})^{\oplus 2}.$$

The *Hodge-Tate period* $\hat{\mathbf{z}} \in \hat{\mathcal{O}}_Y(\mathcal{Y}_x) \subset \mathcal{O}\mathbb{B}_{\mathrm{dR},Y}^+(\mathcal{Y}_x)$ measures the position of the Hodge-Tate filtration

$$\omega^{-1}|_{\mathcal{Y}_x} = \mathfrak{s}^{-1} \cdot \hat{\mathcal{O}}_{\mathcal{Y}_x} \subset \mathcal{H}_{\mathrm{\acute{e}t}}^1(\mathcal{A}) \otimes_{\hat{\mathbb{Z}}_{p,Y}} \hat{\mathcal{O}}\mathbb{B}_{\mathrm{dR},\mathcal{Y}_x}^+ \xrightarrow[\sim]{\alpha_\infty^{-1}} (\hat{\mathcal{O}}\mathbb{B}_{\mathrm{dR},\mathcal{Y}_x}^+ \cdot t^{-1})^{\oplus 2}.$$

Using these periods, and recalling our notation $\mathcal{O}_{\Delta,\mathcal{Y}_x} = \mathcal{O}\mathbb{B}_{\mathrm{dR},\mathcal{Y}_x}^+/(t)$ and defining $\hat{\mathcal{O}}_{\Delta,\mathcal{Y}_x} = \mathcal{O}_{\Delta,\mathcal{Y}_x} \otimes_{\mathcal{O}_{\mathcal{Y}_x}} \hat{\mathcal{O}}_{\mathcal{Y}_x}$, one can construct a Hodge decomposition

$$T_p\mathcal{A} \otimes_{\hat{\mathbb{Z}}_{p,Y}} \hat{\mathcal{O}}_{\Delta,\mathcal{Y}_x} \xrightarrow{\sim} (\omega \otimes_{\mathcal{O}_Y} \hat{\mathcal{O}}_{\Delta,\mathcal{Y}_x}) \oplus (\omega^{-1} \otimes_{\mathcal{O}_Y} \hat{\mathcal{O}}_{\Delta,\mathcal{Y}_x} \cdot t) \qquad (1.9)$$

where the projection onto the first factor is given by $HT_\mathcal{A}$ (i.e., the inclusion of the first factor is a section of $HT_\mathcal{A}$), and so this gives a splitting of the Hodge-Tate filtration.

However, for the purposes of using this splitting to construct a differential operator, there is a technical issue that there is no natural way to define a connection on $\hat{\mathcal{O}}\mathbb{B}_{\mathrm{dR},Y}^+$, precisely because it contains a copy of the p-adically completed structure sheaf $\hat{\mathcal{O}}_Y$: the pullback $\hat{\mathcal{O}}_{\mathcal{Y}}$ is (essentially) the structure sheaf of the perfectoid space $\hat{\mathcal{Y}} \sim \mathcal{Y}$ associated with $\mathcal{Y}$, and there are no nontrivial differentials on perfectoid spaces since differentials are infinitely divisible, and hence 0 in the p-adic completion.

There are two ways to remedy this. One is to instead replace (1.9) with another splitting

$$T_p\mathcal{A}\otimes_{\hat{\mathbb{Z}}_{p,Y}}\mathcal{O}_{\Delta,\mathcal{Y}_x}\xrightarrow{\sim}(\omega\otimes_{\mathcal{O}_Y}\mathcal{O}_{\Delta,\mathcal{Y}_x})\oplus\mathcal{L}, \tag{1.10}$$

where $\mathcal{L}$ is a free $\mathcal{O}_{\Delta,\mathcal{Y}_x}$-module of rank 1. This splitting is constructed by using the natural "horizontal" embedding $\hat{z}\in\hat{\mathcal{O}}_{\mathcal{Y}_x}\subset\mathcal{O}_{\Delta,\mathcal{Y}_x}$. Now the projection onto the first factor is *not* given by $HT_\mathcal{A}$, but instead its kernel is *horizontal* in the sense that

$$\nabla_w(\mathcal{L})\subset\mathcal{L}$$

for any section w of $\Omega_Y^1\otimes_{\mathcal{O}_Y}\mathcal{O}_{\Delta,\mathcal{Y}_x}$. Moreover, (1.10) recovers the usual Hodge-Tate decomposition upon applying the natural map

$$\theta:\mathcal{O}_{\Delta,\mathcal{Y}_x}\twoheadrightarrow\hat{\mathcal{O}}_{\mathcal{Y}_x},$$

where $\hat{\mathcal{O}}_{\mathcal{Y}_x}$ denotes the p-adically completed structure sheaf (and θ is analogous to Fontaine's universal cover $\theta:B_{\mathrm{dR}}^+\twoheadrightarrow\mathbb{C}_p$)

$$T_p\mathcal{A}\otimes_{\hat{\mathbb{Z}}_{p,Y}}\hat{\mathcal{O}}_{\mathcal{Y}_x}\xrightarrow{\sim}(\omega\otimes_{\mathcal{O}_Y}\hat{\mathcal{O}}_{\mathcal{Y}_x})\oplus(\hat{\omega}_{\mathcal{Y}_x}(1)). \tag{1.11}$$

The other approach is to instead construct intermediate period sheaves of "nearly holomorphic coefficients"

$$\mathcal{O}_{\Delta,\mathcal{Y}_x}\subset\mathcal{O}_{\Delta,\mathcal{Y}_x}^\diamond\subset\mathcal{O}_{\Delta,\mathcal{Y}_x}^\dagger\subset\hat{\mathcal{O}}_{\Delta,\mathcal{Y}_x},$$

where $\mathcal{O}_\Delta^\dagger$ contains the Hodge-Tate period $\hat{z}$ and so is large enough to construct a Hodge splitting (essentially one just adjoins a few sections in $\hat{\mathcal{O}}_{\Delta,\mathcal{Y}_x}$ to $\mathcal{O}_{\Delta,\mathcal{Y}_x}$), and on which one can also extend the natural connection on $\mathcal{O}_\Delta$ (namely by declaring that $\nabla(\hat{z})=0$, and showing that this gives rise to a well-defined connection on $\mathcal{O}_{\Delta,\mathcal{Y}_x}^\dagger$ since $\hat{z}$ is transcendental over $\mathcal{O}_\Delta$). One can then define a splitting like (1.10) using $\mathcal{O}_{\Delta,\mathcal{Y}_x}^\dagger$, and consequently construct a p-adic Maass-Shimura operator with coefficients in $\mathcal{O}_{\Delta,\mathcal{Y}_x}^\dagger$; one can even show that this differential operator is defined over the smaller sheaf of coefficients $\mathcal{O}_{\Delta,\mathcal{Y}_x}^\diamond$.

Both approaches have their virtues: while the first approach stays within the smaller period sheaf $\mathcal{O}_{\Delta,\mathcal{Y}_x}$, it requires the use of Fontaine's map θ in order to recover a true Hodge-Tate decomposition, whereas the second requires enlarging (slightly) to $\mathcal{O}_{\Delta,\mathcal{Y}_x}^\diamond$ but then does not require the use of θ. While the p-adic Maass-Shimura operators arising from each approach are different, they satisfy the same algebraicity properties at CM points and are both equal in value (after normalizing by a p-adic period) to the value of the complex Maass-Shimura operator (normalized by a complex period) at CM points. Hence, either Maass-Shimura operator can be used in order to construct our p-adic L-function.

We will follow the first approach in this outline in the introduction. Now we can define a p-adic Maass-Shimura operator d with respect to the splitting (1.10). Since (1.11) recovers the relative Hodge-Tate decomposition, it is induced at CM points by the algebraic CM splitting, and so as in Katz's theory one can show (*using* the horizontalness of (1.10)) that for an algebraic modular form $w \in \omega^{\otimes k}(Y)$, writing

$$w|_{\mathcal{Y}_x} = f \cdot \omega_{\mathrm{can}}^{\otimes k}, \; f \in \mathcal{O}_{\Delta, \mathcal{Y}_x}(\mathcal{Y}_x), \qquad F \cdot (2\pi i dz)^{\otimes k}, \; F \in \mathcal{O}^{\mathrm{hol}}(\mathcal{H}^+),$$

where $\mathcal{O}^{\mathrm{hol}}$ denotes the sheaf of (complex) holomorphic function and $\mathcal{H}^+ \to Y$ the complex universal cover (i.e., the complex upper half-plane), we have that the value

$$(\theta \circ d^j) f(y) / \Omega_p(y)^{k+2j}$$

at a CM point $y \in \mathcal{Y}_x$ is an algebraic number for an appropriate p-adic period $\Omega_p(y)$ (depending on y), and in fact is equal (in $\overline{\mathbb{Q}}$) to the algebraic number

$$\eth^j F(y) / \Omega_\infty(y)^{k+2j}$$

at the same CM point $y \in \mathcal{Y}_x$ for an appropriate complex period $\Omega_\infty(y)$ (only depending on the image of y under the natural projection $\mathcal{Y} \to Y$):

$$(\theta \circ d^j) f(y) / \Omega_p(y)^{k+2j} = \eth^j F(y) / \Omega_\infty(y)^{k+2j}. \tag{1.12}$$

The key fact for proving this algebraicity is that the fiber $\mathcal{O}_{\Delta, \mathcal{Y}_x}(y)$ contains a unique copy of $\overline{\mathbb{Q}}_p$ by Hensel's lemma, and so composition

$$\overline{\mathbb{Q}}_p \subset \mathcal{O}_{\Delta, \mathcal{Y}_x}(y) \overset{\theta}{\twoheadrightarrow} \mathbb{C}_p$$

is the natural inclusion. Then since the specialization $\hat{\mathbf{z}}(y) \in \overline{\mathbb{Q}}_p$, we have

$$\theta(\hat{\mathbf{z}}(y)) = \hat{\mathbf{z}}(y),$$

and so

$$(\theta \circ d^j) f(y) = \theta(d^j f(y)) = d^j f(y),$$

and this latter value is equal (after normalizing by periods) to

$$\eth^j f(y)$$

since both (1.10) and the complex analytic Hodge decomposition are both induced by the CM splitting at y.

It is the algebraicity of $\theta \circ d^j$ at CM points, and moreover the fact that it is equal in value to complex Maass-Shimura derivatives, which makes it applicable to questions regarding interpolation of critical L-values and hence construction of p-adic L-functions. Ultimately, for the construction of the latter, it is necessary to understand the analytic behavior of

$$(\theta \circ d^j)f$$

around CM points y, and here the framework for understanding such analytic properties is provided by q_{dR}-expansions of modular forms, given by a q_{dR}-expansion map

$$\omega^{\otimes k}|_{\mathcal{Y}_x} \xrightarrow[\sim]{\omega_{can}^{\otimes k}} y_{dR}^k \mathcal{O}_{\mathcal{Y}_x} \xrightarrow{q_{dR}-\exp} \hat{\mathcal{O}}_{\mathcal{Y}_x}[\![q_{dR}-1]\!] \subset \mathcal{O}_{\Delta,\mathcal{Y}_x} \xrightarrow[\sim]{\omega_{can}^{\otimes k}} \omega^{\otimes k} \otimes_{\mathcal{O}_Y} \mathcal{O}_{\Delta,\mathcal{Y}_x}. \tag{1.13}$$

A key fact is that on the supersingular locus $\mathcal{Y}^{ss} \subset \mathcal{Y}_x$, (1.13) *coincides* with the natural inclusion

$$\omega|_{\mathcal{Y}^{ss}} \hookrightarrow \omega \otimes_{\mathcal{O}_Y} \mathcal{OB}_{dR,\mathcal{Y}^{ss}}^+ \xrightarrow{\mathrm{mod}\ \iota} \omega \otimes_{\mathcal{O}_Y} \mathcal{O}_{\Delta,\mathcal{Y}^{ss}}, \tag{1.14}$$

which is induced by the composition

$$\mathcal{O}_{\mathcal{Y}} \subset \mathcal{OB}_{dR,\mathcal{Y}}^+ \xrightarrow{\mathrm{mod}\ t} \mathcal{O}_\Delta$$

which turns out to be an inclusion. In fact, recalling that $\hat{\mathcal{O}}_Y$ denotes the p-adic completion of the structure sheaf $\mathcal{O}_Y$, we have a natural inclusion

$$\hat{\mathcal{O}}_{\mathcal{Y}}[\![q_{dR}-1]\!] \subset \mathcal{O}_\Delta$$

which is compatible with the natural connections on each sheaf, and which is in fact an *equality* on $\mathcal{Y}^{ss}$:

$$\hat{\mathcal{O}}_{\mathcal{Y}^{ss}}[\![q_{dR}-1]\!] = \mathcal{O}_{\Delta,\mathcal{Y}^{ss}}.$$

Hence we see that, at least on the supersingular locus $\mathcal{Y}^{ss}$, q_{dR} provides the correct coordinate when viewing a rigid modular form

$$w \in \omega^{\otimes k}(\mathcal{Y}^{ss}) \subset (\omega \otimes_{\mathcal{O}_Y} \mathcal{O}_{\Delta,\mathcal{Y}^{ss}})^{\otimes k}(\mathcal{Y}^{ss}) \xrightarrow[\sim]{\omega_{can}^{\otimes k}} \mathcal{O}_{\Delta,\mathcal{Y}^{ss}}(\mathcal{Y}^{ss})$$

as a "nearly rigid function." The coordinate

$$q_{dR} \in \mathcal{O}_{\Delta,\mathcal{Y}_x}(\mathcal{Y}_x)$$

plays the role analogous to that of the Serre-Tate coordinate, and in fact the q_{dR}-expansion of a modular form recovers the Serre-Tate expansion upon restricting to $\mathcal{Y}^{Ig}$ (due to the fact that $\omega_{can}|_{\mathcal{Y}^{Ig}} = \omega_{can}^{Katz}$).

In fact, one can write down an explicit formula for $\theta \circ d^j$ in terms of q_{dR}-coordinates

$$\theta \circ d^j = \sum_{i=0}^{j} \binom{j}{i} \binom{j+k-1}{i} i! \left(-\frac{\theta(y_{\mathrm{dR}})}{\hat{\mathbf{z}}} \right)^i \theta \circ \left(\frac{q_{\mathrm{dR}}d}{dq_{\mathrm{dR}}} \right)^{j-i}. \tag{1.15}$$

On $\mathcal{Y}^{\mathrm{Ig}}$, as we noted before, $\hat{\mathbf{z}} = \infty$ and so we have

$$(\theta \circ d^j)|_{\mathcal{Y}^{\mathrm{Ig}}} = \theta \circ \left(\frac{q_{\mathrm{dR}}d}{dq_{\mathrm{dR}}} \right)^j \bigg|_{\mathcal{Y}^{\mathrm{Ig}}} = \theta \circ \theta_{\mathrm{AS}}^j = \theta_{\mathrm{AS}}^j$$

where the last equality follows from the fact that

$$\frac{dq_{\mathrm{dR}}}{q_{\mathrm{dR}}} \bigg|_{\mathcal{Y}^{\mathrm{Ig}}} = dz_{\mathrm{dR}}|_{\mathcal{Y}^{\mathrm{Ig}}} = dT|_{\mathcal{Y}^{\mathrm{Ig}}}$$

and that

$$\mathcal{O}_{\mathcal{Y}_x} \subset \mathcal{O}_{\Delta,\mathcal{Y}_x} \xrightarrow{\theta} \hat{\mathcal{O}}_{\mathcal{Y}_x}$$

is the natural completion map. Hence, again restricting to $\mathcal{Y}^{\mathrm{Ig}} \subset \mathcal{Y}_x$, we recover Katz's theory.

In order to construct the p-adic L-function, we consider the image of a modular form

$$w \in \omega^{\otimes k}(Y)$$

under the q_{dR}-expansion map (1.13), and study the growth of the coefficients of its q_{dR}-expansion around supersingular CM points y:

$$\omega^{\otimes k}|_{\mathcal{Y}^{\mathrm{ss}}} \xrightarrow{q_{\mathrm{dR}}-\exp} \hat{\mathcal{O}}_{\mathcal{Y}^{\mathrm{ss}}} \llbracket q_{\mathrm{dR}} - 1 \rrbracket \xrightarrow{\text{stalk at } y} \hat{\mathcal{O}}_{\mathcal{Y}^{\mathrm{ss}},y} \llbracket q_{\mathrm{dR}} - 1 \rrbracket = \hat{\mathcal{O}}_{\mathcal{Y}^{\mathrm{ss}},y} \llbracket q_{\mathrm{dR}}^{1/p^b} - 1 \rrbracket$$

for any $b \in \mathbb{Q}$, where the last equality is just a formal change of variables. By the remarks above involving (1.13) and (1.14), since $y \in \mathcal{Y}^{\mathrm{ss}}$, we see that the above map coincides with the natural map

$$\omega^{\otimes k}|_{\mathcal{Y}^{\mathrm{ss}}} \hookrightarrow \omega^{\otimes k} \otimes_{\mathcal{O}_Y} \mathcal{O}_{\Delta,\mathcal{Y}^{\mathrm{ss}}} \xrightarrow{\text{stalk at } y} \omega^{\otimes k} \otimes_{\mathcal{O}_Y} \mathcal{O}_{\Delta,\mathcal{Y}^{\mathrm{ss}},y}$$

and hence is compatible with the natural connections on all sheaves; in particular, this compatibility shows that the formula (1.15) gives the action of the p-adic Maass-Shimura operator $\theta^j \circ d^j$. One of the main results of Section 6 is that for appropriate

$$b \in \mathbb{Q}$$

(a priori depending on y, but we later show that b is the same for all y in a certain CM orbit), in fact we have that the above map factors through

$$\omega^{\otimes k}|_{\mathcal{Y}^{\mathrm{ss}}} \to \hat{\mathcal{O}}^+_{\mathcal{Y}^{\mathrm{ss}},y}[\![q_{\mathrm{dR}}^{1/p^b} - 1]\!][1/p]. \tag{1.16}$$

Proving the integrality of the q_{dR}^{1/p^b}-expansion involves the consideration of another *Lubin-Tate period* $\mathbf{z}_{\mathrm{LT}}$ coming from the Rapoport-Zink uniformization of the infinite-level supersingular locus $LT_\infty \to \mathcal{Y}^{\mathrm{ss}}$ by the Lubin-Tate space LT_∞ at infinite level. This aforementioned period comes from the Grothendieck-Messing crystalline period map $LT_\infty \to \mathbb{P}^1$ associated to this Rapoport-Zink space. Viewing LT_∞ as an object in the proétale site of Y, one can in fact show that there is a canonical isomorphism

$$\mathcal{O}_{\Delta,LT_\infty} \cong \hat{\mathcal{O}}_{LT_\infty}[\![\mathbf{z}_{\mathrm{LT}} - \bar{\mathbf{z}}_{\mathrm{LT}}]\!], \tag{1.17}$$

where $\bar{\mathbf{z}}_{\mathrm{LT}}$ denotes $\mathbf{z}_{\mathrm{LT}}$ viewed as a section in $\hat{\mathcal{O}}_{LT_\infty}$, which in turn has a natural (horizontal) embedding into $\mathcal{O}_{\Delta,LT_\infty}$. From the above isomorphism, one can show another natural isomorphism

$$\mathcal{O}_{\Delta,\mathcal{Y}^{\mathrm{ss}}} \cong \hat{\mathcal{O}}_{\mathcal{Y}^{\mathrm{ss}}}[\![q_{\mathrm{dR}} - 1]\!] - \hat{\mathcal{O}}_{\mathcal{Y}^{\mathrm{ss}}}[\![q_{\mathrm{dR}}^{1/p^b} - 1]\!]. \tag{1.18}$$

Using a variant of the Dieudonné-Dwork lemma for integrality of power series, one can show that integrality of coefficients (at certain geometric stalks) of the power series expansion (1.17) transfer to integrality in the q_{dR}^{1/p^b}-expansion (1.18).

Given

$$w \in \omega^{\otimes k}(Y),$$

we can construct the p-adic L-function associated with w by considering sums of the images

$$w(q_{\mathrm{dR}}^{1/p^b})_y$$

of w under (1.16), where the subscript y denotes the stalks at various orbits of CM points y on $\mathcal{Y}^{\mathrm{ss}}$, then applying the normalized Maass-Shimura operators

$$p^{bj}\theta \circ d^j$$

using the formula (1.15). By the formula, we see that as long as p-adic valuations

$$|y_{\mathrm{dR}}(y)|, \quad |\hat{\mathbf{z}}(y)|$$

of the specializations

$$y_{\mathrm{dR}}(y), \quad \hat{\mathbf{z}}(y)$$

of the p-adic periods

$$y_{\mathrm{dR}}, \quad \hat{\mathbf{z}}$$

satisfy certain bounds, then images in the stalks

$$p^{bj}(\theta \circ d^j)w(q_{\mathrm{dR}}^{1/p^b})_y$$

"converge" to some p-adic continuous function in $j \in \mathbb{Z}/(p-1) \times \mathbb{Z}_p$, in a sense we now make more precise. Define the *stabilization* by

$$w^\flat(q_{\mathrm{dR}}^{1/p^b})_y = w(q_{\mathrm{dR}}^{1/p^b})_y - \frac{1}{p}\sum_{j=0}^{p-1} w(\zeta_p^j q_{\mathrm{dR}}^{1/p^b})_y. \qquad (1.19)$$

We also denote

$$w^\flat(q_{\mathrm{dR}}^{1/p^b})_y = f^\flat(q_{\mathrm{dR}}^{1/p^b})_y \cdot w_{\mathrm{can},y}^{\otimes k}. \qquad (1.20)$$

One can show directly from (1.15) that

$$p^{bj}(\theta \circ d^j)w^\flat(q_{\mathrm{dR}}^{1/p^b})_y$$

is a p-adic continuous function of $j \in \mathbb{Z}/(p-1) \times \mathbb{Z}_p$. Then for any $j_0 \in \mathbb{Z}_{\geq 0}$ we have

$$\lim_{m \to \infty} p^{b(j_0+p^m(p-1))}(\theta \circ d^{j_0+p^m(p-1)})w(q_{\mathrm{dR}}^{1/p^b})_y = p^{bj_0}(\theta \circ d^{j_0})w^\flat(q_{\mathrm{dR}}^{1/p^b})_y. \quad (1.21)$$

Roughly, summing

$$p^{bj}(\theta \circ d^j)w^\flat(q_{\mathrm{dR}}^{1/p^b})_y$$

against anticyclotomic Hecke characters χ evaluated at ideals corresponding to y for y over an appropriate CM orbit (associated with an order $\mathcal{O}$ of an imaginary quadratic field K) gives the construction of our p-adic continuous L-function. In reality, we will be able to bound the p-adic periods at CM points y' which are related to the natural CM points y associated with $\mathcal{O}$ via

$$y' \cdot \begin{pmatrix} 1 & 0 \\ 0 & q \end{pmatrix} = y,$$

where $q = p$ if $p > 2$ and $q = 8$ if $p = 2$. As a consequence, the $q_{\mathrm{dR}}^{1/q}$-expansions at the $\mathcal{O}$-orbit of y' are well-behaved, and we then construct our p-adic L-function by summing

$$w'_j(q_{\mathrm{dR}}^{1/q})_{y'}$$

against anticyclotomic Hecke characters evaluated at an $\mathcal{O}$ orbit of y', where

$$w_j' = \begin{pmatrix} 1 & 0 \\ 0 & q \end{pmatrix}^* (q^j(\theta \circ d^j)w^\flat)$$

is a certain Hecke translate of the Maass-Shimura derivatives, which allows us to relate our p-adic L-function to period sums over CM orbits of y, and hence obtain our interpolation property using the algebraicity theorem (1.12). Namely, values of the p-adic L-function in a certain range are equal to certain algebraic normalizations of central critical L-values associated with the Rankin-Selberg family (w, χ).

We end this outline with a few remarks on how we obtain the p-adic Waldspurger formula in Section 9, focusing on the case when $k = 2$. A key property of the p-adic Maass-Shimura operator d^j is that it sends p-adic modular forms of weight k in the sense of (1.7) to modular forms of weight $k + 2j$. Hence the limit

$$\lim_{m \to \infty} p^{bp^m(p-1)}(\theta \circ d^{p^m(p-1)})w^\flat(q_{\mathrm{dR}})_y$$

converges to a p-adic modular form of weight 0 on some small affinoid neighborhood of y, for some subgroup

$$\Gamma \subset GL_2(\mathbb{Z}_p).$$

Let K_p denote the p-adic completion of K with respect to a fixed embedding

$$\overline{\mathbb{Q}} \hookrightarrow \overline{\mathbb{Q}}_p.$$

In fact, one can show that on some affinoid

$$\mathcal{U} \supset \mathcal{Y}^{\mathrm{Ig}} \sqcup \mathcal{C},$$

where $\mathcal{C} \subset \mathcal{Y}$ is a locus of CM points associated with K such that

$$\mathrm{Gal}(\mathcal{C}/C) \cong \mathcal{O}_{K_p}^\times \subset GL_2(\mathbb{Z}_p)$$

(induced by some embedding

$$K_p \hookrightarrow M_2(\mathbb{Z}_p);$$

the subadic space $\mathcal{C}$ itself does not depend on this choice of embedding), the limit

$$G := \lim_{m \to \infty} p^{bp^m(p-1)}(\theta \circ d^j)w^\flat(q_{\mathrm{dR}})|_{\mathcal{U}} \tag{1.22}$$

converges to a p-adic modular form of weight 0 on $\mathcal{U}$ for some Γ with

$$B \subset \Gamma \subset GL_2(\mathbb{Z}_p).$$

In particular, by restriction it induces a rigid function on $\mathcal{Y}^{\mathrm{Ig}} \sqcup \mathcal{U}$, which is of weight 0 for B on $\mathcal{Y}^{\mathrm{Ig}}$ and of weight 0 for Γ on $\mathcal{U}$. This means that G descends to a section G on an affinoid open

$$U \subset Y'$$

for some finite étale cover

$$Y' \to Y;$$

here we use the fact that while

$$\mathcal{Y}^{\mathrm{Ig}} \subset \mathcal{Y}$$

is not affinoid open, its image on any finite cover is isomorphic to a copy of the ordinary locus

$$Y^{\mathrm{ord}} \subset Y,$$

which, being the complement of a finite union of residue discs (the supersingular locus), is an (admissible) affinoid open. In particular, G is rigid on U, and one can show using Coleman's theory of integration that on

$$U \cap Y'^{\mathrm{ord}},$$

G is in fact equal to the formal logarithm

$$\log_{w^\flat} \big|_{U \cap Y'^{\mathrm{ord}}}$$

for some p-stabilization $w^\flat$ of the newform w. (Here p-stabilization denotes the image of w under some explicit Hecke operator at p.) Then the rigidity of G on U implies that dG is a rigid 1-form on U, and so by the theory of Coleman integration the rigid primitive G on $\mathcal{U}$ is unique up to constant, which implies

$$G = \log_{w^\flat} \big|_U. \tag{1.23}$$

Since the relevant special value of our p-adic L-function corresponds to evaluating (1.22) on an orbit of the CM point y, one sees that we arrive at our p-adic Waldspurger formula by evaluating (1.23) at an appropriate Heegner point.

1.3   MAIN RESULTS

We now finally state our main results. We adopt the notation of Chapter 8, and the reader should refer to there for precise definitions and assumptions.

Fix an algebraic closure $\overline{\mathbb{Q}}$ of $\mathbb{Q}$, and view all number fields as embedded in $\overline{\mathbb{Q}}$. Let p denote a prime number. Fix an algebraic closure $\overline{\mathbb{Q}}_p$ of $\mathbb{Q}_p$, and fix embeddings

$$i_\infty : \overline{\mathbb{Q}} \hookrightarrow \mathbb{C}, \quad i_p : \overline{\mathbb{Q}} \hookrightarrow \overline{\mathbb{Q}}_p.$$

Fix an imaginary quadratic extension $K/\mathbb{Q}$. Now suppose the prime p is inert or ramified in K, and let w be a new eigenform (i.e., a newform or Eisenstein series) of weight $k \geq 2$ for $\Gamma_1(N)$ and nebentype ϵ_w, where $p \nmid N$, $N \geq 4$. Let $a_p(w)$ denote the Hecke eigenvalue of T_p. Suppose that

1. k is even *or* $p > 2$, and
2. N satisfies the Heegner hypothesis, i.e., that each prime $\ell \mid N$ splits or ramifies in K, and if $\ell^2 \mid N$ then ℓ splits in K.

Let A be a fixed elliptic curve with CM by an order $\mathcal{O}_c \subset \mathcal{O}_K$ of conductor $p \nmid c$ also with $(c, Nd_K) = 1$, let

$$\alpha : \mathcal{O}_{K_p} \xrightarrow{\sim} T_p A$$

be a choice of full p^∞-level structure as in Choice 8.6, let

$$y = (A, \alpha) \in \mathcal{C}(\overline{K}_p, \mathcal{O}_{\overline{K}_p})$$

and let $\Omega_p(y)$ and $\Omega_\infty(y)$ be the associated periods as in Definition 5.45, and also let

$$w|_{\mathcal{H}^+} = F \cdot (2\pi i dz)^{\otimes k}.$$

Then for Hecke characters $\chi \in \Sigma$, in the notation of Chapter 8, Section 8.2, we have that the values $L(F, \chi^{-1}, 0)$ are central critical. On a certain subset $\Sigma_+ \subset \Sigma$, characters satisfy root number conditions so that these central critical L-values are nonvanishing, and so present as candidates for interpolation. Given an algebraic Hecke character χ, let $\check{\chi}$ denote its p-adic avatar, and let $N_K : \mathbb{A}_K^\times \to \mathbb{C}^\times$ denote the norm character, which has infinity type $(1, 1)$. We let $\overline{\check{\Sigma}}$ denote the p-adic closure of the p-adic avatar $\check{\Sigma}_+$ (in the space of functions on $\mathbb{A}_K^{(p\infty)}$, equipped with the uniform convergence topology) of Σ_+: we note that one can naturally view $\check{\Sigma} \subset \overline{\check{\Sigma}}$. Finally, suppose (A, t, α) is a suitable CM point on infinite level (see Choice 8.6), let

$$\theta(\Omega_p)(A, t, \alpha) \in \mathbb{C}_p^\times$$

be the period as in Definition 5.45 (see also Propositions 7.3 and 7.7), so that denoting by $\mathfrak{p}$ the prime of K above p, $\#\kappa = \#\mathcal{O}/\mathfrak{p}$ the order of the residue field at $\mathfrak{p}$, and e the ramification index of $K_\mathfrak{p}/\mathbb{Q}_p$ ($=1$ if p is inert in K, $=2$ if p is ramified in K), we have

$$|\theta(\Omega_p)(A,t,\alpha)| = |2|p^{\frac{1}{p-1} - \frac{1}{e(\#\kappa-1)}} = \begin{cases} |2|p^{\frac{p}{p^2-1}} & p \text{ inert in } K \\ |2|p^{\frac{1}{2(p-1)}} & p \text{ ramified in } K. \end{cases}$$

Let

$$\Omega(A,t) \in \overline{\mathbb{Q}}^\times$$

be the period as in Definition 8.10. We collect our results into one Main Theorem.

Theorem 1.1 (Theorems 8.9, 8.14, 9.10). *There is a p-adic continuous function*

$$\mathcal{L}_{p,\alpha}(w,\cdot) : \overline{\check{\Sigma}}_+ \to \mathbb{C}_p$$

that satisfies the following interpolation property. Let $q = p$ if $p > 2$ and $q = 8$ if $p = 2$. Then for all $\chi \in \Sigma_+$, when w is a newform we have

$$\mathcal{L}_{p,\alpha}(w,\check\chi) = \frac{(q\theta(\Omega_p)(A,t,\alpha))^{k+2j}}{\Omega(A,t)^j} \cdot \Xi_p(w,\chi) \cdot i_p\left(L^{\mathrm{alg}}(F,\chi^{-1},0)\right) \qquad (1.24)$$

and when w is an Eisenstein series $(F = E_k^{\psi_1,\psi_2}$, see Definition 8.3) we have

$$\mathcal{L}_{p,\alpha}(w,\check\chi) = \frac{(q\theta(\Omega_p)(A,t,\alpha))^{k+2j}}{\Omega(A,t)^j} \cdot \Xi_p(w,\chi) \cdot i_p\left(L^{\mathrm{alg}}(E_k^{\psi_1,\psi_2},\chi^{-1},0)\right), \quad (1.25)$$

where $L^{\mathrm{alg}}(F,\chi^{-1},0)$ and $L^{\mathrm{alg}}(E_k^{\psi_1,\psi_2},\chi^{-1},0)$ are certain algebraic normalizations of square roots of the Rankin-Selberg central L-value $L((\pi_w)_K \times \chi^{-1}, 1/2)$ and Hecke central L-value $L^{\mathrm{alg}}(E_k^{\psi_1,\psi_2},\chi^{-1},0)$ as defined in Definition 8.5, and

$$\Xi_p(w,\chi) = \begin{cases} 1 - a_p(w)^2\chi^{-1}(p)\frac{p-1}{p+1} - \frac{1}{p^2} & p \text{ inert in } K \\ 1 - a_p(w)\chi^{-1}(\mathfrak{p})\frac{p-1}{p} - \frac{1}{p^2} & p \text{ ramified in } K. \end{cases}$$

We have the following "p-adic Waldspurger formula": For any $\chi \in \check\Sigma_- \subset \overline{\check{\Sigma}}_+$ of infinity type $(k-(j+1), j+1)$ where $0 \le j \le r := k-2$, that

$$\mathcal{L}_{p,\alpha}(w,\check\chi) = \frac{(q\theta(\Omega_p)(A,t,\alpha))^{k-2(j+1)}}{\Omega(A,t)^{-(j+1)}} \Xi_p(w,\check\chi) \cdot \frac{c^{-j}}{j!} \mathrm{AJ}_F(\Delta_{\chi\mathbb{N}_K^{-1}})(w \wedge \omega_{A_0}^j \eta_{A_0}^{r-j}),$$

where

$$\mathrm{AJ}_F(\Delta_{\chi\mathbb{N}_K^{-1}})(w \wedge \omega_{A_0}^j \eta_{A_0}^{r-j})$$

is the p-adic Abel-Jacobi image of a specific generalized Heegner cycle as defined in Chapter 10, Section 9.2 (which depends on some fixed elliptic curve A_0 with CM by $\mathcal{O}_K$). In particular, we have the following application toward the Beilinson-Bloch conjecture. Let χ take values in E. In particular, if

$$\mathcal{L}_{p,\alpha}(w, \check{\chi}) \neq 0, \tag{1.26}$$

then

$$\epsilon_w \epsilon_\chi \mathrm{N}_K^{-1} \Delta_{\chi \mathrm{N}_K^{-1}} \in \varepsilon_w \varepsilon_{\chi \mathrm{N}_K^{-1}} \mathrm{CH}_0^{r+1}(X_r)(F)_E^{\chi \mathrm{N}_K^{-1}}$$

is nontrivial, where the right-hand side denotes the $\varepsilon_w \varepsilon_{\chi \mathrm{N}_K^{-1}}$-isotypic component of an appropriate Chow group for the underlying (Chow) motive attached to (w, χ^{-1}).

When $k = 2$, we have a simpler statement. Let H_c denote the ring class field associated with the order $\mathcal{O}_c$. For any character $\chi : \mathcal{C}\ell(\mathcal{O}_c) \to \overline{\mathbb{Q}}_p^\times$ with $\check{\mathrm{N}}_K \chi \in \check{\Sigma}_- \subset \check{\overline{\Sigma}}_+$, we have

$$\mathcal{L}_{p,\alpha}(w, \check{\mathrm{N}}_K \chi) = \Omega(A, t) \Xi_p(w, \check{\mathrm{N}}_k \chi) \cdot \log_w P_K(\chi)$$

where $P_K(\chi) \in \mathrm{Jac}(Y_1(N))(H_c)$ is the Heegner point as defined in Section 9.

In particular, if

$$\mathcal{L}_{p,\alpha}(w, \check{\mathrm{N}}_K \chi) \neq 0,$$

then $P_K(\chi)$ projects via a modular parametrization to a non-torsion point in $A_w(H_c)^\chi$, where $A_w/\mathbb{Q}$ is the GL_2-type abelian variety associated uniquely up to isogeny with w. Then by the Gross-Zagier formula and Kolyvagin, we have

$$\mathrm{rank}_{\mathbb{Z}} A_w(H_c)^\chi = \dim_{\mathbb{Q}} A_w = \mathrm{ord}_{s=1} L(A_w, \chi, s).$$

1.4   SOME REMARKS ON OTHER WORKS IN SUPERSINGULAR IWASAWA THEORY

Finally, we point out that there has been much groundbreaking work done in supersingular Iwasawa theory by many other authors. We give a brief summary of some of these results. In our setting of Iwasawa theory for imaginary quadratic fields in which p is inert or ramified, Rubin ([54]), following methods of Katz and invoking the machinery of Coleman power series, succeeded in constructing 1-variable continuous p-adic L-functions in the Lubin-Tate direction (i.e., for characters of type $(k, 0)$ and varying $k \in \mathbb{Z}/(p-1) \times \mathbb{Z}_p$). In [53], he also formulated analogues of supersingular main conjectures assuming certain conjectures on the structure of the Iwasawa module of universal norms of local units. Agboola-Howard ([1]) also developed anticyclotomic main conjectures for imaginary quadratic fields, assuming the aforementioned conjecture on the structure of local units (though even with this assumption, they did not construct an analytic anticyclotomic p-adic L-function, as we do in this book). Schneider-Teitelbaum ([63]), using their p-adic Fourier theory and Coleman power series, also constructed distributions interpolating Hecke L-values over imaginary quadratic fields in which p is inert or ramified.

In the GL_2-setting, particularly for newforms attached to elliptic curves, there has also been great progress, although not directly related to our situation. Previous works have mainly addressed the Iwasawa theory of families of twists $V \otimes \chi$ where the weight of V is greater than the weight of the characters χ. In this case the Galois representations in consideration are supersingular at p exactly when V itself is supersingular at p, since the Hodge-Tate weights of V dominate those of χ. In contrast, we address the case where the weight of the χ's is at least the weight of V, and hence the twists are supersingular precisely when the character χ is supersingular at p (i.e., p is inert or ramified in K), since the Hodge-Tate weights of χ dominate those of V. The former situation, however, already has potent applications to the Birch and Swinnerton-Dyer conjecture. For V attached to elliptic curves over $\mathbb{Q}$, see the fundamental work of Pollack ([50]) who introduced "$+/-$" constructions in order to produce 1-variable (cyclotomic) measures from the classical distributions attached to elliptic curves with good supersingular reduction at p (the construction of which is due to Višik [66], Amice-Vélu [2], and Mazur-Tate-Teitelbaum [47]). Kobayashi, soon after Pollack, gave an algebraic construction of these "$+/-$" p-adic L-functions (see [39]) by defining a suitable "$+/-$" Coleman map and evaluating on Kato's Euler system (see [32]). Later, Sprung ([65]) extended this to the $a_p \neq 0$ case by constructing an appropriate generalization of the "$+/-$" Coleman map. From their algebraic constructions of p-adic L-functions, Kobayashi and Sprung were also able to formulate appropriate "$+/-$" cyclotomic main conjectures in the non-CM case, and use Kato's Euler system to prove one divisibility of these main conjectures. Pollack-Rubin soon afterward formulated and proved the CM analogue of this main conjecture ([51]), building on Kobayashi's construction and Rubin's previous work on the Euler system of elliptic units ([55]). Kim was also able to generalize Kobayashi's constructions to 2 variables in certain height 1 settings ([38]). In more general settings, for elliptic curves there is also the work of Wan ([67]), who proved the supersingular analogue of Skinner-Urban's GL_2 main conjecture ([64]), and of Castella-Wan ([11]), who formulated and proved $+/-$ analogues of Perrin-Riou's main conjecture on Heegner points ([49]). See also the works addressing more general settings such as that of Lei-Loeffler-Zerbes ([45]), Büyükboduk-Lei ([7]), and Castella-Çiperiani-Skinner-Sprung ([10]), who also addressed the setting of a general elliptic modular form.

Chapter Two

Preliminaries: Generalities

2.1 GROTHENDIECK SITES AND TOPOI

We recall some notions of Grothendieck sites and their associated topoi, as we will use this language freely throughout our discussion. Throughout, we will assume knowledge of basic category theory.

Definition 2.1. *Given a category $\mathcal{C}$, we denote its objects by $\mathrm{Ob}(\mathcal{C})$ and its morphisms by $\mathrm{Hom}(\mathcal{C})$. When given two objects $U, V \in \mathrm{Ob}(\mathcal{C})$, we denote the class of all morphisms from U to V by $\mathrm{Hom}_{\mathcal{C}}(U, V)$. We let $\mathcal{C}^{\mathrm{op}}$ denote the opposite category, i.e., the category with the same objects as those of $\mathcal{C}$ but with all the directions of the morphisms between them reversed.*

Later, we will often abuse notation for an object $U \in \mathrm{Ob}(\mathcal{C})$ by simply writing $U \in \mathcal{C}$.

Definition 2.2. *A covering family $\{f_i : U_i \to U\}$ on an object $U \in \mathcal{C}$ is a collection of morphisms $f_i \in \mathrm{Hom}_{\mathcal{C}}(U_i, U)$.*

Definition 2.3 (See [3]). *A Grothendieck pretopology $\mathcal{T}$ on $\mathcal{C}$ consists of a collection of covering families $\{U_i \to U\}$, subject to the following axioms:*

1. *(Existence of fiber products): For any object $U \in \mathrm{Ob}(\mathcal{C})$, for all morphisms $U_i \to U$ which belong to some covering family of U, and for all morphisms $V \to U$, the fiber product $U_i \times_U V$ exists.*
2. *(Base change): For any object $U \in \mathrm{Ob}(\mathcal{C})$, if $\{U_i \to U\}$ is a covering family and $f : V \to U$ is a morphism, then $\{U_i \times_U V \to V\}$ is a covering family.*
3. *(Local characterization): Suppose $\{U_i \to U\}$ is a covering family and $\{U_{ij} \to U_i\}$ is a covering family for all i, then $\{U_{ij} \to U_i \to U\}$ is a covering family.*
4. *(Identity): If $V \to U$ is an isomorphism, then $\{V \to U\}$ is a covering family.*

Given an object $U \in \mathrm{Ob}(\mathcal{C})$, we often call a covering family $\{U_i \to U\}$ a covering of U.

Definition 2.4. *A site consists of a category $\mathcal{C}$ along with a Grothendieck pretopology $\mathcal{T}$ on $\mathcal{C}$. Suppose that $(\mathcal{C}, \mathcal{T})$ and $(\mathcal{D}, \mathcal{T}')$ are sites. A continuous functor*

$f : \mathcal{C} \to \mathcal{D}$ *is a functor* $\mathcal{C} \to \mathcal{D}$ *such that for every sheaf* $\mathcal{F}$ *on* $\mathcal{D}$ *with respect to* $\mathcal{T}'$, *the presheaf* $\mathcal{F} \circ f$ *is a sheaf on* $\mathcal{C}$ *with respect to the topology* $\mathcal{T}$.

Note that for any topological space X, the category of open sets along with open covers on X is a Grothendieck site.

A morphism of sites $(\mathcal{C}, \mathcal{T}) \to (\mathcal{D}, \mathcal{T}')$ consists of a continuous functor

$$\mathcal{D} \to \mathcal{C}$$

such that $f^* \mathcal{F}$ preserves finite limits in $\mathcal{C}$.

Definition 2.5. *Suppose we are given arbitrary categories* $\mathcal{C}, \mathcal{D}$. *A* ($\mathcal{D}$-valued) *presheaf* $\mathcal{F}$ *on* $\mathcal{C}$ *is a contravariant functor* $\mathcal{F} : \mathcal{C} \to \mathcal{D}$.

Now suppose again that $\mathcal{C}$ *admits fiber products, and let* $(\mathcal{C}, \mathcal{T})$ *be a site. A* ($\mathcal{D}$-valued) *sheaf* $\mathcal{F}$ (*with respect to the topology* $\mathcal{T}$) *on* $\mathcal{C}$ *is a* $\mathcal{D}$-*valued presheaf such that for all* $U \in \mathrm{Ob}(\mathcal{C})$ *and all coverings* $\{f_c : U_c \to U\}_{c \in \mathrm{Ob}(\mathcal{C})}$, *the natural map*

$$\mathrm{Hom}_{\mathcal{C}}(\mathrm{Hom}_{\mathcal{C}}(\cdot, U), \mathcal{F}) \to \mathrm{Hom}_{\mathcal{C}}(\{f_c : U_c \to U\}_{c \in \mathrm{Ob}(\mathcal{C})}, \mathcal{F}),$$

which comes from the inclusion $\{f_c : U_c \to U\}_{c \in \mathrm{Ob}(\mathcal{C})} \subset \mathrm{Hom}_{\mathcal{C}}(\cdot, U)$, *is bijective.*

Given a presheaf $\mathcal{F}$ *on* $(\mathcal{C}, \mathcal{T})$ *and a functor* $f : \mathcal{C}' \to \mathcal{C}$, *we denote by* $f^* \mathcal{F}$ *the presheaf on* $(\mathcal{C}', \mathcal{T}')$ *sending*

$$U \mapsto \mathcal{F}(f(U)).$$

A morphism of presheaves and sheaves is simply a natural transformation of functors.

Definition 2.6. *Let* SETs *denote the category of sets. The category* $\tilde{\mathcal{C}}$ *of all* SETs-*valued sheaves on* $\mathcal{C}$ *is called the* topos *on* $(\mathcal{C}, \mathcal{T})$. (*Note that we slightly abuse notation by suppressing the dependence on* $\mathcal{T}$ *in the notation for* $\tilde{\mathcal{C}}$.)

Note that any set S *can be viewed as a sheaf on the point-set topological space* $\{x\}$ *with the discrete topology (i.e., with open sets consisting of* $\emptyset$ *and* $\{x\}$), *by letting* $S(\emptyset) = \emptyset$ *and* $S(\{x\}) = S$. *Hence* SETs *can be viewed as the topos of the Grothendieck site given by the discrete topology on* $\{x\}$.

Definition 2.7. *Note that a continuous functor* $f : \mathcal{C} \to \mathcal{D}$ *induces a functor* $f_* : \tilde{\mathcal{C}} \to \tilde{\mathcal{D}}$ *by sending*

$$\mathcal{F} \mapsto f_* \mathcal{F} := \mathcal{F} \circ f.$$

The functor f_* *is often called the* pushforward *along* f, *and the sheaf* $f_* \mathcal{F}$ *is called the* pushforward *of* $\mathcal{F}$ *along* f. *The functor* f_* *admits a left adjoint* f^* *called the* pullback *along* f, *and a sheaf* $f^* \mathcal{G}$ *is called the* pullback *of* $\mathcal{G}$ *along* f.

Later, given a site $(\mathcal{C}, \mathcal{T})$ we will often abuse notation and write $U \in \mathrm{Ob}(\mathcal{C})$ simply as $U \in (\mathcal{C}, \mathcal{T})$.

We now define the notion of a point in a site $(\mathcal{C}, \mathcal{T})$.

Definition 2.8. *Given topoi* $\tilde{C}$ *and* $\tilde{D}$, *a geometric morphism* $f : \tilde{C} \to \tilde{D}$ *is a pair of adjoint functors* $(f^* : \tilde{D} \to \tilde{C}, f_* : \tilde{C} \to \tilde{D})$ *where* f^* *is left adjoint to* f_* *such that* f^* *preserves finite limits. Hence a morphism of sites induces a morphism of topoi via the usual pullback and pushforward functors constructed above.*

A point x *of a topos* $\tilde{C}$ *is a geometric morphism from* SETs *(viewed as a topos on a one-point discrete topological space) to* $\tilde{C}$. *Given* $\mathcal{F} \in \mathrm{Ob}(\tilde{C})$, *we define the stalk of* $\mathcal{F}$ *at* x *to be* $\mathcal{F}_x := x^*\mathcal{F}$. *We say that the topos* $\tilde{C}$ *has enough points if the collection of functors* $x^* : \tilde{C} \to$ SETs, *for all points* x *is jointly faithful, i.e., a morphism of objects* $\mathcal{F} \to \mathcal{G}$ *in* $\tilde{C}$ *is an isomorphism if and only if for each point* x, *the induced map of sets* $\mathcal{F}_x \to \mathcal{G}_x$ *is an isomorphism.*

In fact, the definition of topos given above (i.e., the category of sheaves on a Grothendieck site) is often called a coherent *topos in the literature in order to distinguish it from a more general abstract notion of topos. A theorem of Deligne ([20]) says that any coherent topos has enough points.*

Given a topological space X *and a Grothendieck pretopology* $\mathcal{C}$ *on* X, *note that any point* $x \subset X$, *viewed as a continuous map* $x \to X$ *(e.g., a geometric point* $x \to X$ *of a locally ringed space* X), *determines a point of the topos* $\tilde{C}$ *on* X *in the following way. We define the skyscraper sheaf functor* $x_* :$ SETs $\to \tilde{C}$ *as follows: Given* $S \in \mathrm{ob}(\text{SETs})$, *we view* S *as a sheaf on* x *in the way described earlier, and define a presheaf*

$$x_*S(U) = S(x^{-1}(U))$$

which one immediately checks is a sheaf (the usual skyscraper sheaf at x). *Given* $\mathcal{F} \in \mathrm{Ob}(\tilde{C})$, *one defines* $x^* = \mathcal{F}_x$ *as the usual stalk at* x

$$x^*\mathcal{F} = \mathcal{F}_x = \varinjlim_{U \in \mathcal{C}, x \in |U|} \mathcal{F}(U)$$

which is a set. It is standard knowledge that (x^*, x_*) *is an adjoint pair of functors.*

2.2   PRO-CATEGORIES

Throughout this section, let $\mathcal{C}$ denote a category. Recall that a category is *small* if both its class of underlying objects and its class of underlying morphisms are sets. We call a category I *cofiltered* if

1. $\mathrm{Ob}(\mathcal{C}) \neq \emptyset$,
2. for all $i, j \in \mathrm{Ob}(I)$, there exists $k \in \mathrm{Ob}(I)$ and morphisms $f : k \to i, g : k \to j \in \mathrm{Hom}(I)$, and
3. for any $f, g : i \to j \in \mathrm{Hom}(I)$, there exists $h : k \to i \in \mathrm{Hom}(I)$ such that $f \circ h = g \circ h$.

A *(small) cofiltered limit* is a limit of a functor $F : I \to \mathcal{C}$ where I is a (small) cofiltered category.

Definition 2.9. *Given a category $\mathcal{C}$, we define the category pro-$\mathcal{C}$ as that with the following objects and morphism:*

1. $\mathrm{Ob}(pro\text{-}\mathcal{C}) = \{functors \;\; F : I \to \mathcal{C} : I \;\; small \; cofiltered \; category\},$

2. $\mathrm{Hom}(pro\text{-}\mathcal{C}) = \varinjlim_i \varprojlim_j \mathrm{Hom}_{\mathcal{C}}(F(i), G(j)).$

Equivalently, letting SETs denote the category of sets, and

$$\mathcal{C}^* = \mathrm{Funct}(\mathcal{C}, SETs)^{\mathrm{op}}$$

where $\mathrm{Funct}(\mathcal{D}, \mathcal{E})$ denotes the category of functors between two categories $\mathcal{D}$ and $\mathcal{E}$, we have a fully faithful embedding

$$\mathcal{C} \to \mathcal{C}^*, \quad x \mapsto \mathrm{Hom}(x, \cdot) \tag{2.1}$$

called the Yoneda completion. Recall that an object $F \in \mathrm{Ob}(\mathcal{C}^*)$ is called representable *if it is in the image of (2.1). Then*

$$pro\text{-}\mathcal{C} \subset \mathcal{C}^*$$

is the full subcategory which are small cofiltered inverse limits of representable objects.

2.3   ADIC SPACES

We recall some basic notions of Huber's theory of adic spaces which will be used throughout our discussion, following [30], [59, Section 2].

Definition 2.10. *Given a commutative topological ring R (i.e., R is a topological space whose ring operations are continuous with respect to the topology on R), recall that an ideal $I \subset R$ defines the topology on R if and only if $\{I^n\}$ is a basis of open neighborhoods of 0, in which case R is called an (I-)adic ring.*
    We say that R is Huber[1] *if there exists an open subring $R_0 \subset R$ such that the subspace topology on R_0 is defined by a finitely generated ideal.*
    We say that R is a Tate ring *if it is Huber and if there exists an invertible element in $r \in R$ which is topologically nilpotent, that is,*

$$r^n \to 0 \quad as \quad \mathbb{Z}_{\geq 0} \ni n \to \infty.$$

---

[1] *Huber's original terminology is f-adic ([30, Chapter 1]). Here, we adopt the terminology proposed in [61].*

We also say that a set $S \subset R$ is bounded *if for all open neighborhoods $0 \in U$, there exists an open neighborhood $0 \in V$ such that $SV \subset U$. An element $r \in R$ is called* power bounded *if $\{r^n \mid n \in \mathbb{Z}_{\geq 0}\}$ is bounded. We let $R^0 \subset R$ denote the set of power bounded elements of R.*

Assumption 2.11. *We note that Huber makes the following assumption on Huber rings in [30, Assumption 1.1.1]: all Huber rings satisfy one of the following:*

1. the completion $\hat{R}$ of R has a noetherian ring of definition,
2. R is a strongly noetherian Tate ring,
3. the topology of $\hat{R}$ is discrete.

We make this Assumption until our discussion on pre-adic spaces (Section 2.4). All results in this chapter on Huber adic spaces which we use later in the text will be applied to adic spaces satisfying (1) or (2) in the above Assumption, (e.g., modular curves and Lubin-Tate spaces). Later, our discussion will involve objects in the proétale site of such adic spaces which will have proétale presentations consisting of inverse systems of étale morphisms between adic spaces satisfying the above Assumption.

Example 2.12. *For example, $\mathbb{Z}_p$ has the p-adic topology defined by (p), and $R = \mathbb{Q}_p$ is Huber taking $R_0 = \mathbb{Z}_p$, and is also Tate taking the invertible element p, which is topologically nilpotent as $p^n \to 0$ as $n \to \infty$ in the p-adic topology.*
Another example is given by the Tate algebra

$$\mathbb{Z}_p\langle T\rangle = \{f = \sum_{n=0}^{\infty} a_n T^n \in \mathbb{Z}_p[\![T]\!] : |a_n| \to 0 \ \ as \ \ n \to \infty\}, \quad \mathbb{Q}_p\langle T\rangle = \mathbb{Z}_p\langle T\rangle[1/p].$$

Again one can take $R_0 = \mathbb{Z}_p\langle T\rangle$ to make $R = \mathbb{Q}_p\langle T\rangle$ into a Huber ring, and the element p makes it further into a Tate ring.

Definition 2.13. *Let R be a commutative topological ring, and let $(\Gamma, \cdot)$ denote a totally ordered (multiplicative) abelian group. Let $\{0\}$ be a symbol, and consider the set $\Gamma_0 = \Gamma \cup \{0\}$, which we endow with a total ordering by declaring that 0 is a minimal element of Γ_0, and the rest of Γ_0 inherits the ordering from Γ. We also extend the operation $\cdot$ on Γ to Γ_0 by declaring that $0 \cdot \gamma = 0$ for all $\gamma \in \Gamma$.*

1. A valuation *(or absolute value) on R is a (not necessarily surjective) map*

$$|\cdot| : R \to \Gamma_0$$

which satisfies the following axioms:

a) $|ab| = |a| \cdot |b|$,
b) $|a + b| \leq \max(|a|, |b|)$,
c) $|0| = 0, |1| = 1$.

2. *Let $\Gamma_{|\cdot|}$ denote the value group of $|\cdot|$, which is the smallest subgroup of Γ containing $|R|$ (all the values of $|\cdot|$ on R). We further say that $|\cdot|$ is continuous if for all $\gamma \in \Gamma_{|\cdot|} \cup \{0\}$, there exists an open neighborhood $0 \in U \subset R$ such that the image*

$$|U| \subset \{\gamma' \in \Gamma_0 : \gamma' < \gamma\}.$$

3. *Two valuations $|\cdot|$ and $|\cdot|'$, with corresponding value groups Γ and Γ' are called* equivalent *if there exists an order-preserving isomorphism $\iota : \Gamma \to \Gamma'$ such that for all $r \in R$,*

$$|\iota(x)| = |x|'.$$

Given a valuation $|\cdot|$ on R, the *support of $|\cdot|$* is given by

$$\mathrm{supp}(|\cdot|) = \{r \in R : |r| = 0\} \subset R.$$

It is a prime ideal of R, and $|\cdot|$ is easily seen to induce a valuation on $K = \mathrm{Frac}(R/\mathrm{supp}(|\cdot|))$. One also observes that $|\cdot|$ and $|\cdot|'$ are equivalent if and only if $\mathrm{supp}(|\cdot|) = \mathrm{supp}(|\cdot|')$ and they define the same valuation rings in K.

Definition 2.14. *A* Huber pair[2] *is a pair (R, R^+), where R is a Huber ring and $R^+ \subset R^0$ is an open integrally closed subring of R. A morphism of Huber rings*

$$f : (R, R^+) \to (S, S^+)$$

is a continuous morphism $f : R \to S$ such that $f(R^+) \subset S^+$.

Definition 2.15. *Suppose (R, R^+) is a Huber pair. Then define its spectrum by*

$$\mathrm{Spa}(R, R^+) := \{ \text{continuous valuations } |\cdot| : R \to \Gamma_0 : |R^+| \leq 1 \}/\sim$$

where $\sim$ is equivalence of valuations. We endow $\mathrm{Spa}(R, R^+)$ with the topology generated by

$$\{|\cdot| \in \mathrm{Spa}(R, R^+) : |a| \leq |b| \neq 0\}$$

for any $a, b \in R$. We can further define a natural basis of this topology by considering rational subsets

$$U\left(\frac{f_1, \ldots, f_n}{g}\right) = \{|\cdot| \in \mathrm{Spa}(R, R^+) : |f_i| \leq |g| \neq 0, \quad i = 1, \ldots, n\}$$

---

[2] *Huber's original terminology in [30, Chapter 1] for such a pair (R, R^+) is affinoid ring. We adopt the terminology of [61] here.*

where $\{f_i\}_{i=1,\ldots,r}$ generates an open ideal of R. The set of rational subsets form a basis of the topology on $\mathrm{Spa}(R, R^+)$ which is stable under finite intersection. If R is Tate, so that R is the only open ideal of R, the stipulation "$\neq 0$" is not needed.

The construction Spa defines a functor

$$\{\text{Huber pairs}\} \to \{\text{topological spaces}\}.$$

Now we define the structure presheaf on $X = \mathrm{Spa}(R, R^+)$. Suppose

$$U = U\left(\frac{f_1,\ldots,f_n}{g}\right)$$

is a rational subset. Then let

$$A = R\left[\frac{f_1}{g},\ldots,\frac{f_n}{g}\right]$$

and let A^+ denote the integral closure of $R^+\left[\frac{f_1}{g},\ldots,\frac{f_n}{g}\right]$ in A. Let

$$\hat{A} = R\left\langle\frac{f_1}{g},\ldots,\frac{f_n}{g}\right\rangle$$

denote the completion of A, and let $\hat{A}^+$ denote the completion of A^+. Then there is a map of Huber pairs

$$\phi: (R, R^+) \to (\hat{A}, \hat{A}^+)$$

with the following universal property: for any complete Huber pair (S, S^+) and map

$$\psi: (S, S^+) \to (R, R^+)$$

with

$$\mathrm{Spa}(\psi)(\mathrm{Spa}(R, R^+)) \subset U,$$

then there is a unique morphism $\rho: (R, R^+) \to (\hat{A}, \hat{A}^+)$ such that $\psi = \psi \circ \rho$.
Hence, $(\hat{A}, \hat{A}^+)$ depends only on U.

Definition 2.16. *We now define presheaves on X by defining them on rational subsets U as above by*

$$\mathcal{O}_X(U) = \hat{A}, \quad \mathcal{O}_X^+(U) = \hat{A}^+,$$

and then for an arbitrary open $V \subset X$, defining

$$\mathcal{O}_X(V) = \varprojlim_{U \subset V,\, U \subset X \text{ rational subset}} \mathcal{O}_X(U), \quad \mathcal{O}_X^+(V) = \varprojlim_{U \subset V,\, U \subset X \text{ rational subset}} \mathcal{O}_X^+(U).$$

It is apparent that $\mathcal{O}_X$ and $\mathcal{O}_X^+$ are presheaves of complete topological rings, and are respectively called the structure presheaf *and* integral structure presheaf *of X.*

We have

$$\mathcal{O}_X^+(V) = \{ f \in \mathcal{O}_X(V) : x(f) = |f(x)| \leq 1 \}.$$

Given $x \in X$, we also define the stalk by

$$\mathcal{O}_{X,x} = \varinjlim_{x \in U,\, U \subset X \text{ rational subset}} \mathcal{O}_X(U), \quad \mathcal{O}_{X,x}^+ = \varinjlim_{x \in U,\, U \subset X \text{ rational subset}} \mathcal{O}_X^+(U).$$

In fact, as shown in [16, Proposition 14.3.1], we have

$$\mathcal{O}_{X,x}^+ = \{ f \in \mathcal{O}_{X,x} : x(f) = |f(x)| \leq 1 \}. \tag{2.2}$$

Moreover, it is shown in Section 14 of loc. cit. that any $x \in U$ for $U \subset X$ a rational subset can be extended to $x : \mathcal{O}_{X,x} \to \Gamma$, and subsequently $\mathcal{O}_{X,x}$ is a local ring with maximal ideal given by

$$\mathfrak{m}_x = \{ f \in \mathcal{O}_{X,x} : x(f) = |f(x)| = 0 \}.$$

In certain cases, the structure presheaves defined above can be upgraded to sheaves.

Definition 2.17. *Recall that a topological ring R is strongly noetherian if for all $n \in \mathbb{Z}_{\geq 0}$, $R\langle T_1, \ldots, T_n \rangle$ is noetherian.*

Theorem 2.18 ([31], Theorem 2.2). *Suppose R is strongly noetherian. Then for $X = \operatorname{Spa}(R, R^+)$, $\mathcal{O}_X$ and $\mathcal{O}_X^+$ are sheaves.*

Definition 2.19 (See [61], Definition 3.2.1). *We define the category (V) as follows. Objects are triples $(X, \mathcal{O}_X, (|\cdot(x)|_{x \in X}))$, where X is a topological space, $\mathcal{O}_X$ is a sheaf of topological rings, and for each $x \in X$, $|\cdot(x)|$ is an equivalence class of continuous valuations on $\mathcal{O}_{X,x}$ (which also uniquely defines $\mathcal{O}_{X,x}^+$). The morphisms are maps of topologically ringed spaces $f : X \to Y$ that make the following diagram commute*

$$\begin{array}{ccc} \mathcal{O}_{Y,f(x)} & \longrightarrow & \mathcal{O}_{X,x} \\ \downarrow & & \downarrow \\ \Gamma_{f(x)} \cup \{0\} & \longrightarrow & \Gamma_x \cup \{0\}. \end{array}$$

An adic space is an object $(X, \mathcal{O}_X, (|\cdot(x)|)_{x \in X})$ *of (V) that admits a covering by spaces* $\{U_i\}$ *such that* $(U_i, \mathcal{O}_X|_{U_i}, (|\cdot(x)|))_{x \in X}$ *is isomorphic to* $\mathrm{Spa}(A_i, A_i^+)$ *for a sheafy Huber pair* (A_i, A_i^+).

For a sheafy Huber pair (A, A^+), *the topological space* $X = \mathrm{Spa}(A, A^+)$ *equipped with its structure sheaf and continuous valuations is an* affinoid adic space. *Often, we simply write the affinoid adic space as* $\mathrm{Spa}(A, A^+)$.

If we are given an adic space X *(encoding a triple in the sense of above), we denote its underlying topological space by* $|X|$.

Example 2.20. *Let* k *be any nonarchimedean field. Then any affinoid* R-*algebra, i.e., any quotient of the Tate algebra*

$$k\langle T_1, \ldots, T_n \rangle = \left\{ \sum_{i_1, \ldots, i_n = 0}^{\infty} a_{i_1, \ldots, i_n} T_1^{i_1} \cdots T_n^{i_n} : |a_{i_1, \ldots, i_n}| \to 0 \text{ as } i_1 + \cdots + i_n \to \infty \right\}$$

by a closed ideal, is Huber. In fact, all ideals are closed in this situation.

Theorem 2.21. *If* X *is an adic space and* $\mathrm{Spa}(S, S^+)$ *is an affinoid adic space, then there is a canonical bijection of sets*

$$\mathrm{Hom}_{adic\ spaces}(X, \mathrm{Spa}(S, S^+)) \cong \mathrm{Hom}_{Huber\ pairs}((\hat{S}, \hat{S}^+), (\mathcal{O}_X(X), \mathcal{O}_X^+(X))).$$

Definition 2.22 (Definition 1.1.13 in [30]). *A topological space* X *is called* spectral *if* X *is quasi-compact (i.e., every open cover of* X *has a finite subcover), and has a basis of quasi-compact open subsets stable under finite intersection, and if every irreducible closed subset is the closure of a unique point. (In particular, affinoid rigid analytic spaces and affinoid adic spaces are examples of spectral spaces.) A topological space is called* locally spectral *if every point of* X *has an open neighborhood that is spectral. (In particular, rigid analytic spaces and adic spaces are examples of locally spectral spaces.) A locally spectral space is spectral if and only if* X *is quasi-separated (i.e., the intersection of two quasi-compact open subsets of* X *is quasi-compact) and is quasi-compact.*

Let $f : X \to Y$ *be a continuous mapping between locally spectral spaces. Then*

1. f *is called* quasi-separated *if, for every quasi-separated open subset* $V \subset Y$, $f^{-1}(V)$ *is quasi-separated, and*
2. f *is called* quasi-compact *if, for every quasi-compact open subset* $V \subset Y$, $f^{-1}(V)$ *is quasi-compact.*

Moreover, there is an "adicification functor" from rigid spaces to adic spaces over a complete nonarchimedean field k.

Theorem 2.23. *There is a functor*

$$\{rigid\ spaces\ over\ k\} \rightarrow \{adic\ spaces\ over\ \mathrm{Spa}(k, k^0)\}$$

arising from gluing the functors

$$\mathrm{Max}(A) \mapsto \mathrm{Spa}(A, A^0).$$

This functor is fully faithful and induces an equivalence

$$\{quasi\text{-}separated\ rigid\ spaces\ over\ k\}$$
$$\cong \{quasi\text{-}separated\ adic\ spaces\ locally\ of\ finite\ type\ over\ \mathrm{Spa}(k, k^0)\}.$$

(For the definitions of quasi-separated and finite type, see [30, Chapter 0].)

We finally end this by recalling the notion of étale morphisms of adic spaces. Henceforth, assume that R is Tate.

Definition 2.24 ([30], Chapter 1). *A morphism of Huber pairs $f: (R, R^+) \rightarrow (S, S^+)$ is called a* quotient map *if $f: R \rightarrow S$ is surjective, continuous and open, and S^+ is the integral closure of $f(R^+)$ in S.*

We call f of topologically finite type *if it factorizes as*

$$(R, R^+) \rightarrow (R\langle T_1, \ldots, T_n\rangle, R^+\langle T_1, \ldots, T_n\rangle) \xrightarrow{\phi} (S, S^+)$$

where ϕ is a quotient map.

Definition 2.25 ([30], Chapter 1). *Suppose $f: X \rightarrow Y$ is a morphism of adic spaces.*

1. *We say f is* quasi-compact *if Y can be covered by open affinoid subspaces $\{V_i\}$ such that each $f^{-1}(V_i)$ is quasi-compact as a topological space; that is, every open cover of $f^{-1}(V_i)$ has a finite subcover.*
2. *We say f is* locally of weak finite type *if for all $x \in X$, there exist $x \in U \subset X$ and $V \subset X$ open such that $f(U) \subset V$ and such that the induced morphism of Huber rings*

$$\mathcal{O}_Y(V) \rightarrow \mathcal{O}_X(U)$$

is of topologically finite type.
3. *We say f is* locally of finite type *if for all $x \in X$, there exist open affinoid subspaces $x \in U \subset X$ and $V \subset Y$ with $f(U) \subset V$ such that the induced morphism*

of Huber pairs

$$(\mathcal{O}_Y(V), \mathcal{O}_Y^+(V)) \to (\mathcal{O}_X(U), \mathcal{O}_X^+(U))$$

is of topologically finite type.
4. f is of finite type *if it is quasi-compact and locally of finite type.*

Proposition 2.26 ([30], Chapter 1). *If $f : X \to Z$ and $g : Y \to Z$ are morphisms of adic spaces where f is locally of finite type, then one can define the fiber product of f and g.*

Definition 2.27. *Suppose $f : X \to Y$ is a morphism of adic spaces which is locally of finite type. Then*

1. f is separated *if $\Delta(X) \subset X \times_Y X$ is closed, where*

$$\Delta : X \to X \times_Y X$$

is the diagonal morphism.
2. f is universally closed *if it is locally of weakly finite type, and for all adic morphisms $Z \to Y$, the induced morphism*

$$X \times_Y Z \to Z$$

is closed.
3. f is proper *if it is of finite type, separated and universally closed.*

We have the following definitions of unramified, smooth and étale morphisms of adic spaces, which are analogous to the usual notions in algebraic geometry.

Definition 2.28 ([30], Chapter 1). *1. A morphism of Huber pairs $f : (R, R^+) \to (S, S^+)$ is* finite *if it is of topologically finite type, $R \to S$ is finite, and S^+ is the integral closure of $f(R^+)$.*
2. A morphism of adic spaces $f : X \to Y$ is finite *if for all $y \in Y$, there exists an affinoid open $y \in V \subset$ such that $U = f^{-1}(V)$ is affinoid and*

$$(\mathcal{O}_Y(U), \mathcal{O}_Y^+(U)) \to (\mathcal{O}_X(U), \mathcal{O}_X^+(U))$$

is finite.
3. f is locally quasi-finite *if for any $y \in Y$, the fiber $f^{-1}(y)$ is discrete.*
4. f is quasi-finite *if f is quasi-compact and locally quasi-finite.*

Definition 2.29 ([30], Chapter 1). *A morphism of adic spaces $f : X \to Y$ that is locally of finite type is called* unramified (*resp.* smooth, *resp.* étale) *if for all Huber pairs (R, R^+) and all ideals $I \subset R$ such that $I^2 = 0$ and maps $g :$*

$\mathrm{Spa}(R, R^+) \to Y$, *we have the following "infinitesimal lifting property": the natural map*

$$\mathrm{Hom}_Y(\mathrm{Spa}(R, R^+), X) \to \mathrm{Hom}_Y(\mathrm{Spa}(R/I, R^+/I), X)$$

is injective (resp. surjective, resp. bijective).

An arbitrary morphism $f : X \to Y$ is **unramified** (resp. **smooth**, resp. **étale**) *at $x \in X$ if there exist open subsets $x \in U \subset X$ and $V \subset Y$ such that $f(U) \subset V$ and $f|_U : U \to V$ are unramified (resp. smooth, resp. étale). Finally, f is said to be* **unramified** (resp. **smooth**, resp. **étale**) *if it is unramified (resp. smooth, resp. étale) at all $x \in X$.*

Suppose $f : X \to Y$ is locally of finite type. Let $\Omega^1_{X/Y}$ be the $\mathcal{O}_X$-module (which is a sheaf) of relative differentials as defined in [30, Chapter 1].

Proposition 2.30 ([30], Chapter 1). *Suppose $f : X \to Y$ is a morphism of adic spaces which is locally of finite type.*

1. *f is unramified if and only if $\Omega^1_{X/Y} = 0$.*
2. *If f is smooth, $\Omega^1_{X/Y}$ is a locally free $\mathcal{O}_x$-module.*
3. *If (R, R^+) is a Huber pair where R is Tate and letting $Y = \mathrm{Spa}(R, R^+)$, then $f : X \to Y$ is smooth if and only if for all $x \in X$, there exists $x \in U \subset X$ open such that*

where h is the natural morphism and g is étale.

We have the following local characterization of étale morphisms.

Proposition 2.31 ([30], Proposition 1.6.10). *Suppose $f : X \to Y = \mathrm{Spa}(R, R^+)$ is a morphism of affinoid adic spaces where R is Tate. Then the following are equivalent:*

1. *f is étale.*
2. *There exist $n \in \mathbb{Z}_{\geq 0}$ and $f_1, \ldots, f_n \in R\langle T_1, \ldots, T_n \rangle$ such that if $I = (f_1, \ldots, f_n)$, then*

$$\det\left(\frac{\partial f_i}{\partial T_j}\right)_{1 \leq i,j \leq n}$$

is invertible in $R\langle T_1, \ldots, T_n \rangle / I$, and X is isomorphic as an adic space over Y to $\mathrm{Spa}(R\langle T_1, \ldots, T_n \rangle / I, R^+\langle T_1, \ldots, T_n \rangle / I)$.

3. *There exist $n \in \mathbb{Z}_{\geq 0}$ and $f_1, \ldots, f_n \in R[T_1, \ldots, T_n]$ such that if $I = (f_1, \ldots, f_n)$,*
   then

$$\det \left(\frac{\partial f_i}{\partial T_j} \right)_{1 \leq i,j \leq n}$$

is invertible in $R\langle T_1, \ldots, T_n \rangle / I$, and X is isomorphic as an adic space over Y to $\mathrm{Spa}(R\langle T_1, \ldots, T_n \rangle / I, R^+ \langle T_1, \ldots, T_n \rangle / I)$.

2.4   (PRE-)ADIC SPACES AND (PRE)PERFECTOID SPACES

In this section, we review the generalized notion of adic spaces introduced by Scholze-Weinstein in [62, Section 2], in which the (analytic) structure presheaf is no longer required to be a sheaf, as they are in Huber's sense of adic spaces. Our discussion follows Section 2 of loc. cit. very closely. In certain situations, projective limits of Huber adic spaces are not Huber adic spaces but rather give rise to so-called pre-adic spaces. Examples of such include objects that we will consider later, such as the infinite-level modular curve $\mathcal{Y} \in Y_{\text{proét}}$, the infinite-level Lubin-Tate tower $LT_\infty \in \mathbb{P}^1_{\text{proét}}$ and the infinite-level Drinfeld tower $\Sigma_\infty \in \mathbb{P}^1_{\text{proét}}$. In fact, these are *preperfectoid spaces*, and hence after p-adic completion give rise to perfectoid spaces which *are* Huber adic spaces.

Definition 2.32 (Definition 2.1.5 of [62]). *Let Aff denote the category of Huber pairs, and let CAff be the category of complete Huber pairs (A, A^+).[3] Following loc. cit., we give $\mathrm{CAff}^{\mathrm{opp}}$ the structure of a site by defining the following Grothendieck pretopology on $\mathrm{CAff}^{\mathrm{opp}}$: We define a cover of $(A, A^+) \in \mathrm{CAff}$ to be a family of morphisms $\{(A, A^+) \to (A_i, A_i^+)\}$ such that $(A_i, A_i^+) = (\mathcal{O}_Y(U_i), \mathcal{O}_Y^+(U_i))$ for a covering $\{U_i \to Y\}$ by rational subsets $U_i \subset Y$ of the affinoid adic space $Y = \mathrm{Spa}(A, A^+)$. Note that this site has fiber products, by [30, Proposition 1.2.2 b]. The other axioms are easily verified.*

Let $\widetilde{\mathrm{CAff}^{\mathrm{opp}}}$ be the associated topos. A (not necessarily complete) Huber pair $(A, A+)$ gives rise to a presheaf

$$(B, B^+) \mapsto \mathrm{Hom}_{\mathrm{Aff}}((A, A^+), (B, B^+))$$

and we let $\mathrm{Spa}(A, A^+)$ denote the associated sheaf. Letting $(\hat{A}, \hat{A}^+)$ denote the completion of (A, A^+), note that $\mathrm{Spa}(A, A^+) = \mathrm{Spa}(\hat{A}, \hat{A}^+)$. We call $\mathrm{Spa}(A, A^+)$ an affinoid pre-adic space.[4]

We say $\mathcal{F} \to \mathrm{Spa}(A, A^+)$ is an open immersion if there is an open subset $U \subset Y = \mathrm{Spa}(A, A^+)$ such that

---

[3] *This notation is based on older terminology.*

[4] *Here we follow the terminology of [61]; in loc. cit. Huber affinoid adic spaces are simply called affinoid adic spaces.*

$$\mathcal{F} = \varinjlim_{V \subset U, V \ rational \ subset} \mathrm{Spa}(\mathcal{O}_Y(V), \mathcal{O}_Y^+(V)).$$

Given a morphism $f : \mathcal{F} \to \mathcal{G}$ in $\widetilde{\mathrm{CAff}}^{\mathrm{opp}}$, we say that f is an open immersion if for all (A, A^+) and all morphisms $\mathrm{Spa}(A, A^+) \to \mathcal{G}$, the fiber product

$$\mathcal{F} \times_{f, \mathcal{G}} \mathrm{Spa}(A, A^+) \to \mathrm{Spa}(A, A^+)$$

(where the fiber product here is defined just as the abstract pushout of sheaves) is an open immersion. In particular, open immersions are injective, and often one says simply that $\mathcal{F} \subset \mathcal{G}$ is open.

We call $\mathcal{F} \in \widetilde{\mathrm{CAff}}^{\mathrm{opp}}$ a pre-adic space if

$$\mathcal{F} = \varinjlim_{\mathrm{Spa} \subset \mathcal{F} \ open} \mathrm{Spa}(A, A^+)$$

and we denote the category of pre-adic spaces by Adic. *In particular, the category of adic spaces (in Huber's sense) is a full subcategory of* Adic. *When we want to emphasize that a pre-adic space is in fact an adic space in Huber's sense, we will write "Huber adic space."*

We have the following result relating pre-adic spaces to usual adic spaces. For just the following proposition, following loc. cit., we use the following notation to delineate affinoid spaces in Huber's sense from the generalized sense above: If (A, A^+) is a Huber pair, we let

$$\mathrm{Spa}_{\mathrm{naive}}(A, A^+) := \{X, \mathcal{O}_X^\#, \{v_x : x \in X\}\},$$

where X is the previously defined topological space

$$X = \{\text{set of equivalence classes of continuous valuations}$$
$$\text{on } A \text{ satisfying } |A^+| \leq 1\},$$

with opens generated by rational subsets, $\mathcal{O}_X^\#$ is the sheafification of the presheaf $\mathcal{O}_X$ (so $\mathcal{O}_X^\# = \mathcal{O}_X$ if (A, A^+) is sheafy), and v_x is the continuous valuation corresponding to $x \in X$. Let Adic^h denote the category of Huber adic spaces for the next proposition (called "honest adic spaces" in loc. cit.).

Proposition 2.33 (Proposition 2.1.6 of [62])**.** *1. Let (A, A^+) be a Huber pair, and assume that (A, A^+) is complete and sheafy (i.e., its structure presheaf is a sheaf). Then*

$$\mathrm{Hom}_{\mathrm{Adic}}(\mathrm{Spa}(A, A^+), \mathrm{Spa}(B, B^+)) = \mathrm{Hom}_{\mathrm{Aff}}((B, B^+), (A, A^+)).$$

In particular, the functor $(A, A^+) \mapsto \text{Spa}(A, A^+)$ *is fully faithful on the full subcategory of sheafy complete Huber pairs.*

2. *The functor* $(A, A^+) \mapsto \text{Spa}_{\text{naive}}(A, A^+)$ *factors through* $(A, A^+) \mapsto \text{Spa}$ (A, A^+).

3. *The functor*

$$\text{Spa}(A, A^+) \mapsto \text{Spa}_{\text{naive}}(A, A^+)$$

from part (2) extends to a functor

$$X \mapsto X_{\text{naive}} : \text{Adic} \rightarrow (V),$$

where the target is the category (V) of Definition 2.19. In particular, any adic space X has an associated topological space X, given by the first component of X_{naive}.

4. *The full subcategory of adic spaces that are covered by* $\text{Spa}(A, A^+)$ *with* (A, A^+) *sheafy is equivalent to the category of* Adic^h *under* $X \mapsto X_{\text{naive}}$. *Under this identification, Adic^h is a full subcategory of Adic.*

5. *Suppose X is a (Huber) adic space, and let (B, B^+) be a Huber pair. Then*

$$\text{Hom}_{\text{Adic}}(X, \text{Spa}(B, B^+)) = \text{Hom}((B, B^+), \mathcal{O}_X(X), \mathcal{O}_X^+(X)),$$

where the right-hand side is the set of continuous ring homomorphisms $B \rightarrow \mathcal{O}_X(X)$ which map $B^+ \rightarrow \mathcal{O}_X^+(X)$.

Henceforth, we will use the notation of $\text{Spa}(A, A^+)$ for an affinoid pre-adic space. When (A, A^+) gives rise to an honest affinoid adic space, the above proposition shows there is no information lost in viewing it as a pre-adic space.

We now define the notion of perfectoid and preperfectoid spaces, the former notion belonging to the category of usual adic spaces (where the structure presheaf is a sheaf) and the latter belonging to the category of pre-adic spaces Aff.

Definition 2.34 (Definition 2.3.1-2 in [62][5]). *We say K is a perfectoid field if it is a complete nonarchimedean field with nondiscrete (rank 1) valuation, and such that the Frobenius morphism $\mathcal{O}_K/p \rightarrow \mathcal{O}_K/p$ is surjective. A perfectoid K algebra is a complete Banach K-algebra A such that:*

1. *the subalgebra A^0 of power-bounded elements is bounded in A, and*
2. *the Frobenius map $A^0/p \rightarrow A^0/p$ is surjective.*

---

[5]We note that [62] uses a more modern definition of perfectoid space in which the perfectoid space is no longer required to be defined over a perfectoid field. However, since we will use results from [62] and [57], which seem to require perfectoid spaces to be defined over perfectoid fields, we will adopt the more traditional definition in this text.

A perfectoid affinoid $(K, \mathcal{O}_K)$-algebra is an affinoid $(K, \mathcal{O}_K)$-algebra (A, A^+) such that A is a perfectoid K-algebra.

Theorem 2.35 (Theorem 6.3 in [59]). *If (A, A^+) is a perfectoid affinoid $(K, \mathcal{O}_K)$-algebra, then the structure presheaf on $\mathrm{Spa}(A, A^+)$ is a sheaf. Mroeover, if $U \subset Y = \mathrm{Spa}(A, A^+)$ is a rational subset, then the pair $(\mathcal{O}_Y(U), \mathcal{O}_Y^+(U))$ is also a perfectoid affinoid $(K, \mathcal{O}_K)$-algebra.*

Definition 2.36 (Definition 2.3.3 of [62]). *A perfectoid space over $\mathrm{Spa}(K, \mathcal{O}_K)$, where K is a perfectoid field, is a (usual) adic space that is locally isomorphic to $\mathrm{Spa}(A, A^+)$ for a perfectoid affinoid $(K, \mathcal{O}_K)$-algebra (A, A^+).*[6]

Example 2.37. *Let K be a perfectoid field, and let ϖ denote the maximal ideal of $\mathcal{O}_K$. Let*

$$\mathbb{T} = \mathrm{Spa}(K\langle T^{\pm 1}\rangle, \mathcal{O}_K\langle T^{\pm 1}\rangle),$$

where

$$\mathcal{O}_K\langle T^{\pm 1}\rangle = \varprojlim_n \mathcal{O}_K[T^{\pm 1}]/(\varpi^n), \qquad K\langle T^{\pm 1}\rangle = \mathcal{O}_K\langle T^{\pm 1}\rangle[1/p]$$

are the usual Tate algebras. Then

$$\tilde{\mathbb{T}} := \mathrm{Spa}(A, A^+) = \mathrm{Spa}(K\langle T^{\pm 1/p^\infty}\rangle, \mathcal{O}_K\langle T^{\pm 1/p^\infty}\rangle)$$

is a perfectoid space, where $A^+ := \mathcal{O}_K\langle T^{\pm 1/p^\infty}\rangle$ is the ϖ-adic completion of $\varinjlim_i \mathcal{O}_K\langle T^{\pm 1/p^i}\rangle$, and $A = A^+[1/p]$. Here, A^+ is given the structure of a $\mathcal{O}_K$-Banach space by the Gauss norm

$$\left\| \sum_{n \in \mathbb{Z}} a_n T^n \right\| = \sup_{n \in \mathbb{Z}} |a_n|,$$

where $|\cdot|$ denote the valuation (or rather absolute value) on K. Then A is given the unique K-Banach structure which makes A^+ the subring of elements of norm ≤ 1 (i.e., letting the norm $\|p\| = |p|$).

This is not the most basic (nontrivial) example of a perfectoid space, but it is one which we will make great use of later in the text. A more basic example of

---

[6] *We adopt this notion of perfectoid space to follow loc. cit., [56] and [60], which are frequently cited throughout our discussion. Moreover, the proofs of key results in these latter references use the notion of perfectoid space defined over a perfectoid field, though we presume that they do not fundamentally use this definition over a perfectoid field and can be generalized. For a more modern definition of perfectoid space, which is not required to be defined over a perfectoid field, see [61]. See also [37] for a comprehensive discussion of different notions of perfectoid and preperfectoid spaces.*

a perfectoid space is given by $\mathrm{Spa}(K\langle T^{1/p^{\infty}}\rangle, \mathcal{O}_K\langle T^{1/p^{\infty}}\rangle)$, *with the analogous notations and definitions as in the previous example.*

Definition 2.38. *Suppose Y is a pre-adic space over* $\mathrm{Spa}(K, \mathcal{O}_K)$*, with K a perfectoid field. Then Y is called* preperfectoid *if there is a cover by open affinoid pre-adic $U_i = \mathrm{Spa}(A_i, A_i^+)$ such that $(\hat{A}_i, \hat{A}_i^+)$ is a perfectoid affinoid $(K, \mathcal{O}_K)$-algebra, where $\hat{A}_i^+$ denotes the p-adic completion of A_i^+, and $\hat{A}_i = \hat{A}_i^+[1/p]$. We call $(\hat{A}_i, \hat{A}_i^+)$ the* strong completion *of (A_i, A_i^+).*

Example 2.39 (Remark 2.3.5 of [62]). *Let K be a perfectoid field. An example of a preperfectoid space is given by the nonreduced rigid analytic point $Y = \mathrm{Spa}(K[X], X^2, \mathcal{O}_K + KX)$. Note that $(K[X], X^2, \mathcal{O}_K + KX)$ is already complete with respect to the natural X-adic topology. Taking the p-adic completion results in $(K, \mathcal{O}_K)$, which is trivially a $(K, \mathcal{O}_K)$-algebra.*

Proposition 2.40 (Proposition 2.3.6 of [62]). *Let Y be a preperfectoid space over* $\mathrm{Spa}(K, \mathcal{O}_K)$*, where K is a perfectoid field. Then there exists a perfectoid space $\hat{Y}$ over* $\mathrm{Spa}(K, \mathcal{O}_K)$ *with the same underlying topological space, and such that for any open subset $U = \mathrm{Spa}(A, A^+) \subset Y$ with $(\hat{A}, \hat{A}^+)$ a perfectoid $(K, \mathcal{O}_K)$-algebra, the corresponding open subset of $\hat{Y}$ is given by $\hat{U} = \mathrm{Spa}(\hat{A}, \hat{A}^+)$. Moreover, $\hat{Y}$ is universal for morphisms from perfectoid spaces over* $\mathrm{Spa}(K, \mathcal{O}_K)$ *to Y. We call $\hat{Y}$ the strong completion of Y.*

Proposition 2.41 (Proposition 2.3.7 of [62]). *Suppose Y is a preperfectoid space over* $\mathrm{Spa}(K, \mathcal{O}_K)$*, where K is a perfectoid field, and suppose $U \subset Y$ is a locally closed subspace. Then U is preperfectoid.*

2.5 SOME COMPLEMENTS ON INVERSE LIMITS OF (PRE-)ADIC SPACES

Certain subtleties arise upon taking inverse limits of (pre-)adic spaces. For example, a direct limit of Huber pairs need not be Huber. In this section, we recall some results from [62] on the nice behavior of certain inverse limits of pre-adic spaces.

Definition 2.42 (Definition 2.4.1 of [62]). *Suppose Y_i is a filtered inverse system of pre-adic spaces with quasi-compact and quasi-separated transition maps, let Y be a pre-adic space, and let $f_i : Y \to Y_i$ be a compatible family of morphisms. We write*

$$Y \sim \varprojlim_i Y_i$$

if the map of underlying topological spaces $|Y| \to \varprojlim_i |Y_i|$ is a homeomorphism (see Definition 2.19 for the general notion of underlying topological space), and *if there is an open cover of Y by generalized affinoid $(A, A^+) \subset Y$ such that the map*

$$\varinjlim_{\mathrm{Spa}(A_i, A_i^+) \subset Y} A_i \to A$$

has dense image, where the direct limit above is taken over all open affinoid pre-adic spaces

$$\mathrm{Spa}(A_i, A_i^+) \subset Y_i$$

over which $\mathrm{Spa}(A, A^+) \subset Y \to Y_i$ factors.

Proposition 2.43 (Proposition 2.4.2 of [62]). *Suppose (A_i, A_i^+) is a direct system of complete Huber pairs, and assume that there are rings of definitions $A_{i,0} \subset A_i$ compatible for all i, and finitely generated ideals of definition $I_i \subset A_{i,0}$ such that $I_j = I_i A_{j,0} \subset A_{j,0}$ for all $j \geq i$. Let*

$$(A, A^+) = \varinjlim_i (A_i, A_i^+),$$

equipped with the topology making $A_0 = \varinjlim_i A_{i0}$ and open adic subring with ideal of definition $I = \varinjlim_i I_i$. Then

$$\mathrm{Spa}(A, A^+) \sim \varprojlim_i \mathrm{Spa}(A_i, A_i^+).$$

Proposition 2.44 (Proposition 2.4.3 of [62]). *In the situation of Definition 2.42, suppose $Y_i \to X_i$ is an open immersion of pre-adic spaces, and let*

$$Y_j = Y_i \times_{X_i} X_j \to X_j$$

for all $j \geq i$ be the pullback, as well as

$$Y = Y_i \times_{X_i} X \to X.$$

Then

$$Y \sim \varprojlim_{j \geq i} Y_j.$$

Proposition 2.45 (Proposition 2.4.4 of [62]). *Let K be a perfectoid field and Y a preperfectoid space over $\mathrm{Spa}(K, \mathcal{O}_K)$. Then*

$$\hat{Y} \sim Y.$$

We will use the following results ubiquitously and implicitly throughout our discussion.

Proposition 2.46 (Proposition 2.4.5 of [62]). *Suppose K is a perfectoid field, and X_i is an inverse system of pre-adic spaces over $\mathrm{Spa}(K, \mathcal{O}_K)$ with quasi-compact, quasi-separated (qcqs) transition maps, and assume that there is a perfectoid space X over $\mathrm{Spa}(K, \mathcal{O}_K)$ such that $X \sim \varprojlim_i X_i$. Then for any perfectoid affinoid $(K, \mathcal{O}_K)$-algebra (B, B^+), we have*

$$X(B, B^+) = \varprojlim_i X_i(B, B^+).$$

In particular, if Y is a perfectoid space over $\mathrm{Spa}(K, \mathcal{O}_K)$ with a compatible system of maps $Y \to X_i$, then Y factors over X uniquely, making X unique up to unique isomorphism.

Theorem 2.47 (Theorem 2.4.7 of [62], cf. Theorem 7.17 of [59]). *Let K be a perfectoid field, and assume that all X_i are strongly noetherian adic spaces (and so, in particular, quasi-compact and quasi-separated) over $\mathrm{Spa}(K, \mathcal{O}_K)$, and that X is a perfectoid space over $\mathrm{Spa}(K, \mathcal{O}_K)$ with $X \sim \varprojlim_i X_i$. Then the étale topos of X is the projective limit of the étale topos of the X_i.*

2.6 THE PROÉTALE SITE OF AN ADIC SPACE

Throughout this section, we will follow [57, Section 3]. We let Y denote a locally (strongly) noetherian adic space, which means it is locally of the form $\mathrm{Spa}(R, R^+)$ where R is *strongly noetherian* (see Definition 2.17).

Definition 2.48. *Let K denote a nonarchimedean field, and let Y denote a locally (strongly) noetherian adic space over $\mathrm{Spa}(K, \mathcal{O}_K)$. Let $Y_{\text{ét}}$ denote the small étale site of the adic space Y over $\mathrm{Spa}(K, \mathcal{O}_K)$. Hence, objects ("opens") of $Y_{\text{ét}}$ are étale maps $U \to Y$, coverings are étale coverings, which for an open $V \to Y$ consist of $\{f_i : U_i \to U\}_{i \in I}$ for a set I, such that the f_i are étale (and so the images $|U_i|$ are open in U in the adic topology), and we have $\bigcup_{i \in I} f(|U_i|) = U$. It is well-known that $Y_{\text{ét}}$ satisfies all the properties of a Grothendieck site.*

Consequently, $Y_{\text{ét}}$ has a notion of points (also corresponding to equivalence classes of valuations) and an associated topos and consequently a structure sheaf $\mathcal{O}$, along with an integral structure sheaf given by

$$\mathcal{O}^+(U) = \{f \in \mathcal{O}(U) : x(f) = |f(x)| \le 1 \quad \forall x \in U\}.$$

Definition 2.49. *The category pro-$Y_{\text{ét}}$ consists of functors $F : I \to Y_{\text{ét}}$ from small cofiltered index categories I in which morphisms are given by*

$$\mathrm{Hom}_{\mathrm{pro}-Y_{\text{ét}}}(F, G) := \varprojlim_J \varinjlim_I \mathrm{Hom}_{Y_{\text{ét}}}(F(i), G(j)).$$

A morphism $U \to V$ in pro-$Y_{ét}$ is called étale (resp. finite étale) if there is an étale (resp. a finite étale) morphism $U_0 \to V_0$ in $Y_{ét}$ and $V \to V_0$ in pro-$Y_{ét}$ such that $U \to Y$ is equal to the projection

$$U \times_{V_0} V \to V.$$

Definition 2.50. Let $Y_{fét}$ denote the site with objects $U \to Y$ finite étale, and with coverings $\{f_i : U_i \to U\}_{i \in I}$ such that the f_i are finite étale and $\bigcup_{i \in I} f(|U_i|) = |U|$. (It is again standard to check that $Y_{fét}$ is a site.) For any $U = \varprojlim_i U \in$ pro $- Y_{fét}$, we have on topological spaces $|U| = \varprojlim_i |U_i|$. We define a site $Y_{pro\text{-}fét}$ with underlying category pro-$Y_{fét}$, and with coverings $\{U_i \to U\}_{i \in I}$ where the f_i are open and $|U| = \bigcup_{i \in I} f_i(|U_i|)$.

Suppose now that Y is connected. As is in the case for $Y_{ét}$, there is an equivalent way to describe $Y_{pro\text{-}fét}$ using the profinite étale fundamental group.

Definition 2.51 (Definition 3.9 of [57]). *We now define the* proétale site $Y_{proét}$ *as defined in [57, Section 3], whose objects form a full subcategory*

$$Y_{proét} \subset \text{pro} - Y_{ét}.$$

Explicitly, objects of $Y_{proét}$ are proétale maps $U \to Y$, which means that U can be written as $\varprojlim_i U_i$, where U_i are objects of pro-$Y_{ét}$, such that each $U_i \to Y$ is an object in pro-$Y_{ét}$, and $U_i \to U_j$ is a finite étale map in pro-$Y_{ét}$ in the sense of loc. cit. for all large $i > j$. Coverings are given by proétale coverings, i.e., collections of proétale morphisms $\{U_i \to U\}_i$ such that $\bigcup_i |U_i| = |U|$ such that each $U_i \to U$ satisfies a condition, which we now describe. We require that each $U_i \to U$ can be written as an inverse limit $U_i = \lim_{\mu < \lambda} U_\mu \to U$ of $U_\mu \in \text{pro} - Y_{ét}$, where the terminal object in the inverse system $U_0 = U$, and where the limit is along the set of ordinals μ less than some ordinal λ, with the following property: for all $\mu < \lambda$, $U_\mu \to U_{<\mu} := \lim_{\mu' < \mu} U_{\mu'}$ is the pullback of an étale, surjective map in $Y_{ét}$. That is, there exist $U_{\mu,0}, U_{\mu' < \mu,0} \in Y_{ét}$ and a morphism $U_{<\mu} \to U_{<\mu,0}$ in pro-$Y_{ét}$ such that $U_\mu = U_{\mu,0} \times_{U_{<\mu,0}} U_{<\mu}$ and $U_\mu \to U_{<\mu}$ is given by the natural projection. We note that if the projective limit $U_i - \lim_{\mu < \lambda} U_\mu \to U$ is countable (i.e., $\lambda = \aleph_0$), then this imposes no additional condition on the proétale morphism $U_i \to U$.

Given a $U \in Y_{proét}$, a presentation

$$U = \varprojlim_i U_i$$

as above is called a proétale presentation. *Recall that each of the U_i is an object of pro-$Y_{ét}$, and is not necessarily an object of $Y_{ét}$.*

It is shown in [57, Lemma 3.10] that $Y_{proét}$ is a site. Hence, $Y_{proét}$ has a notion of points (also corresponding to equivalence classes of valuations) and an associated topos.

Remark 2.52. In defining covers on the proétale site, we follow the modification of the original definition in [57, Definition 3.9] given in [58].

There is a natural projection of sites

$$\nu : Y_{\text{proét}} \to Y_{\text{ét}}, \quad \nu(U) = \varprojlim U$$

where the maps $U \to U$ in the projective system above are just the identity. (Recall that a morphism of sites induces a functor on underlying category of opens going in the opposite direction.) Let $\tilde{Y}_{\text{ét}}$ and $\tilde{Y}_{\text{proét}}$ denote the topoi attached to $Y_{\text{ét}}$ and $Y_{\text{proét}}$. Then there is a natural pullback functor on the associated topoi

$$\nu^* : \tilde{Y}_{\text{ét}} \to \tilde{Y}_{\text{proét}},$$

$\nu^* \mathcal{F} =$ sheafification of the presheaf given on qcqs objects by

$$\varprojlim_i U_i \mapsto \varinjlim_i \varinjlim_{j \in J_i} \mathcal{F}(\Gamma_i(j)) \tag{2.3}$$

where $U_i = (F_i : J_i \to Y_{\text{ét}}) \in \text{pro} - Y_{\text{ét}}$, with J_i a small cofiltered index category, is quasi-compact and quasi-separated (qcqs). Any object of pro-$Y_{\text{ét}}$ has a covering by qcqs objects.

The structure sheaf $\mathcal{O}_Y$ on $Y_{\text{proét}}$ is defined as the pullback $\nu^* \mathcal{O}_{Y_{\text{ét}}}$ of the étale structure sheaf $\mathcal{O}_{Y_{\text{ét}}}$. It has an integral structure sheaf given by

$$\mathcal{O}_Y^+(U) = \{f \in \mathcal{O}(U) : x(f) = |f(x)| \leq 1 \ \forall x \in U\}.$$

Proposition 2.53 (cf. Lemma 4.2 of [57]). *We have the following characterization of the proétale stalk of the integral proétale structure sheaf:*

$$\mathcal{O}_{Y,y}^+ = \{f \in \mathcal{O}_{Y,y} : |f(y)| \leq 1\}. \tag{2.4}$$

Proof. The inclusion $\subset$ is obvious. For $\supset$, let $f \in \mathcal{O}_{Y,y}$ come from a proétale neighborhood $U \to Y$ of $y \in Y$. Then since affinoid perfectoids form a basis of the proétale topology, without loss of generality we can "shrink" U to assume it is affinoid perfectoid. Then the subset

$$U' = \{y \in U : |f(y)| = 1\}$$

is a rational subset of U, and hence itself affinoid perfectoid by [59, Theorem 6.3(ii)], and so $f|_{U'} \in \mathcal{O}_Y^+(U)$ by (2.4), and is a representative of the stalk element f, i.e., $f \in \mathcal{O}_{Y,y}^+$. $\square$

Given any extension $K/\mathbb{Q}_p$, we can form the base change

$$Y_K = Y \times_{\mathrm{Spa}(\mathbb{Q}_p, \mathbb{Z}_p)} \mathrm{Spa}(K, \mathcal{O}_K).$$

Henceforth, given any separable extension $K/\mathbb{Q}_p$, we can define an object also denoted by $Y_K \in Y_{\mathrm{proét}}$ via

$$Y_K = \varprojlim_{\mathbb{Q}_p \subset L \subset K,\, L/\mathbb{Q}_p \text{ finite}} Y_{L_i}.$$

Given any object $U \in Y_{\mathrm{proét}}$, let $Y_{\mathrm{proét}}/U$ denote the localized site, consisting of objects $V \in Y_{\mathrm{proét}}$ with a map $V \to U$. As we remarked before, if Y_K denotes the base change of Y to $\mathrm{Spa}(K, \mathcal{O}_K)$, then there is a natural equivalence of sites

$$Y_{K, \mathrm{proét}} \cong Y_{\mathrm{proét}}/Y_K.$$

Definition 2.54. *An* affinoid perfectoid *is an object $U \in Y_{proét}/Y_K$ with a proétale presentation $\varprojlim_i U_i$ where $U_i = \mathrm{Spa}(R_i, R_i^+)$ is an affinoid such that if R^+ denotes the p-adic completion of $\varinjlim_i R_i^+$ and $R = R^+[1/p]$, then (R, R^+) is a perfectoid $(K, \mathcal{O}_K)$-algebra. A* perfectoid *object in $Y_{proét}$ is one with an open covering by affinoid perfectoids.*

We recall the key fact that, after base change to a perfectoid field K, affinoid perfectoids form a basis of $Y_{K, \mathrm{proét}}$ ([57]).

Proposition 2.55 ([57], Corollary 4.7). *Let K be a perfectoid field. Then the set of $U \in Y_{K,proét}$ which are affinoid perfectoid form a basis for the proétale topology.*

Convention 2.56. Given any sheaf $\mathcal{F} \in \tilde{Y}_{\mathrm{proét}}$ and any object $U \in Y_{\mathrm{proét}}$, we will often let

$$\mathcal{F}_U := \mathcal{F}|_U$$

denote the induced sheaf on $Y_{\mathrm{proét}}/U$. If the sheaf is written as $\mathcal{F}_Y$, we will often write

$$\mathcal{F}_U := \mathcal{F}_Y|_U$$

for brevity.

Definition 2.57. *Using rational subsets as in Section 2.3, one can define the structure sheaf $\mathcal{O}_Y$ on $Y_{proét}$ (which is also equal to $\nu^* \mathcal{O}$, where $\mathcal{O}$ is the structure sheaf on $Y_{ét}$). Similarly, one can define the integral structure sheaf $\mathcal{O}_Y^+$. Let*

$$\hat{\mathcal{O}}_Y^+ := \varprojlim_n \mathcal{O}_Y^+/p^n$$

denote the integral p-adically completed structure sheaf, and let

$$\hat{\mathcal{O}}_Y := \hat{\mathcal{O}}_Y^+[1/p]$$

denote the p-adically completed structure sheaf.

2.7   PERIOD SHEAVES

We will recall construction of the period sheaves and $\mathcal{O}\mathbb{B}_{\mathrm{dR},Y}^+$ and $\mathcal{O}\mathbb{B}_{\mathrm{dR},Y}$ on $Y_{\mathrm{proét}}$, defined as in [57, Section 6] (with corrected definition in [58]), as well as some of their key properties. Let K denote any perfectoid field containing $\mathbb{Q}_p$, let K^+ denote its ring of integral elements, let

$$K^{\flat,+} = \varprojlim_{x \mapsto x^p} K^+/p = \varprojlim_{x \mapsto x^p} K^{+1}, \qquad K^\flat = K^{\flat,+}[1/p]$$

denote the corresponding tilts, and let Y_K denote the base change

$$Y \times_{\mathrm{Spa}(\mathbb{Q}_p,\mathbb{Z}_p)} \mathrm{Spa}(K,\mathcal{O}_K).$$

Since $Y_{K,\mathrm{proét}} \cong Y_{\mathrm{proét}}/Y_K$, and affinoid perfectoids form a basis of $Y_{K,\mathrm{proét}}$, then it suffices to define the sections of the above period sheaves on affinoid perfectoids in $Y_{K,\mathrm{proét}}$ for any perfectoid field K containing $\mathbb{Q}_p$. Recall the tilted structure sheaf

$$\hat{\mathcal{O}}_{Y_K^\flat}^+ := \varprojlim_{x \mapsto x^p} \mathcal{O}_{Y_K}^+/p = \varprojlim_{x \mapsto x^p} \hat{\mathcal{O}}_{Y_K}^+$$

and the period sheaves

$$\mathbb{A}_{\mathrm{inf},Y_K} := W(\hat{\mathcal{O}}_{Y_K^\flat}^+) \qquad \mathbb{B}_{\mathrm{inf},Y_K} := \mathbb{A}_{\mathrm{inf}}[p^{-1}] \qquad \mathcal{O}\mathbb{B}_{\mathrm{inf},Y_K} := \mathcal{O}_{Y_K} \otimes_{W(\kappa)} \mathbb{B}_{\mathrm{inf},Y_K}$$

which are initially defined naturally on affinoid perfectoids (see [57, Section 6]). We have natural maps

$$\theta : \mathbb{A}_{\mathrm{inf},Y_K} \twoheadrightarrow \hat{\mathcal{O}}_{Y_K}^+, \qquad \theta : \mathbb{B}_{\mathrm{inf},Y_K} \twoheadrightarrow \hat{\mathcal{O}}_{Y_K} \qquad\qquad (2.5)$$

given by

$$\theta \left(\sum_{k \gg -\infty} p^k [(x_k, x_k^{1/p}, \ldots)] \right) = \sum_{k \gg -\infty} p^k x_k$$

where

$$(x_k, x_k^{1/p}, \ldots) \qquad \forall k \gg 0$$

are sections of $\hat{\mathcal{O}}^+_{Y^\flat_K}$, and

$$[\cdot] : \hat{\mathcal{O}}^+_{Y^\flat_K} \to W(\hat{\mathcal{O}}^+_{Y^\flat_K})$$

is the usual Teichmüller lift.

There is a natural connection

$$\nabla : \mathcal{O}\mathbb{B}_{\mathrm{inf},Y_K} \to \mathcal{O}\mathbb{B}_{\mathrm{inf},Y_K} \otimes_{W(\kappa)} \Omega^1_{Y_K} \tag{2.6}$$

given by $\otimes_{W(\kappa)}\mathbb{B}_{\mathrm{inf}}$-linearly extending $d : \mathcal{O}_{Y_K} \to \Omega^1_{Y_K}$.

We let

$$\mathbb{B}^+_{\mathrm{dR},Y_K} = \varprojlim_i \mathbb{B}_{\mathrm{inf},Y_K}/(\ker\theta)^i, \qquad \mathbb{B}_{\mathrm{dR},Y_K} := \mathbb{B}^+_{\mathrm{dR},Y_K}[1/t]$$

where t is any generator of $(\ker\theta)\mathbb{B}^+_{\mathrm{dR},Y_K}$. The map

$$\theta : \mathbb{B}_{\mathrm{inf},Y_K} \twoheadrightarrow \hat{\mathcal{O}}_{Y_K}$$

extends $\mathcal{O}_{Y_K}$-linearly to a map

$$\theta : \mathcal{O}\mathbb{B}_{\mathrm{inf},Y_K} \twoheadrightarrow \hat{\mathcal{O}}_{Y_K}.$$

We have a natural filtration on $\mathbb{B}^+_{\mathrm{dR},Y_K}$ given by

$$\mathrm{Fil}^i \mathbb{B}^+_{\mathrm{dR},Y_K} = (\ker\theta)^i \mathbb{B}^+_{\mathrm{dR},Y_K}$$

as well as a filtration on $\mathbb{B}_{\mathrm{dR},Y_K}$ given by

$$\mathrm{Fil}^i \mathbb{B}_{\mathrm{dR},Y_K} = \sum_{j \in \mathbb{Z}} t^{-j} \mathrm{Fil}^{i+k} \mathbb{B}^+_{\mathrm{dR},Y_K}$$

where t is any element that generates $\mathrm{Fil}^1 \mathbb{B}^+_{\mathrm{dR},Y_K}$. (It is a consequence of the following proposition that such t exists *locally* on $Y_{K,\mathrm{pro\acute{e}t}}$ and is not a zero divisor, and so gives a well-defined filtration.)

It can be shown that $\ker(\theta : \mathbb{A}_{\mathrm{inf}} \to \hat{\mathcal{O}}^+)$ is locally generated by some section ξ, which we do in the following lemma.

Proposition 2.58 ([57], Lemma 6.3). *For any perfectoid affinoid (K, K^+)-algebra (R, R^+) (where K is of characteristic 0), there is an element $\xi \in \mathbb{A}_{\mathrm{inf}}(R, R^+)$ generating $\ker(\theta : \mathbb{A}_{\mathrm{inf}}(R, R^+) \to R^+)$ which is not a zero divisor in $\mathbb{A}_{\mathrm{inf}}(R, R^+)$.*

Proof. We give the proof of [57, Lemma 6.3] nearly verbatim. Choose an element $\pi \in \mathcal{O}_{K^\flat}$ such that $\pi^\#/p \in (K^+)^\times$, where $\pi^\# \in K^+$ is any element lifting the first coordinate of π. Hence we can write

$$\theta([\pi]) = \sum_{k=1}^{\infty} p^k \theta([x_k])$$

with $x_k \in (K^\flat)$. Now let

$$\xi := [\pi] - \sum_{k=1}^{\infty} p^k [x_k].$$

We now show that ξ is not a zero divisor in $W(R^{\flat,+})$. Suppose

$$y = \sum_{k=0}^{\infty} p^k [y_k] \in W(R^{\flat,+})$$

with $y \neq 0$ and

$$\xi y = 0. \tag{2.7}$$

Write $y = p^r y'$ where $r \in \mathbb{Z}_{\geq 0}$ is minimal such that $y_r \neq 0$. Since $W(R^{\flat,+})$ is flat over $\mathbb{Z}_p$, we have that multiplication by p is injective, and hence p^r is not a zero divisor. Hence we may assume, without loss of generality, that $y_0 \neq 0$. Reducing (2.7) modulo p, we get

$$\pi y_0 = 0,$$

which, since $R^{\flat,+}$ is flat over $K^{\flat,+}$ (and so multiplication by π is injective), implies $y_0 = 0$, a contradiction.

Finally, we show that ξ generates $\ker \theta$. Suppose again that $y = \sum_{k=0}^{\infty} p^k [y_k] \in \ker \theta$, then we need to show that $\xi | y$. Since R^+ is flat over $\mathbb{Z}_p$, so that multiplication by p is injective, we may assume without loss of generality that $y_0 \neq 0$. Now note that

$$W(R^{\flat,+})/(\xi, p) = R^{\flat,+}/\pi = R^+/p.$$

Since $\xi \in \ker \theta$, then θ factors through

$$\theta : W(R^{\flat,+})/(\xi) \to R^+$$

and so

$$\theta(y) = 0 \implies y \equiv 0 \pmod{(\xi, p)}$$

which means there exists $z_0 \in W(R^{\flat,+})$ with

$$p | y - z_0 \xi.$$

Now assume that we have shown the existence of $z_0, z_1, \ldots, z_{i-1} \in W(R^{\flat,+})$ with

$$p^i \,\big|\, y - \left(\sum_{k=0}^{i-1} z_k p^k\right) \xi.$$

Applying the same argument as above to

$$\frac{1}{p^{i-1}}\left(y - \left(\sum_{k=0}^{i-1} z_k p^k\right)\xi\right),$$

noting that

$$\theta\left(\frac{1}{p^{i-1}}\left(y - \left(\sum_{k=0}^{i-1} z_k p^k\right)\xi\right)\right) = \frac{1}{p^{i-1}}\left(\theta(y) - \theta\left(\sum_{k=0}^{i-1} z_k p^k\right)\theta(\xi)\right) = 0,$$

we see that there exists $z_i \in W(R^{\flat,+})$ with

$$p \,\Big|\, \frac{1}{p^{i-1}}\left(y - \left(\sum_{k=0}^{i} z_k p^k\right)\xi\right) - z_i\xi$$

and so

$$p^{i+1} \,\big|\, y - \left(\sum_{k=0}^{i} z_k p^k\right)\xi.$$

Since $W(R^{\flat,+})$ is p-adically complete, we hence have that

$$y = \left(\sum_{k=0}^{\infty} z_k p^k\right)\xi$$

and we are done. $\qquad\square$

We note that we could use ξ in lieu of the element $t \in \mathrm{Fil}^1\mathbb{B}^+_{\mathrm{dR},Y_K}$ used in defining $\mathbb{B}_{\mathrm{dR},Y_K}$ and its filtration above.

Definition 2.59. *We now define the period sheaf $\mathcal{OB}^+_{\mathrm{dR},Y}$, which will be the most relevant to our discussion. We note that the original definition of $\mathcal{OB}^+_{\mathrm{dR},Y}$ given in [57, Section 6] is slightly incorrect, and has been corrected in the subsequent corrigendum [58], and we include the definition here from the latter, following the discussion there nearly verbatim. Let*

$$U = \varprojlim_{i} U_i \in Y_{K,pro\acute{e}t}$$

be any affinoid perfectoid, where

$$U_i = \mathrm{Spa}(R_i, R_i^+)$$

with

$$(R, R^+) = (\widehat{\varinjlim_i R_i^+}[1/p], \widehat{\varinjlim_i R_i^+})$$

perfectoid affinoid. We define $\mathcal{O}\mathbb{B}_{\mathrm{dR},Y_K}^+$ as the sheafification of the presheaf sending

$$U \mapsto \varinjlim_i \varprojlim_n \left(R_i^+ \hat{\otimes}_{W(\kappa)} \mathbb{A}_{\mathrm{inf}}(R, R^+)\right)[1/p]/(\ker \theta)^n \qquad (2.8)$$

where the completed tensor product is the p-adic completion of the tensor product, and

$$\theta : \left(R_i^+ \hat{\otimes}_{W(\kappa)} \mathbb{A}_{\mathrm{inf}}(R, R^+)\right)[1/p] \to R$$

is the tensor product over $\otimes_{W(\kappa)}$ of the map $R_i^+ \to R^+$ and the map

$$\theta : \mathbb{A}_{\mathrm{inf}}(R, R^+) \to R^+$$

from (2.5).

Then we also define

$$\mathcal{O}\mathbb{B}_{\mathrm{dR},Y_K} := \mathcal{O}\mathbb{B}_{\mathrm{dR},Y_K}^+[t^{-1}]$$

where $t \in \mathrm{Fil}^1\mathbb{B}_{\mathrm{dR},Y_K}^+$ is any generator. It is a fact that the composition

$$\mathcal{O}_{Y_K} \to \mathcal{O}\mathbb{B}_{\mathrm{dR},Y_K}^+ \to \mathcal{O}\mathbb{B}_{\mathrm{dR},Y_K}^+/\ker\theta = \hat{\mathcal{O}}_{Y_K} \qquad (2.9)$$

is the natural map $\mathcal{O}_{Y_K} \to \hat{\mathcal{O}}_{Y_K}$.

Crucially, $\mathcal{O}\mathbb{B}_{\mathrm{dR},Y_K}^+$ is equipped with a natural integrable $\mathbb{B}_{\mathrm{dR},Y}^+$-linear connection

$$\nabla : \mathcal{O}\mathbb{B}_{\mathrm{dR},Y_K}^+ \to \mathcal{O}\mathbb{B}_{\mathrm{dR},Y_K}^+ \otimes_{\mathcal{O}_{Y_K}} \Omega_{Y_K}^1 \qquad (2.10)$$

extending the trivial connection $d : \mathcal{O}_{Y_K} \to \Omega_{Y_K}^1$. This is defined using the fact that the connection

$$\nabla : \mathcal{O}\mathbb{B}_{\mathrm{inf},Y_K} \to \mathcal{O}\mathbb{B}_{\mathrm{inf},Y_K} \otimes_{\mathcal{O}_{Y_K}} \Omega_{Y_K}^1$$

from (2.6) satisfies *Griffiths transversality* with respect to the filtration

$$\mathrm{Fil}^i\mathcal{O}\mathbb{B}_{\mathrm{inf},Y_K} = (\ker\theta)^i\mathcal{O}\mathbb{B}_{\mathrm{infy},Y_K}$$

on $\mathcal{OB}_{\mathrm{inf},Y_K}$, that is:

$$\nabla(\mathrm{Fil}^i\mathcal{OB}_{\mathrm{inf},Y_K}) \subset \mathrm{Fil}^{i-1}\mathcal{OB}_{\mathrm{inf},Y_K} \otimes_{\mathcal{O}_{Y_K}} \Omega^1_{Y_K}$$

and we can extend to a $\mathbb{B}^+_{\mathrm{dR},Y_K}$-linear connection

$$\nabla : \mathcal{OB}^+_{\mathrm{dR},Y_K} \to \mathcal{OB}^+_{\mathrm{dR},Y_K} \otimes_{\mathcal{O}_{Y_K}} \Omega^1_{Y_K}$$

by passing to the limit in (2.8). Since ∇ is $\mathbb{B}^+_{\mathrm{dR},Y_K}$-linear, it also extends to a connection

$$\nabla : \mathcal{OB}_{\mathrm{dR},Y_K} \to \mathcal{OB}_{\mathrm{dR},Y_K} \otimes_{\mathcal{O}_{Y_K}} \Omega^1_{Y_K}.$$

The sheaf $\mathcal{OB}_{\mathrm{dR},Y_K}$ has a natural filtration given by

$$\mathrm{Fil}^i\mathcal{OB}^+_{\mathrm{dR},Y_K} = (\ker\theta)^i\mathcal{OB}^+_{\mathrm{dR},Y_K},$$

and $\mathcal{OB}_{\mathrm{dR},Y_K}$ has a filtration given by

$$\mathrm{Fil}^i\mathcal{OB}_{\mathrm{dR},Y_K} = \sum_{j\in\mathbb{Z}} t^j \,\mathrm{Fil}^{i-j}\mathcal{OB}^+_{\mathrm{dR},Y_K}.$$

By construction of ∇, one sees that it satisfies Griffiths transversality on either $\mathcal{OB}^+_{\mathrm{dR},Y_K}$ or $\mathcal{OB}_{\mathrm{dR},Y_K}$ with respect to the above filtrations:

$$\nabla(\mathrm{Fil}^i\mathcal{OB}^{(+)}_{\mathrm{dR},Y_K}) \subset \mathrm{Fil}^{i-1}\mathcal{OB}^{(+)}_{\mathrm{dR},Y_K} \otimes_{\mathcal{O}_{Y_K}} \Omega^1_{Y_K}.$$

Finally, we define a "p-adically completed $\mathcal{OB}^+_{\mathrm{dR},Y}$"

$$\hat{\mathcal{OB}}_{\mathrm{dR},Y} := \mathcal{OB}^+_{\mathrm{dR},Y} \otimes_{\mathcal{O}_Y} \hat{\mathcal{O}}_Y$$

which is a priori a presheaf (and not a sheaf). We will not need to know whether this is a sheaf, or consider its sheafification, in our discussion. Note that by tensoring $\theta : \mathcal{OB}_{\mathrm{dR},Y} \twoheadrightarrow \hat{\mathcal{O}}_Y$ with $\otimes_{\mathcal{O}_Y}\hat{\mathcal{O}}_Y$, we get

$$\theta : \hat{\mathcal{OB}}^+_{\mathrm{dR},Y} \twoheadrightarrow \hat{\mathcal{O}}_Y.$$

Hence, $\hat{\mathcal{OB}}^+_{\mathrm{dR},Y}$ inherits a natural filtration from $\mathcal{OB}^+_{\mathrm{dR},Y}$:

$$\mathrm{Fil}^i\hat{\mathcal{OB}}^+_{\mathrm{dR},Y} := (\ker\theta)^i\hat{\mathcal{OB}}^+_{\mathrm{dR},Y}.$$

However, we note that there is no (natural) nontrivial connection on $\hat{\mathcal{O}}_Y$, since $\hat{\mathcal{O}}_{\mathcal{Y}_K} = \mathcal{O}_{\hat{\mathcal{Y}}_K}$, being the structure sheaf of a perfectoid space $\hat{\mathcal{Y}}_K$, has no nontrivial differentials.

2.8 THE PROÉTALE "CONSTANT SHEAF" $\hat{\mathbb{Z}}_{p,Y}$

By the previous discussion, since K was an arbitrary perfectoid field, we have defined all period sheaves and their associated objects as above on $Y_{\text{proét}}$.

Henceforth, given an adic space Y with proétale site $Y_{\text{proét}}$, let

$$\hat{\mathbb{Z}}_{p,Y} = \nu^*(\mathbb{Z}_p)_Y = \varprojlim_n \nu^*(\mathbb{Z}/p^n)_Y$$

be the pullback under ν of the usual constant sheaf $(\mathbb{Z}_p)_Y$ on $Y_{\text{ét}}$, which is equivalently the inverse limit of the pullbacks under ν of the constant sheaves $(\mathbb{Z}/p^n)_Y$ on $Y_{\text{ét}}$. Hence, on a proétale $U = \varprojlim_i U_i \in Y_{\text{proét}}$, we have by (2.3)

$$\hat{\mathbb{Z}}_{p,Y}(U) = \varprojlim_n \nu^*(\mathbb{Z}/p^n)_Y(U) = \varprojlim_n \varinjlim_i \nu^*(\mathbb{Z}/p^n)_Y(U_i) = \varinjlim_i \text{Hom}_{\text{cts}}(U_i, \varprojlim_n \mathbb{Z}/p^n)$$

$$= \varinjlim_i \varprojlim_n \text{Hom}_{\text{cts}}(U_i, \mathbb{Z}/p^n).$$

We then see that

$$\hat{\mathbb{Z}}_{p,Y}(U) = \text{Hom}_{\text{cts}}(U, \mathbb{Z}_p)$$

where $\mathbb{Z}_p$ is given the *p-adic topology* and *not* the discrete topology.

On the other hand, letting $\mathbb{Z}_{p,Y}$ denote the constant sheaf associated with $\mathbb{Z}_p$ on $Y_{\text{ét}}$, we have

$$\nu^*\mathbb{Z}_{p,Y}(U) = \varinjlim_i \nu^*\mathbb{Z}_{p,Y}(U_i) = \varinjlim_i \text{Hom}_{\text{cts}}(U_i, \mathbb{Z}_p)$$

where $\mathbb{Z}_p$ has the discrete topology. We then see that

$$\nu^*\mathbb{Z}_{p,Y}(U) = \{\text{locally constant functions } U \to \mathbb{Z}_p\}$$

which is a strict subset of $\hat{\mathbb{Z}}_{p,Y}(U)$.

We also define the *Tate twist* $\hat{\mathbb{Z}}_{p,Y}(1)$ by

$$\hat{\mathbb{Z}}_{p,Y}(1) = \varprojlim_n \nu^*(\mu_{p^n})_Y$$

where $\nu^*(\mu_{p^n})_Y$ is the pullback under ν of the usual constant sheaf $(\mu_{p^n})_Y$ on $Y_{\text{ét}}$. For any non-negative integer n, we let

$$\hat{\mathbb{Z}}_{p,Y}(n) = \hat{\mathbb{Z}}_{p,Y}(1)^{\otimes n}$$

(where $\otimes$ is taken over $\hat{\mathbb{Z}}_{p,Y}$), and we let

$$\hat{\mathbb{Z}}_{p,Y}(-n) = \mathrm{Hom}_{\hat{\mathbb{Z}}_{p,Y}}(\hat{\mathbb{Z}}_{p,Y}(n), \hat{\mathbb{Z}}_{p,Y}).$$

Given any $\hat{\mathbb{Z}}_{p,Y}$-module $\mathcal{F}$ and $n \in \mathbb{Z}$, we let

$$\mathcal{F}(n) := \mathcal{F} \otimes_{\hat{\mathbb{Z}}_{p,Y}} \hat{\mathbb{Z}}_{p,Y}(n).$$

2.9 $\mathbb{B}^+_{\mathrm{dR},Y}$-LOCAL SYSTEMS, $\mathcal{O}\mathbb{B}^+_{\mathrm{dR},Y}$-MODULES WITH CONNECTION, AND THE GENERAL DE RHAM COMPARISON THEOREM

Suppose now in this section that Y is locally noetherian smooth adic space over $\mathrm{Spa}(k, \mathcal{O}_k)$ where k is a complete discretely valued nonarchimedean field extension of $\mathbb{Q}$ with perfect residue field κ.

Definition 2.60 ([57], Definition 7.1). *1. A $\mathbb{B}^+_{\mathrm{dR},Y}$-local system is a sheaf of $\mathbb{B}^+_{\mathrm{dR},Y}$-modules $\mathbb{M}$ that on $Y_{\mathrm{pro\acute{e}t}}$ is locally free of finite rank.*
2. An $\mathcal{O}\mathbb{B}^+_{\mathrm{dR},Y}$-module with integral connection is a sheaf of $\mathcal{O}\mathbb{B}^+_{\mathrm{dR},Y}$-modules $\mathcal{M}$ that on $Y_{\mathrm{pro\acute{e}t}}$ is locally free of finite rank, along with an integrable connection

$$\nabla_{\mathcal{M}} : \mathcal{M} \to \mathcal{M} \otimes_{\mathcal{O}_Y} \Omega^1_Y$$

which satisfies the Leibniz rule with respect to the connection $\nabla : \mathcal{O}\mathbb{B}^+_{\mathrm{dR},Y} \to \mathcal{O}\mathbb{B}^+_{\mathrm{dR},Y} \otimes_{\mathcal{O}_Y} \Omega^1_Y$ as defined in Section 2.7.

We recall the fundamental equivalence between the category of $\mathbb{B}^+_{\mathrm{dR},Y}$-local systems and $\mathcal{O}\mathbb{B}^+_{\mathrm{dR},Y}$-modules.

Theorem 2.61 ([57], Theorem 7.2). *The functor*

$$\mathbb{M} \mapsto (\mathcal{M}, \nabla_{\mathcal{M}}), \quad \mathcal{M} = \mathbb{M} \otimes_{\mathbb{B}^+_{\mathrm{dR},Y}} \mathcal{O}\mathbb{B}^+_{\mathrm{dR},Y}, \quad \nabla_{\mathbb{M}} = \mathrm{id} \otimes \nabla$$

induces an equivalence of categories between the category of $\mathbb{B}^+_{\mathrm{dR},Y}$-local systems and the category of $\mathcal{O}\mathbb{B}^+_{\mathrm{dR},Y}$-modules with integrable connection. The inverse functor is given by

$$(\mathcal{M}, \nabla_{\mathcal{M}}) \mapsto \mathbb{M} = \mathcal{M}^{\nabla_{\mathcal{M}}=0}.$$

We also have the following compatibility between vector bundles on Y for different topologies.

Proposition 2.62 ([57], Lemma 7.3). *Let Y_{ad} denote the usual site (i.e., topology) of opens on the adic space Y. The following categories are equivalent.*

1. *The category of $\mathcal{O}_{Y_{ad}}$-modules on Y_{ad} that are locally free of finite rank.*
2. *The category of $\mathcal{O}_{Y_{\acute{e}t}}$-modules on $Y_{\acute{e}t}$ that are locally free of finite rank.*
3. *The category of $\mathcal{O}_{Y_{pro\acute{e}t}}$-modules on $Y_{pro\acute{e}t}$ that are locally free of finite rank.*

Definition 2.63 ([57], Definition 7.4). *A filtered $\mathcal{O}_Y$-module with integrable connection is a locally free $\mathcal{O}_Y$-module on Y (in the adic topology) with a decreasing exhaustive filtration $\mathrm{Fil}^i\mathcal{E}, i \in \mathbb{Z}$, which is separable (i.e., $\bigcap_{i \in \mathbb{Z}} \mathrm{Fil}^i\mathcal{E} = 0$) and an integrable connection*

$$\nabla : \mathcal{E} \to \mathcal{E} \otimes_{\mathcal{O}_Y} \Omega^1_Y$$

satisfying Griffiths transversality with respect to $\mathrm{Fil}^i\mathcal{E}$, that is,

$$\nabla(\mathrm{Fil}^i\mathcal{E}) \subset \mathrm{Fil}^{i-1}\mathcal{E} \otimes_{\mathcal{O}_Y} \Omega^1_Y.$$

Definition 2.64 ([57], Definition 7.5). *A filtered $\mathcal{O}\mathbb{B}^+_{dR,Y}$-module with integrable connection $\mathcal{M}$ and a filtered $\mathcal{O}_Y$-module with integrable connection $\mathcal{E}$ are called associated if there is an isomorphism of sheaves on $Y_{pro\acute{e}t}$*

$$\mathcal{M} \otimes_{\mathcal{O}\mathbb{B}^+_{dR,Y}} \mathcal{O}\mathbb{B}_{dR,Y} \cong \mathcal{E} \otimes_{\mathcal{O}_Y} \mathcal{O}\mathbb{B}_{dR,Y}.$$

Theorem 2.65 ([57], Theorem 7.6). *Suppose $\mathcal{M}$ is a filtered $\mathcal{O}\mathbb{B}^+_{dR,Y}$-module with integrable connection associated with $\mathcal{E}$ in the sense of Definition 2.64. Consider the $\mathbb{B}^+_{dR,Y}$-local system $\mathbb{M} = \mathcal{M}^{\nabla_\mathcal{M}=0}$. Then*

$$\mathbb{M} = \mathrm{Fil}^0(\mathcal{E} \otimes_{\mathcal{O}_Y} \mathcal{O}\mathbb{B}_{dR,Y})^{\nabla=0}. \tag{2.11}$$

Moreover, for any filtered $\mathcal{O}_Y$-module with integrable connection $\mathcal{E}$, $\mathbb{M}$ as defined in (2.11) is a $\mathbb{B}^+_{dR,Y}$-local system such that $\mathcal{E}$ is associated with $\mathcal{M} = \mathbb{M} \otimes_{\mathbb{B}^+_{dR,Y}} \mathcal{O}\mathbb{B}^+_{dR,Y}$.

In particular, the property of being associated as in Definition 2.64 defines a fully faithful functor

$$\mathcal{E} \mapsto \mathcal{M}$$

from the category of filtered $\mathcal{O}_Y$-modules with integrable connection to the category of filtered $\mathcal{O}\mathbb{B}^+_{dR,Y}$-modules with connection. By Theorem 2.61, this also defines a fully faithful functor into the category of $\mathbb{B}^+_{dR,Y}$-local systems

$$\mathcal{M} \mapsto \mathbb{M}.$$

Definition 2.66 ([57], Definition 8.1). *Suppose Y is a locally noetherian adic space. A lisse $\mathbb{Z}_p$-sheaf $\mathbb{L}_\bullet$ on $Y_{\text{ét}}$ is an inverse system of sheaves of $(\mathbb{Z}/p^n)_Y$-modules $\mathbb{L}_n$ on $Y_{\text{ét}}$, such that each $\mathbb{L}_n$ is a locally constant sheaf on $Y_{\text{ét}}$ associated to a finitely generated $\mathbb{Z}/p^n$-module, and such that this inverse system is isomorphic in pro-$Y_{\text{ét}}$ to an inverse system with $\mathbb{L}_{n+1}/p^n \cong \mathbb{L}_n$.*

Let $\hat{\mathbb{Z}}_{p,Y}$ be as in Section 2.8. Then a lisse $\hat{\mathbb{Z}}_{p,Y}$-sheaf on $Y_{\text{proét}}$ is a sheaf $\mathbb{L}$ of $\hat{\mathbb{Z}}_{p,Y}$-modules such that $\mathbb{L}$ is locally on $Y_{\text{proét}}$ isomorphic to $\hat{\mathbb{Z}}_p \otimes_{\mathbb{Z}_p} M$, where M is a finitely generated $\mathbb{Z}_p$-module.

Definition 2.67 ([57], Definition 8.3). *Let k be a complete discretely valued nonarchimedean extension of $\mathbb{Q}_p$, with perfect residue field κ. Suppose Y is a proper smooth adic space over $\text{Spa}(k, \mathcal{O}_k)$. A lisse $\hat{\mathbb{Z}}_{p,Y}$-sheaf $\mathbb{L}$ is called de Rham if the associated $\mathbb{B}_{\text{dR},Y}^+$-local system $\mathbb{M} = \mathbb{L} \otimes_{\hat{\mathbb{Z}}_{p,Y}} \mathbb{B}_{\text{dR},Y}^+$ is associated to some filtered module with integrable connection $(\mathcal{E}, \nabla, \text{Fil}^\bullet)$.*

Given a filtered $\mathcal{O}_Y$-module with integrable connection $(\mathcal{E}, \nabla, \text{Fil}^\bullet)$, we associate the following $\mathbb{B}_{\text{dR},Y}^+$-local system. Consider the $\mathcal{O}\mathbb{B}_{\text{dR},Y}^+$-module

$$\mathcal{M}_0 = \mathcal{E} \otimes_{\mathcal{O}_Y} \mathcal{O}\mathbb{B}_{\text{dR},Y}^+$$

with integrable connection $\nabla_{\mathcal{M}_0}$ given by extending the connections on $\mathcal{E}$ and $\mathcal{O}\mathbb{B}_{\text{dR},Y}^+$ by the Leibniz rule. Then we let

$$\mathbb{M}_0 = \mathcal{M}_0^{\nabla_{\mathcal{M}_0} = 0}.$$

We endow $\mathbb{M}_0$ with the natural filtration induced by the filtration on $\mathbb{B}_{\text{dR},Y}^+$.

By Theorem 2.61, we have

$$\mathbb{M}_0 \otimes_{\mathbb{B}_{\text{dR},Y}^+} \mathcal{O}\mathbb{B}_{\text{dR},Y}^+ \cong \mathcal{M}_0$$

and so

$$\mathcal{E} \otimes_{\mathcal{O}_Y} \mathcal{O}\mathbb{B}_{\text{dR},Y}^+ \cong \mathbb{M}_0 \otimes_{\mathbb{B}_{\text{dR},Y}^+} \mathcal{O}\mathbb{B}_{\text{dR},Y}^+. \tag{2.12}$$

Reducing (2.12) modulo $\ker(\theta : \mathcal{O}\mathbb{B}_{\text{dR},Y}^+ \to \hat{\mathcal{O}}_Y)$, we get an isomorphism

$$\mathcal{E} \otimes_{\mathcal{O}_Y} \hat{\mathcal{O}}_Y \cong \text{gr}^0 \mathbb{M}_0.$$

Similarly, multiplying (2.12) through by $\ker(\theta : \mathcal{O}\mathbb{B}_{\text{dR},Y}^+ \to \hat{\mathcal{O}}_Y)^i$ and reducing modulo $\ker(\theta : \mathcal{O}\mathbb{B}_{\text{dR},Y}^+ \to \hat{\mathcal{O}}_Y)^{i+1}$, we get

$$\mathcal{E} \otimes_{\mathcal{O}_Y} \hat{\mathcal{O}}_Y(i) \cong \text{gr}^i \mathbb{M}_0.$$

Theorem 2.68 ([57], Proposition 7.9). *The $\mathbb{B}_{dR,Y}^+$-local system $\mathbb{M}$ associated with $(\mathcal{E}, \nabla, \mathrm{Fil}^\bullet)$ has*

$$\mathbb{M} \subset \mathbb{M}_0 \otimes_{\mathbb{B}_{dR,Y}^+} \mathbb{B}_{dR,Y},$$

and has

$$(\mathbb{M} \cap \mathrm{Fil}^i \mathbb{M}_0)/(\mathbb{M} \cap \mathrm{Fil}^{i+1} \mathbb{M}_0) = \mathrm{Fil}^{-i} \mathcal{E} \otimes_{\mathcal{O}_Y} \hat{\mathcal{O}}_Y(i) \subset \mathrm{gr}^i \mathbb{M}_0 \cong \mathcal{E} \otimes_{\mathcal{O}_Y} \hat{\mathcal{O}}_Y(i). \tag{2.13}$$

In fact, $\mathbb{M}$ is the unique submodule of $\mathbb{M}_0 \otimes_{\mathbb{B}_{dR,Y}^+} \mathbb{B}_{dR,Y}$ satisfying (2.13). In particular, we get

$$\mathbb{M}_0 \subset \mathbb{M} \tag{2.14}$$

and

$$\mathbb{M}_0 \otimes_{\mathbb{B}_{dR,Y}^+} \mathbb{B}_{dR,Y} \cong \mathbb{M} \otimes_{\mathbb{B}_{dR,Y}^+} \mathbb{B}_{dR,Y}.$$

In particular, we have the following immediate corollary of Theorem 2.65.

Corollary 2.69. *Suppose we are in the setting of Theorem 2.65. The associations*

$$\mathcal{E} \mapsto \mathbb{M}_0, \quad \mathcal{E} \mapsto (\mathbb{M}_0 \subset \mathbb{M})$$

are functors from the category of filtered $\mathcal{O}_Y$-modules with integrable connection to the category of $\mathbb{B}_{dR,Y}^+$-local systems, and to the category of inclusions of $\mathbb{B}_{dR,Y}^+$- local systems, respectively. By Theorem 2.61, this gives functors

$$\mathcal{E} \mapsto \mathbb{M}_0 \otimes_{\mathbb{B}_{dR,Y}^+} \mathcal{O}\mathbb{B}_{dR,Y}^+, \quad \mathcal{E} \mapsto (\mathbb{M}_0 \otimes_{\mathbb{B}_{dR,Y}^+} \mathcal{O}\mathbb{B}_{dR,Y}^+ \subset \mathbb{M} \otimes_{\mathbb{B}_{dR,Y}^+} \mathcal{O}\mathbb{B}_{dR,Y}^+)$$

from the category of filtered $\mathcal{O}_Y$-modules with integrable connection to the category of filtered $\mathcal{O}\mathbb{B}_{dR,Y}^+$-modules with integrable connection, and to the category of inclusion of filtered $\mathcal{O}\mathbb{B}_{dR,Y}^+$-modules with integrable connection where the inclusion does not necessarily respect filtrations.

Definition 2.70. *Given an $\mathcal{O}_X$-module $\mathcal{E}$ with integrable connection. Let*

$$R^i \pi_{dR,*} \mathcal{E} := R^i \pi_* (0 \to \mathcal{E} \xrightarrow{\nabla} \mathcal{E} \otimes_{\mathcal{O}_X} \Omega_{X/Y}^1 \xrightarrow{\nabla} \cdots \xrightarrow{\nabla} \mathcal{E} \otimes \Omega_{X/Y}^n \to 0)$$

where

$$0 \to \mathcal{E} \xrightarrow{\nabla} \mathcal{E} \otimes_{\mathcal{O}_X} \Omega_{X/Y}^1 \xrightarrow{\nabla} \cdots \xrightarrow{\nabla} \mathcal{E} \otimes \Omega_{X/Y}^n \to 0$$

is the usual de Rham complex (which is indeed a complex by the integrality of ∇) in the category of $\pi^{-1}\mathcal{O}_Y$-modules.

Now assume that $\pi : X \to Y$ is a proper smooth morphism of adic spaces over a complete discretely valued nonarchimedean field extension k of $\mathbb{Q}_p$ with residue field κ. We have the following de Rham comparison theorem for the family $\pi : X \to Y$.

Theorem 2.71. *Let* $\mathbb{L}$ *be a lisse* $\hat{\mathbb{Z}}_{p,X}$-*sheaf on* $X_{pro\acute{e}t}$, *and let*

$$\mathbb{M} = \mathbb{L} \otimes_{\hat{\mathbb{Z}}_{p,X}} \mathbb{B}^+_{dR,X}$$

be the associated $\mathbb{B}^+_{dR,X}$-*local system. Assume that* $R^i \pi_* \mathbb{L}$ *is a lisse* $\hat{\mathbb{Z}}_{p,Y}$-*sheaf on* $Y_{pro\acute{e}t}$.[7] *Then we have the following.*

1. *There is a canonical isomorphism*

$$R^i \pi_* \mathbb{M} \cong R^i \pi_* \mathbb{L} \otimes_{\hat{\mathbb{Z}}_{p,Y}} \mathbb{B}^+_{dR,Y}.$$

   In particular, $R^i \pi_* \mathbb{M}$ *is a* $\mathbb{B}^+_{dR,Y}$-*local system on* $Y_{pro\acute{e}t}$ *which is associated with* $R^i \pi_* \mathbb{L}$.
2. *Suppose further that* $\mathbb{L}$ *is de Rham with respect to a filtered* $\mathcal{O}_X$-*module with integrable connection* $(\mathcal{E}, \nabla, \mathrm{Fil}^\cdot)$. *Then* $R^i \pi_* \mathbb{L}$ *is de Rham, with associated filtered* $\mathcal{O}_Y$-*module with integrable connection given by* $R^i \pi_* \mathcal{E}$. *In particular, we have an isomorphism*

$$R^i \pi_* \mathbb{L} \otimes_{\mathbb{B}^+_{dR,Y}} \mathcal{O}\mathbb{B}_{dR,Y} \cong R^i \pi_{dR,*} \mathcal{E} \otimes_{\mathcal{O}_Y} \mathcal{O}\mathbb{B}_{dR,Y}$$

   compatible with filtrations and connections.

From (2.14), tensoring by $\otimes_{\mathbb{B}^+_{dR,Y}} \mathcal{O}\mathbb{B}^+_{dR,Y}$, we hence get an inclusion

$$R^i \pi_* \mathbb{L} \otimes_{\hat{\mathbb{Z}}_{p,Y}} \mathcal{O}\mathbb{B}^+_{dR,Y} = \mathbb{M}_0 \otimes_{\mathbb{B}^+_{dR,Y}} \mathcal{O}\mathbb{B}^+_{dR,Y} \subset \mathbb{M} \otimes_{\mathbb{B}^+_{dR,Y}} \mathcal{O}\mathbb{B}^+_{dR,Y}$$
$$= R^i \pi_{dR,*} \mathcal{E} \otimes_{\mathcal{O}_Y} \mathcal{O}\mathbb{B}^+_{dR,Y}$$

of sheaves on $Y_{pro\acute{e}t}$, compatible with connections, but not necessarily filtrations.

We now specialize the above to the situation $\mathcal{E} = \mathcal{O}_X, \mathbb{L} = \hat{\mathbb{Z}}_{p,X}$.

Theorem 2.72 ([57], Theorem 8.8 specialized to $\mathcal{E} = \mathcal{O}_X, \mathbb{L} = \hat{\mathbb{Z}}_{p,X}$). *Suppose* $R^i \pi_* \hat{\mathbb{Z}}_{p,X}$ *is a lisse* $\hat{\mathbb{Z}}_{p,Y}$-*sheaf on* $Y_{pro\acute{e}t}$ *for all* $i \geq 0$.[8] *Then for all* $i \geq 0$, $R^i \pi_* \hat{\mathbb{Z}}_{p,S}$ *is de Rham with associated filtered* $\mathcal{O}_X$-*module with integrable connection* $R^i \pi_*$

---

[7] *This condition is known to hold whenever* π *is algebraizable, which is the case in our applications by [41].*

[8] *Again, as in Theorem 2.71, this holds whenever* π *is algebraizable, as is the case in our applications. Moreover, as stated in [9, Theorem 2.2.2], this is known to be true if* $R^i \pi_* \mathbb{F}_p$

$\mathcal{O}_X$, endowed with the Hodge filtration and the Gauss-Manin connection. Hence, there is an isomorphism

$$R^i\pi_*\hat{\mathbb{Z}}_{p,X} \otimes_{\hat{\mathbb{Z}}_{p,Y}} \mathcal{O}\mathbb{B}_{\mathrm{dR},Y} \cong R^i\pi_{\mathrm{dR},*}\mathcal{O}_X \otimes_{\mathcal{O}_Y} \mathcal{O}\mathbb{B}_{\mathrm{dR},Y}$$

of sheaves on $Y_{pro\acute{e}t}$, compatible with filtrations and connections.

From (2.14), tensoring by $\otimes_{\mathbb{B}_{\mathrm{dR},Y}^+} \mathcal{O}\mathbb{B}_{\mathrm{dR},Y}^+$, we hence get an inclusion

$$R^i\pi_*\hat{\mathbb{Z}}_{p,X} \otimes_{\hat{\mathbb{Z}}_{p,Y}} \mathcal{O}\mathbb{B}_{\mathrm{dR},Y}^+ = \mathrm{M}_0 \otimes_{\mathbb{B}_{\mathrm{dR},Y}^+} \mathcal{O}\mathbb{B}_{\mathrm{dR},Y}^+ \subset \mathrm{M} \otimes_{\mathbb{B}_{\mathrm{dR},Y}^+} \mathcal{O}\mathbb{B}_{\mathrm{dR},Y}^+$$
$$= R^i\pi_{\mathrm{dR},*}\mathcal{O}_X \otimes_{\mathcal{O}_Y} \mathcal{O}\mathbb{B}_{\mathrm{dR},Y}^+$$

of sheaves on $Y_{pro\acute{e}t}$, compatible with connections.

Finally, we end this section with some p-adically completed versions of some of the notions above.

Definition 2.73. *We call a $\hat{\mathcal{O}}_Y$-module $\hat{\mathcal{E}}$ a completed filtered $\mathcal{O}_Y$-module with integrable connection if it is of the form*

$$\hat{\mathcal{E}} = \mathcal{E} \otimes_{\mathcal{O}_Y} \hat{\mathcal{O}}_Y$$

where $\mathcal{E}$ is a filtered $\mathcal{O}_Y$-module with integrable connection. Note that $\hat{\mathcal{E}}$ inherits a canonical filtration from $\mathcal{E}$:

$$\mathrm{Fil}^i\hat{\mathcal{E}} = \mathrm{Fil}^i\mathcal{E} \otimes_{\mathcal{O}_Y} \hat{\mathcal{O}}_Y.$$

Hence, a morphism of completed filtered $\mathcal{O}_Y$-modules is a map of $\hat{\mathcal{O}}_Y$-modules $\hat{\mathcal{E}} \to \hat{\mathcal{F}}$ which preserves the inherited filtrations.

Similarly, we call a $\hat{\mathcal{O}}\mathbb{B}_{\mathrm{dR},Y}^+$-module $\hat{\mathcal{M}}$ a completed filtered $\mathcal{O}\mathbb{B}_{\mathrm{dR},Y}^+$-module with integrable connection if it is of the form

$$\hat{\mathcal{M}} = \mathcal{M} \otimes_{\mathcal{O}\mathbb{B}_{\mathrm{dR},Y}^+} \hat{\mathcal{O}}\mathbb{B}_{\mathrm{dR},Y}^+ = \mathcal{M} \otimes_{\mathcal{O}_Y} \hat{\mathcal{O}}_Y$$

where $\mathcal{M}$ is a filtered $\mathcal{O}\mathbb{B}_{\mathrm{dR},Y}^+$-module with integrable connection. Again, we have a natural filtration

$$\mathrm{Fil}^i\hat{\mathcal{M}} = \mathrm{Fil}^i\mathcal{M} \otimes_{\mathcal{O}_Y} \hat{\mathcal{O}}_Y.$$

Now from Corollary 2.69, we immediately get the following functoriality.

---

is locally free on $Y_{pro\acute{e}t}$ for all $i \geq 0$, which as mentioned in the footnote in Theorem 2.2.2 in loc. cit., in turn has been announced by Gabber.

Corollary 2.74. *The association*

$$\hat{\mathcal{E}} \mapsto (\mathbb{M}_0 \otimes_{\mathbb{B}_{\mathrm{dR},Y}^+} \hat{\mathcal{O}}\mathbb{B}_{\mathrm{dR},Y}^+ \subset \mathbb{M} \otimes_{\mathbb{B}_{\mathrm{dR},Y}^+} \hat{\mathcal{O}}\mathbb{B}_{\mathrm{dR},Y}^+)$$

is a functor from completed $\mathcal{O}_Y$-modules with integrable connection into the functor of inclusions of completed $\hat{\mathcal{O}}\mathbb{B}_{\mathrm{dR},Y}^+$-modules

$$\hat{\mathcal{M}}_0 \subset \hat{\mathcal{M}}$$

where the inclusion does not necessarily respect filtrations.

Chapter Three

Preliminaries: Geometry of the infinite-level modular curve

3.1 THE INFINITE-LEVEL MODULAR CURVE

Henceforth, fix an algebraic closure $\overline{\mathbb{Q}}$ of $\mathbb{Q}$, and view all number fields (finite extensions of $\mathbb{Q}$) as embedded in $\overline{\mathbb{Q}}$. Henceforth, fix embeddings

$$i_p : \overline{\mathbb{Q}} \hookrightarrow \overline{\mathbb{Q}}_p, \qquad i_\infty : \overline{\mathbb{Q}} \hookrightarrow \mathbb{C}. \tag{3.1}$$

Let $\mathbb{C}_p$ denote the p-adic completion of $\overline{\mathbb{Q}}_p$. For every field $L \subset \overline{\mathbb{Q}}_p$, let L_p denote the p-adic completion of L in $\overline{\mathbb{Q}}_p$. Let $|\cdot|$ denote the unique p-adic valuation on $\mathbb{C}_p$ normalized with $|p| = 1/p$.

Given an integer N, we consider the following congruence subgroups:

$$\Gamma_0(N) := \left\{ \gamma \in GL_2(\prod_{\ell \mid N} \mathbb{Z}_\ell) : \gamma \equiv \begin{pmatrix} * & * \\ 0 & * \end{pmatrix} \pmod{N \prod_{\ell \mid N} \mathbb{Z}_\ell} \right\},$$

$$\Gamma_1(N) := \left\{ \gamma \in GL_2(\prod_{\ell \mid N} \mathbb{Z}_\ell) : \gamma \equiv \begin{pmatrix} 1 & * \\ 0 & * \end{pmatrix} \pmod{N \prod_{\ell \mid N} \mathbb{Z}_\ell} \right\},$$

$$\Gamma(N) := \left\{ \gamma \in GL_2(\prod_{\ell \mid N} \mathbb{Z}_\ell) : \gamma \equiv \begin{pmatrix} 1 & 0 \\ 0 & 1 \end{pmatrix} \pmod{N \prod_{\ell \mid N} \mathbb{Z}_\ell} \right\},$$

where $\ell \mid N$ ranges over finite primes dividing N. We can naturally view $\Gamma = \Gamma_0(N)$, $\Gamma_1(N)$ or $\Gamma(N)$ as compact open subgroups of $GL_2(\mathbb{A}_{\mathbb{Q}}^{(\infty)})$,[1] where $\mathbb{A}_{\mathbb{Q}}^{(\infty)}$ denotes the finite adèles, via the embedding

$$\Gamma \mapsto \Gamma \cdot GL_2(\prod_{\ell \nmid N} \mathbb{Z}_\ell) \subset GL_2(\hat{\mathbb{Z}}).$$

---

[1] We make the above conventions as we will often use Γ to denote a subgroup of $GL_2(\mathbb{Z}_p)$, rather than $GL_2(\hat{\mathbb{Z}})$.

Given any compact open $\Gamma \subset GL_2(\mathbb{A}_{\mathbb{Q}}^{(\infty)})$, let $Y(\Gamma)$ denote the adicification of the associated (open) modular curve (see [69, Introduction Section 1.2.1, Chapter 3.1], [8, Section 4], [13], or [56, Chapter III.1], though the definition of Γ_0 is slightly nonstandard in loc. cit.). On $\mathbb{C}$-points, we have

$$Y(\Gamma)(\mathbb{C}) = GL_2(\mathbb{Q}) \backslash \mathcal{H}^{\pm} \times GL_2(\mathbb{A}_{\mathbb{Q}}^{(\infty)})/\Gamma$$

where $\mathcal{H}^{\pm} = \mathbb{C} \backslash \mathbb{R}$, the union of the upper and lower complex half-planes. Unless specifically mentioned, we will not work with the compactification, or cusps, throughout this text. For Γ sufficiently small, $Y(\Gamma)$ has its usual interpretation as the solution the moduli problem classifying isomorphism classes of elliptic curves together with Γ-level structure: $Y(\Gamma)$ classifies isomorphism classes of pairs (E, P) consisting of elliptic curves E, and $P \subset E[N]$ a cyclic subgroup of order N if $\Gamma = \Gamma_0(N)$, and $P \in E[N]$ a point of exact order N if $\Gamma = \Gamma_1(N)$. If $\Gamma = \Gamma(N)$, $Y(\Gamma)$ classifies isomorphism classes of pairs (E, P) consisting of an elliptic curve E and a basis P of $E[N] \cong \mathbb{Z}/N \times \mathbb{Z}/N$ if $\Gamma = \Gamma(N)$. For $N \geq 4$, the algebraic scheme giving rise to $Y(\Gamma_1(N) \cap \Gamma)$ is a fine moduli space over $\mathbb{Q}$ for any compact open $\Gamma \subset GL_2(\mathbb{A}_{\mathbb{Q}}^{(\infty)})$ (see [8, Section 4]). Hence we have an adification of the universal object

$$\mathcal{A} \rightarrow Y(\Gamma_1(N) \cap \Gamma).$$

Note that given any compact open Γ, Γ' as above, we can consider their intersection $\Gamma \cap \Gamma'$ in $GL_2(\mathbb{A}_{\mathbb{Q}}^{(\infty)})$, which is also compact open.

Let p be a prime. Fix a positive integer $N \geq 4$ which is coprime with p. Let

$$Y = Y(\Gamma_1(N) \cap \Gamma) \rightarrow Y_1(N)$$

denote (the adification of) any modular curve parametrizing isomorphism classes of elliptic curves with $\Gamma_1(N) \cap \Gamma$-level structure, for some finite-index congruence subgroup $\Gamma \subset GL_2(\mathbb{Z}_p)$. It is an (adic) étale cover of the adic modular curve $Y_1(N)$ parametrizing isomorphism classes of elliptic curves with $\Gamma_1(N)$-level structure (i.e., a point of exact order N on the elliptic curve). We will consider these modular curves as adic spaces over $\mathrm{Spa}(\mathbb{Q}_p, \mathbb{Z}_p)$.

Definition 3.1. *Let*

$$\mathcal{Y} := \varprojlim_{n} Y(\Gamma_1(N) \cap \Gamma(p^n)) \in Y_{pro\acute{e}t},$$

which we call the infinite level modular curve *with* $\Gamma_1(N) \cap \Gamma(p^{\infty})$-level struc-ture. One can view $\mathcal{Y}$ naturally as a profinite étale object of the proétale site $Y_{pro\acute{e}t}$ as defined in [57]. Upon p-adically completing the structure sheaf, one obtains the infinite-level perfectoid modular curve constructed by Scholze in [56].*

If $K \supset \mathbb{Q}_p$ is a perfectoid field that is the completion of an algebraic extension of $\mathbb{Q}_p$, Y_K is an object in $Y_{\text{proét}}$ (see [57, Discussion after Remark 6.9]). There is a natural equivalence of sites

$$Y_{\text{proét},K} \cong Y_{\text{proét}}/Y_K,$$

where the latter denotes $Y_{\text{proét}}$ localized at Y_K.

It is a somewhat subtle point that we work on $\mathcal{Y}$, instead of its associated "strong completion" in the sense of Proposition 2.40

$$\hat{\mathcal{Y}} \sim \mathcal{Y},$$

which is a true perfectoid space in the sense of [56] ($\hat{\mathcal{Y}}$, first constructed in loc. cit., is defined over $\text{Spa}(\mathbb{Q}_p, \mathbb{Z}_p)$, see [13, Theorem 2.8]) and of [59]. It is also a perfectoid space over $\text{Spa}(\mathbb{Q}_p, \mathbb{Z}_p)$ in the sense of Kedlaya-Liu, [37]. Here, the symbol "$\sim$" is defined in Definition 2.42. In particular, we have a natural homeomorphism between the underlying topological spaces:

$$|\hat{\mathcal{Y}}| \cong |\mathcal{Y}|.$$

The reason for this is that we need to work with differentials in $\Omega^1_{\hat{\mathcal{Y}}}$ in order to define our differential operators which arise from extensions of the Gauss-Manin connection

$$\nabla : \mathcal{H}^1_{\text{dR}}(\mathcal{A}) \to \mathcal{H}_{\text{dR}}(\mathcal{A}) \otimes_{\mathcal{O}_Y} \Omega^1_{\mathcal{Y}}.$$

(See Definition 3.16 for definitions of the notations.) It is a fact that

$$\Omega^1_{\hat{\mathcal{Y}}} = 0,$$

and more generally that all perfectoid spaces X have

$$\Omega^i_X = 0$$

for all $i \geq 1$ (for example, see [12, Section 4.4]). On the other hand, as $\mathcal{Y}$ is an object of $Y_{\text{proét}}$, we can work on the proétale site and in particular with the proétale sheaf $\Omega^1_{\mathcal{Y}} = \Omega^1_Y|_{\mathcal{Y}}$.

Remark 3.2. Note that we do *not* claim that $\mathcal{Y}$ equipped with $\mathcal{O}_{\mathcal{Y}}$ is an adic or pre-adic space. Rather, it is an object of $Y_{\text{proét}}$ and $\mathcal{O}_{\mathcal{Y}}$ is the restriction $\mathcal{O}_Y|_{\mathcal{Y}}$. Only after completing to $\hat{\mathcal{Y}}$ (using, for example, Proposition 2.43) do we obtain an adic space with structure sheaf $\mathcal{O}_{\hat{\mathcal{Y}}} = \hat{\mathcal{O}}_{\mathcal{Y}}$. All of our calculations on $\mathcal{Y}$ should therefore be viewed as calculations on the localized site $Y_{\text{proét}}/\mathcal{Y}$.

The $(\mathbb{C}_p, \mathcal{O}_{\mathbb{C}_p})$-points of $\mathcal{Y}$ have the following moduli-theoretic interpretation (see [56, Section III.3]): a $(\mathbb{C}_p, \mathcal{O}_{\mathbb{C}_p})$-point on $\mathcal{Y}$ corresponds to a triple (A, t, α)

consisting of an elliptic curve $A/\mathbb{C}_p$, a Γ-level structure t, and a trivialization $\alpha : \mathbb{Z}_p^{\oplus 2} \xrightarrow{\sim} T_p A$ of its Tate module. (This trivialization is equivalent to a $\Gamma(p^\infty)$-level structure on A.) Recall that by our assumptions, we have a universal object $\mathcal{A} \to Y$. Let

$$\mathcal{A}_\infty := \mathcal{A} \times_Y \mathcal{Y}.$$

We also will write

$$\mathcal{A}_\infty = (\mathcal{A}, \alpha_\infty)$$

where α_∞ denotes the canonical $\Gamma(p^\infty)$-level structure on $\mathcal{A}$ over $\mathcal{Y}$ (see [56, Proof of Lemma III.3.4]), i.e., a trivialization

$$\alpha_\infty : \hat{\mathbb{Z}}_{p,\mathcal{Y}}^{\oplus 2} \xrightarrow{\sim} T_p \mathcal{A}$$

of the universal Tate module $T_p \mathcal{A}$ (where $\hat{\mathbb{Z}}_{p,\mathcal{Y}}$ is the "p-adic constant sheaf corresponding to $\mathbb{Z}_p$, to be defined in Section 2.8). From now on, we will often suppress the Γ-level structure t in our notation for simplicity (and also because it doesn't affect any of the following).

3.2   RELATIVE ÉTALE COHOMOLOGY AND THE WEIL PAIRING

Henceforth, we will make free use of the canonical principal polarization

$$\mathcal{A} \cong \check{\mathcal{A}}$$

where $\check{\mathcal{A}}$ denotes the dual of an abelian variety $\mathcal{A}$.

Definition 3.3. *Let*

$$\mathcal{H}^1_{\text{ét}}(\mathcal{A}) := R^1 \pi_* \hat{\mathbb{Z}}_{p,\mathcal{A}}$$

is a $\hat{\mathbb{Z}}_{p,\mathcal{Y}}$-local system of rank 2 which gives the relative étale cohomology *of the family $\pi : \mathcal{A} \to Y$. We denote*

$$T_p \mathcal{A} := \text{Hom}_{\hat{\mathbb{Z}}_{p,Y}} (\mathcal{H}^1_{\text{ét}}(\mathcal{A}), \hat{\mathbb{Z}}_{p,Y})$$

which gives the relative Tate module *of the same family.*

Recall the Weil pairings given by using the principal polarization $\mathcal{A} \cong \check{\mathcal{A}}$ on $\mathcal{A}$

$$\langle \cdot, \cdot \rangle_n : \mathcal{A}[p^n] \times \mathcal{A}[p^n] \to \mu_{p^n} \tag{3.2}$$

which, taking the inverse limit over n, form a nondegenerate pairing

$$\langle \cdot, \cdot \rangle : T_p \mathcal{A} \times T_p \mathcal{A} \to \hat{\mathbb{Z}}_{p,Y}(1) \tag{3.3}$$

as an inverse limit of constant sheaves on $Y_{\text{proét}}$. This induces a canonical isomorphism

$$T_p \mathcal{A} \cong \mathcal{H}^1_{\text{ét}}(\mathcal{A})(1). \tag{3.4}$$

3.3 THE $GL_2(\mathbb{Q}_p)$-ACTION ON $\mathcal{Y}$ (AND $\hat{\mathcal{Y}}$)

Definition 3.4. *Henceforth, we denote the canonical homogeneous coordinates on*

$$\mathbb{P}^1 = \mathbb{P}(\mathbb{Q}_p^2)$$

(where for a vector space V, $\mathbb{P}(V)$ denotes the Grothendieck $\mathbb{P}^1$) by

$$x, y \in \mathcal{O}_{\mathbb{P}^1}(1)(\mathbb{P}^1).$$

Henceforth, let

$$\mathbb{P}^1_x = \{x \neq 0\}, \qquad \mathbb{P}^1_y = \{y \neq 0\}$$

be the standard affine cover of $\mathbb{P}^1$.

We now recall the right $GL_2(\mathbb{Q}_p)$-action on $\mathcal{Y}$, which acts on $(\mathbb{C}_p, \mathcal{O}_{\mathbb{C}_p})$-points (A, α) in the following way (as recalled in [13, Section 2.2]). Recall $GL_2(\mathbb{Q}_p)$ has a left action on $\mathbb{C}_p^{\oplus 2}$ in the standard way (viewing elements of the latter as column vectors), and thus if we denote the contragredient by

$$g^\vee := g^{-1} \det(g),$$

we get a *right* action

$$L \cdot g = g^\vee(L)$$

on $\mathbb{Z}_p^{\oplus 2}$. Now fix $n \in \mathbb{Z}$ such that

$$p^n g \in M_2(\mathbb{Z}_p) \quad \text{but} \quad p^{n-1} g \notin M_2(\mathbb{Z}_p).$$

Then for all $m \in \mathbb{Z}_{>0}$ sufficiently large, the kernel of $p^n g^\vee$ (mod p^m) stabilizes to some subgroup H of $A[p^m]$. Then we put

$$(A, \alpha) \cdot g := (A/H, \alpha'), \tag{3.5}$$

where α' is defined as the composition

$$\mathbb{Z}_p^{\oplus 2} \xrightarrow{p^n g} \mathbb{Q}_p^{\oplus 2} \xrightarrow{\alpha} V_p A \xrightarrow{(\check{\phi}_*)^{-1}} V_p(A/H), \qquad (3.6)$$

where $p^n g$ acts in the usual way on the left of $\mathbb{Z}_p^{\oplus 2}$ (viewing elements of the latter space as column vectors), where

$$\phi : A \to A/H$$

is natural isogeny given by projection, and where ϕ_* is the induced map $T_p A \to T_p(A/H)$ (extended by linearity to $V_p A \to V_p(A/H)$). Note that if

$$g = \begin{pmatrix} a & b \\ c & d \end{pmatrix} \in GL_2(\mathbb{Z}_p),$$

then

$$(A, \alpha) \cdot g = (A, \alpha')$$

where

$$\alpha'(e_1) = a\alpha(e_1) + c\alpha(e_2) \quad \text{and} \quad \alpha'(e_2) = b\alpha(e_1) + d\alpha(e_2).$$

In fact, one can check that the following diagram is commutative:

$$\begin{array}{ccc} \mathbb{Z}_p^{\oplus 2} & \xrightarrow{\alpha} & T_p A \\ \downarrow{\scriptstyle p^n g^\vee} & & \downarrow{\scriptstyle \phi_*} \\ \mathbb{Z}_p^{\oplus 2} & \xrightarrow{\alpha'} & T_p(A/H) \end{array} \qquad (3.7)$$

where in the arrow on the left $p^n g^\vee$ acts on $\mathbb{Z}_p^{\oplus 2}$ via left multiplication (viewing the latter as column vectors).

One also has the standard *right* $GL_2(\mathbb{Q}_p)$-action on $\mathbb{P}^1$, given by an element g acting on lines $\mathcal{L}$ (viewed as being spanned by column vectors) via

$$\mathcal{L} \cdot g = g^\vee \mathcal{L}.$$

Finally, we remark that given a geometric point y of Y, the automorphism group of the fiber of y under the natural profinite étale map $\mathcal{Y} \to Y$ is $GL_2(\mathbb{Z}_p)$, since $GL_2(\mathbb{Z}_p)$ acts simply transitively on this geometric fiber by changing the basis $\alpha : \mathbb{Z}_p^{\oplus 2} \xrightarrow{\sim} T_p A$.

Since we have an isomorphism of topological spaces $|\hat{\mathcal{Y}}| \cong |\mathcal{Y}|$, as seen in Section 3.1, the same definitions and notions can be defined for $\hat{\mathcal{Y}}$, replacing $\mathcal{Y}$ with $\hat{\mathcal{Y}}$ above.

3.4   THE HODGE-TATE PERIOD AND THE HODGE-TATE PERIOD MAP

In this section, we recall the Hodge-Tate period map $\pi_{HT} : \hat{\mathcal{Y}} \to \mathbb{P}^1$ following [56] and [9], which induces a map of topological spaces $\pi_{HT} : |\mathcal{Y}| \to |\mathbb{P}^1|$. One can also recover our definition of π_{HT} from the more general construction of period morphisms on moduli of p-divisible groups given by Scholze-Weinstein [62].

Definition 3.5. *Recall the* Hodge bundle *defined by*

$$\omega := \pi_* \Omega^1_{\mathcal{A}/Y}.$$

It has Serre dual given by the universal Lie algebra

$$\omega^{-1} := R^1 \pi_* \mathcal{O}_{\mathcal{A}}.$$

We denote

$$\hat{\omega} := \omega \otimes_{\mathcal{O}_Y} \hat{\mathcal{O}}_Y, \quad \hat{\omega}^{-1} := \omega^{-1} \otimes_{\mathcal{O}_Y} \hat{\mathcal{O}}_Y.$$

Definition 3.6. *Now, we define the* Hodge-Tate map

$$HT_{\mathcal{A}} : T_p \mathcal{A} \otimes_{\hat{\mathbb{Z}}_{p,Y}} \hat{\mathcal{O}}_Y \to \hat{\omega} \tag{3.8}$$

to be the following: the truncated Weil pairings (3.2) are compatible in the sense that the following diagram commutes

$$\begin{array}{ccccc}
\mathcal{A}[p^n] & \times & \mathcal{A}[p^n] & \xrightarrow{\langle \cdot, \cdot \rangle_n} & \mu_{p^n} \\
{\scriptstyle p}\uparrow & & \downarrow & & \downarrow \\
\mathcal{A}[p^{n+1}] & \times & \mathcal{A}[p^{n+1}] & \xrightarrow{\langle \cdot, \cdot \rangle_{n+1}} & \mu_{p^{n+1}},
\end{array} \tag{3.9}$$

and hence they can be put together (taking an inverse limit along the left first vertical arrows, and a direct limit along the middle and right vertical arrows) to get a pairing

$$T_p \mathcal{A} \times \mathcal{A}[p^\infty] \to \mu_{p^\infty}. \tag{3.10}$$

Hence, we have a natural identification

$$T_p \mathcal{A} \cong \mathrm{Hom}_{p-\mathrm{div}}(\mathcal{A}[p^\infty], \mu_{p^\infty})$$

(where the Hom *is taken in the category of p-divisible groups over $\mathbb{Z}_p$). Using this identification, we define*

$$HT_{\mathcal{A}}(\alpha) = \alpha^* \frac{dT}{T} \in \pi_* \Omega^1_{\mathcal{A}[p^\infty]/Y} \otimes_{\mathcal{O}_Y} \hat{\mathcal{O}}_Y = \pi_* \Omega^1_{\mathcal{A}/Y} \otimes_{\mathcal{O}_Y} \hat{\mathcal{O}}_Y = \hat{\omega}. \qquad (3.11)$$

By nondegeneracy of the Weil pairing, we know that $HT_{\mathcal{A}}$ is not the zero map. Let

$$\mathcal{H}^{0,1} \subset T_p \mathcal{A} \otimes_{\hat{\mathbb{Z}}_{p,Y}} \hat{\mathcal{O}}_Y$$

denote its kernel.

Pulling back 3.11 along the proétale cover $\mathcal{Y} \to Y$, we get a section

$$HT_{\mathcal{A}}(\alpha_{\infty,2}) \in \omega(\mathcal{Y}).$$

Let $\alpha_{\infty,1} := \alpha_\infty|_{\hat{\mathbb{Z}}_{p,\mathcal{Y}} \oplus \{0\}}$ and $\alpha_{\infty,2} := \alpha_\infty|_{\{0\} \oplus \hat{\mathbb{Z}}_{p,\mathcal{Y}}}$. Define the following affinoid subdomains of $\hat{\mathcal{Y}}$:

$$\hat{\mathcal{Y}}_x = \{HT_{\mathcal{A}}(\alpha_{\infty,2}) \neq 0\} \subset \hat{\mathcal{Y}}, \qquad \hat{\mathcal{Y}}_y := \{HT_{\mathcal{A}}(\alpha_{\infty,1}) \neq 0\} \subset \hat{\mathcal{Y}}.$$

Using the identification $|\hat{\mathcal{Y}}| \cong |\mathcal{Y}|$, we can define corresponding affinoid subdomains (cf. [13, Theorem 2.8])

$$\mathcal{Y}_x \subset \mathcal{Y}, \qquad \mathcal{Y}_y \subset \mathcal{Y}$$

characterized by $|\mathcal{Y}_x| = |\hat{\mathcal{Y}}_x|, \quad |\mathcal{Y}_y| = |\hat{\mathcal{Y}}_y|$. For brevity, denote

$$\mathcal{O}_{\mathcal{Y}} := \mathcal{O}_Y|_{\mathcal{Y}}, \qquad \hat{\mathcal{O}}_{\mathcal{Y}} := \hat{\mathcal{O}}_Y|_{\mathcal{Y}}.$$

Definition 3.7. *We now define the* Hodge-Tate period

$$\hat{\mathbf{z}} \in \hat{\mathcal{O}}_{\mathcal{Y}}(\mathcal{Y}_y)$$

via the relation

$$HT_{\mathcal{A}}(\alpha_{\infty,2}) = \hat{\mathbf{z}} \cdot HT_{\mathcal{A}}(\alpha_{\infty,1}). \qquad (3.12)$$

Since $HT_{\mathcal{A}}$ is not the zero map, we have that

$$\mathcal{Y}_x \cup \mathcal{Y}_y = \mathcal{Y}.$$

Using the trivialization

$$\alpha_\infty^{-1} : T_p \mathcal{A}|_{\mathcal{Y}} \xrightarrow{\sim} \hat{\mathbb{Z}}_{p,\mathcal{Y}}^{\oplus 2},$$

we now see that the line

$$\mathcal{H}^{0,1}|_{\mathcal{Y}} \subset T_p \mathcal{A} \otimes_{\hat{\mathbb{Z}}_{p,\mathcal{Y}}} \hat{\mathcal{O}}_{\mathcal{Y}} \xrightarrow[\sim]{\alpha_\infty^{-1}} \hat{\mathcal{O}}_{\mathcal{Y}}^{\oplus 2} \qquad (3.13)$$

is the sub-$\mathcal{O}_\mathcal{Y}$-module obtained by gluing together the sheaves

$$\mathcal{H}^{0,1}|_{\mathcal{Y}_x} := (\alpha_{\infty,1} - \frac{1}{\hat{z}}\alpha_{\infty,2})\hat{\mathcal{O}}_\mathcal{Y} \quad \text{and} \quad \mathcal{H}^{0,1}|_{\mathcal{Y}_y} := (\hat{z} \cdot \alpha_{\infty,2} - \alpha_{\infty,2})\hat{\mathcal{O}}_\mathcal{Y}$$

along the affinoid subdomain $\mathcal{Y}_x \cap \mathcal{Y}_y \subset \mathcal{Y}$ (and on which one easily sees that the two sheaves coincide, by the definition of $\hat{z}$).

Definition 3.8. *The inclusion (3.13) defines a map of adic spaces over* Spa $(\mathbb{Q}_p, \mathbb{Z}_p)$

$$\pi_{\mathrm{HT}} : \hat{\mathcal{Y}} \to \mathbb{P}^1$$

which is called the Hodge-Tate *period map ([57, Theorem III.1.2] over base* Spa$(\mathbb{C}_p, \mathcal{O}_{\mathbb{C}_p})$, *see also [13, Theorem 1.1] for a version over base* Spa$(\mathbb{Q}_p, \mathbb{Z}_p)$). *Using the homeomorphism* $|\mathcal{Y}| \cong |\hat{\mathcal{Y}}|$, *we get a map of topological spaces*

$$\pi_{\mathrm{HT}} : |\mathcal{Y}| \to |\mathbb{P}^1|.$$

It is straightforward to check that π_{HT} *is equivariant with respect to the* $GL_2(\mathbb{Q}_p)$*-actions defined in Section 3.3. Furthermore, since* $\hat{\mathcal{Y}}$ *is a perfectoid space (once base changed to any perfectoid field), the map* π_{HT} *is étale on the supersingular locus*

$$\hat{\mathcal{Y}}^{\mathrm{ss}} \sim \mathcal{Y}^{\mathrm{ss}} := \varprojlim_n Y(\Gamma_1(N) \cap \Gamma(p^n))^{\mathrm{ss}}$$

and maps to the Drinfeld upper half-plane $\Omega = \mathbb{P}^1(\mathbb{C}_p) \setminus \mathbb{P}^1(\mathbb{Q}_p)$. *(For the étaleness of*

$$\pi_{\mathrm{HT}} : \hat{\mathcal{Y}}^{\mathrm{ss}} \to \Omega \subset \mathbb{P}^1,$$

see also [62, Remark 7.1.2 and Theorem 7.2.3].) The stratification $\mathcal{Y} = \mathcal{Y}^{\mathrm{ord}} \sqcup \mathcal{Y}^{\mathrm{ss}}$ *is in fact given by*

$$\hat{\mathcal{Y}}^{\mathrm{ss}} = \pi_{\mathrm{HT}}^{-1}(\Omega), \qquad \hat{\mathcal{Y}}^{\mathrm{ord}} = \pi_{\mathrm{HT}}^{-1}(\mathbb{P}^1(\mathbb{Q}_p)).$$

Again, using the identification of topological spaces $|\hat{\mathcal{Y}}^{\mathrm{ss}}| = |\mathcal{Y}^{\mathrm{ss}}|$, we get a corresponding stratification using $\pi_{\mathrm{HT}} : |\mathcal{Y}| \to \mathbb{P}^1$

$$\mathcal{Y}^{\mathrm{ss}} = \pi_{\mathrm{HT}}^{-1}(\Omega), \quad \mathcal{Y}^{\mathrm{ord}} = \pi_{\mathrm{HT}}^{-1}(\mathbb{P}^1(\mathbb{Q}_p)).$$

Then clearly

$$|\hat{\mathcal{Y}}^{\mathrm{ss}}| = |\mathcal{Y}^{\mathrm{ss}}| \quad \text{and} \quad |\hat{\mathcal{Y}}^{\mathrm{ord}}| = |\mathcal{Y}^{\mathrm{ord}}|.$$

Finally, we note that pulling back the coordinate $z = -x/y$ on $\mathbb{P}^1$, we also obtain the Hodge-Tate period via

$$\hat{\mathbf{z}} = \pi_{\text{HT}}^* z \tag{3.14}$$

on $\mathcal{Y}$. Moreover,

$$HT_{\mathcal{A}}(\alpha_{\infty,1}) = \pi_{\text{HT}}^* x, \qquad HT_{\mathcal{A}}(\alpha_{\infty,2}) = \pi_{\text{HT}}^*(-y).$$

By (3.7), $\hat{\mathbf{z}}$ satisfies the transformation property

$$\begin{pmatrix} a & b \\ c & d \end{pmatrix}^* \hat{\mathbf{z}} = \frac{d\hat{\mathbf{z}} + b}{c\hat{\mathbf{z}} + a} \tag{3.15}$$

for any $\begin{pmatrix} a & b \\ c & d \end{pmatrix} \in GL_2(\mathbb{Q}_p)$.

3.5   THE LUBIN-TATE PERIOD ON THE SUPERSINGULAR LOCUS

Throughout this section, let $Y_i = Y(\Gamma_1(N) \cap \Gamma(p^i))$. Let F denote the p-adic completion of the maximal unramified extension of $\mathbb{Q}$, and $\mathcal{O}_F = W(\overline{\mathbb{F}}_p)$ its ring of integers.

Definition 3.9. *Fix a 1-dimensional p-divisible group H over $\overline{\mathbb{F}}_p$ of height n. (Recall there is a unique such p-divisible group, up to isomorphism over $\overline{\mathbb{F}}_p$, by Lubin-Tate theory.) Let $M(H)$ denote the Dieudonné module of H, which is a free $\mathcal{O}_F$-module of rank n.*

For each open compact subgroup $K \subset GL_n(\mathbb{Q}_p)$, there is a rigid analytic Lubin-Tate space LT_K, which parametrizes isomorphism classes of triples (G, ρ, α) consisting of a quasi-isogeny class of deformations (G, ρ) of H and a level structure α trivializing some part of the Tate module $T_p G$ (here K is exactly the subgroup of $GL_2(\mathbb{Z}_p)$ which fixes α). Let D be the ramified quaternion algebra over $\mathbb{Q}_p$. Recall that we have

$$\text{End}_{\mathbb{Q}_p}(H) \cong D.$$

Hence each LT_K carries an action of $\mathcal{O}_D^\times$ (which does not act on the level structure K). Each LT_K can be viewed as a Rapoport Zink space ([52, Chapter 6] or [62, Section 6]) and is equipped finite étale Grothendieck-Messing crystalline period map

$$\pi_{\text{GM},K} : LT_i \to \mathbb{P}(M(H)[1/p])$$

which is an étale map of rigid spaces over F (and hence induces an étale map of adic spaces over $\mathrm{Spa}(F, \mathcal{O}_F)$) with fiber set $GL_n(\mathbb{Q}_p)/K$, and which is compatible for varying K. More precisely, if $K' \subset K$, then we have a factorization

$$\pi_{\mathrm{GM},K'} : LT'_K \to LT_K \xrightarrow{\pi_{GM,K}} \mathbb{P}(M(H)[1/p]).$$

Moreover, $\pi_{\mathrm{GM},K}$ is equivariant with respect to the action of $\mathcal{O}_D^\times$, where $\mathcal{O}_D^\times$ acts on $\mathbb{P}(M(H)[1/p])$ after viewing $\mathbb{P}(M(H)[1/p])$ as the Brauer-Severi variety for $D/\mathbb{Q}_p$ (which in particular splits over F).

After fixing a basis of the $\mathcal{O}_F$-module, so that we get trivializations

$$M(H) \cong \mathcal{O}_F^n, \quad M(H)[1/p] \cong F^n,$$

then we get a (non-canonical) isomorphism

$$\mathbb{P}(M(H)[1/p]) \cong \mathbb{P}_F^{n-1}. \tag{3.16}$$

In particular, the Grothendieck-Messing period maps are non canonical if we view them as maps to $\mathbb{P}_F^{n-1}$, unlike the global and local Hodge-Tate period maps constructed in [56] and [62].

The Grothendieck-Messing period map in the Lubin-Tate case was originally constructed by Gross-Hopkins ([26]). By the results of loc. cit., the maps $\pi_{GM,K}$ are surjective. See also [68] for a p-adic Hodge-theoretic discussion of the Grothendieck-Messing period map for Lubin-Tate spaces. Note that this implies that the rigid analytic space $\mathbb{P}(M(H)[1/p]) \cong \mathbb{P}_F^{n-1}$ has nontrivial étale covers, and in fact étale fundamental group (in the sense of [19]) with $GL_n(\mathbb{Q}_p)$ as a quotient. In particular, it is very far from being "simply connected."

The LT_K's form a tower

$$LT_\infty := \varprojlim_{K \subset GL_n(\mathbb{Q}_p)} LT_K$$

which can be viewed as a preperfectoid space over $\mathrm{Spa}(F, \mathcal{O}_F)$, equipped with a (right) action of $GL_n(\mathbb{Q}_p)$ acting on LT_K by the induced isomorphisms

$$g : LT_K \xrightarrow{\sim} LT_{gKg^{-1}}.$$

Such data also form what is called a local Shimura variety of PEL type.

Let us give a more detailed description of LT_i, following the perfectoid description given in [62, Definition 6.1.3]. One can define a functor from affinoid $(F, \mathcal{O}_F)$-algebras to sets of isomorphism classes of deformations of H with $\Gamma(p^i)$-level structure as follows. Given an affinoid $(F, \mathcal{O}_F)$-algebra (R, R^+), let

$$\mathcal{M}_i(R, R^+) = \{(G, \rho, \alpha_i)/(R, R^+)\}/\sim$$

where G/R^+ is a p-divisible group,

$$\rho : H \otimes_{\overline{\mathbb{F}}_p} R^+/p \to G \otimes_{R^+} R^+/p$$

is a quasi-isogeny, and

$$\alpha : (\mathbb{Z}/p^i)^n \xrightarrow{\sim} G[p^i]$$

is an isomorphism. Here $\sim$ denotes "modulo isomorphism," where an isomorphism of pairs $(G, \rho, \alpha) \to (G', \rho', \alpha')$ is an isomorphism $G \cong G'$ over R^+ respecting all the above data. It is proven in loc. cit. that $\mathcal{M}_i$ is representable by the adic space LT_i over $\mathrm{Spa}(F, \mathcal{O}_F)$.

Definition 3.10. *We give a more detailed description of the period maps*

$$\pi_i := \pi_{GM, \Gamma(p^i)} : LT_i \to \mathbb{P}(M(H)[1/p]),$$

from which it is apparent that the fibers of $\pi_i : LT_i \to \mathbb{P}(M(H)[1/p])$ are isogeny classes of (isomorphism classes of) triples (G, ρ, α). (A more general analogous statement holds for all $\pi_{GM,K}$; in particular, the period maps $\pi_{GM,K}$ "forget" the level structures α, and their values depend only on the isogeny class of (G, ρ).)

Let (R, R^+) be an affinoid $(F, \mathcal{O}_F)$ algebra. Suppose we have a (R, R^+) point $(G, \rho, \alpha) \in LT_i(R, R^+)$. Let $\check{G}$ denote the Cartier dual of G. Then the Dieudonné module $D(G/R^+)$ has a Hodge-Tate exact sequence

$$0 \to \Omega^1_{\check{G}} \to D(G/R^+)(R^+ \to R^+/pR^+)^\vee \to \mathrm{Lie}\, G \to 0 \qquad (3.17)$$

where here $D(\cdot)$ denotes the Messing crystal ([48, Chapter 4]), and the superscript $\vee$ denotes the R^+-linear dual. The quasi-isogeny

$$\rho : H \otimes_{\overline{\mathbb{F}}_p} R^+/pR^+ \to G \otimes_{R^+} R^+/pR^+$$

induces an isomorphism of F-vector spaces

$$\rho : D(H \otimes_{\overline{\mathbb{F}}_p} R^+/pR^+)(R^+ \to R^+/pR^+)^\vee[1/p]$$
$$\xrightarrow{\sim} D(G \otimes_{R^+} R^+/pR^+)(R^+ \to R^+/pR^+)^\vee[1/p].$$

By the base change properties of the Messing crystal, the above isomorphism is equivalent to an isomorphism

$$\rho : M(H)^\vee \otimes_{\mathcal{O}_F} R \xrightarrow{\sim} D(G/R^+)^\vee \otimes_{R^+} R.$$

Now using (3.17) and the above isomorphism, we have

$$\Omega^1_{\tilde{G}} \otimes_{R^+} R \subset D(G/R^+)^{\vee} \otimes_{R^+} R \xrightarrow[\sim]{\rho^{-1}} M(H)^{\vee} \otimes_{\mathcal{O}_F} R$$

which identifies a hyperplane in $M(H)^{\vee} \otimes_{\mathcal{O}_F} R$, and hence corresponds to a point in $\mathbb{P}(M(H)[1/p])(R)$, which is the value $\pi_{\mathrm{GM}}(G, \rho, \alpha)$. It is clear from the above construction that the value $\pi_{\mathrm{GM}}(R, \rho, \alpha)$ does not depend on α and only depends on the isogeny class of (G, ρ).

Definition 3.11. *Fix a uniformizer ϖ of $\mathcal{O}_D$. Let*

$$\Sigma_i \to \Sigma_0 = \Omega$$

denote Drinfeld's i^{th} covering of the Drinfeld plane $\Omega = \mathbb{P}^1_{\mathbb{C}_p} \setminus \mathbb{P}^1(\mathbb{Q}_p)$. (Drinfeld even constructed formal models of Σ_i over $\mathrm{Spf}(\mathcal{O}_F)$.) As a rigid analytic space, Σ_i parametrizes formal $\mathcal{O}_D$-modules G of height n with level structure $(\mathbb{Z}/p^i)^n \cong G[\varpi^i]$. Each Σ_i is a finite étale covering of Ω (viewed either as a rigid or adic space) which carries a natural action of $GL_n(\mathbb{Z}_p)$ (acting on the level structure), which can be extended to an action of $GL_2(\mathbb{Q}_p)$. We have isomorphisms $\mathrm{Gal}(\Sigma_i/\Omega) \cong (\mathcal{O}_D/\varpi^i\Omega_D)^{\times}$.

In parallel with LT_∞, we have a tower

$$\Sigma_\infty := \varprojlim_i \Sigma_i,$$

Σ_∞ is a preperfectoid space over $\mathrm{Spa}(F, \mathcal{O}_F)$, equipped with an action of $D^{\times}$.

It is a classical theorem due to Faltings and Fargues-Genestier-Lafforgue ([25]) that there is a $GL_2(\mathbb{Q}_p) \times D^{\times}$-equivariant isomorphism of pro-rigid spaces over F (or of pro-adic spaces over $\mathrm{Spa}(F, \mathcal{O}_F)$)

$$LT_\infty \cong \Sigma_\infty. \tag{3.18}$$

In fact, (3.18) is a special case of a more general theorem due to Faltings ([23]) and Scholze-Weinstein ([62, Theorem 7.2.3]) of the *duality isomorphism* between Rapoport-Zink states with dual Rapoport-Zink data. Scholze-Weinstein formulate (3.18) as an isomorphism of preperfectoid spaces over $\mathrm{Spa}(F, \mathcal{O}_F)$. In fact, the induced isomorphism of strong (p-adic) completions

$$\widehat{LT}_\infty \cong \hat{\Sigma}_\infty$$

is a $GL_2(\mathbb{Q}_p) \times D^{\times}$-equivariant isomorphism of perfectoid spaces over $\mathrm{Spa}(F, \mathcal{O}_F)$, which can be viewed naturally inside the supersingular locus of the perfectoid infinite-level modular curve (base changed to $\mathrm{Spa}(F, \mathcal{O}_F)$)

$$\hat{\Sigma}_\infty \cong \widehat{LT}_\infty \subset \dot{\mathcal{Y}}_F^{\mathrm{ss}}.$$

Again, we have

$$\widehat{LT}_\infty \sim LT_\infty, \quad \hat{\Sigma}_\infty \sim \Sigma_\infty, \quad |\widehat{LT}_\infty| \cong |LT_\infty|, \quad |\hat{\Sigma}_\infty| \cong |\Sigma_\infty|.$$

Let $\mathbb{A}_{\mathbb{Q}}^{(\infty)}$ denote the finite adèles, and let

$$K_p(p^i) \subset GL_2(\mathbb{Q}_p), \quad K^p(N) \subset GL_2(\mathbb{A}_{\mathbb{Q}}^{(p\infty)})$$

denote the usual subgroups corresponding to $\Gamma(p^i)$ and $\Gamma_1(N)$. Base change Y to $\mathrm{Spa}(F, \mathcal{O}_F)$, recalling that we denote this by Y_F. Suppose now that $n = 2$. We recall the Rapoport-Zink uniformization ([52, Theorem 6.30]), which gives a profinite étale map

$$D^\times(\mathbb{Q}) \backslash (LT_i \times GL_2(\mathbb{A}_{\mathbb{Q}}^{(p\infty)})) / K^p(N) \to Y_{i,F}^{\mathrm{ss}}$$

over $\mathrm{Spa}(F, \mathcal{O}_F)$. The left-hand side can alternatively be written as a (countable) disjoint union

$$\bigsqcup_j \Gamma_j \backslash LT_i$$

where Γ_j acts discretely and continuously on LT_i, and so we have profinite étale maps

$$LT_i \to Y_{i,F}^{\mathrm{ss}}$$

of adic spaces over $\mathrm{Spa}(\mathbb{Q}_p, \mathbb{Z}_p)$. Putting these maps together, we get a proétale map

$$LT_\infty = \varprojlim_i LT_i \to \varprojlim_i Y_{i,F}^{\mathrm{ss}} = \mathcal{Y}_F^{\mathrm{ss}}$$

in the sense of Definition 2.49. Hence we can view LT_∞ as an object of $Y_{\mathrm{proét}}/\mathcal{Y}^{\mathrm{ss}} \subset Y_{\mathrm{proét}}$.

Definition 3.12. *For simplicity, from now on let*

$$\pi_{\mathrm{GM}} := \pi_{\mathrm{GM},0} : LT_0 \to \mathbb{P}(M(H)[1/p]).$$

Now fix a basis of $M(H)$, so that as in (3.16), we can view π_{GM} as a map

$$\pi_{\mathrm{GM}} : LT_0 \to \mathbb{P}_F^1.$$

Let

$$x, y \in \mathcal{O}_{\mathbb{P}^1_F}(1)(\mathbb{P}^1_F)$$

be the canonical global sections (which can be viewed as elements

$$x, y \in \mathcal{O}_{\mathbb{P}^1_F}(1)(\mathbb{P}^1_F) = \mathcal{O}_{\mathbb{P}(M(H)[1/p])}(1)(\mathbb{P}(M(H)[1/p]))$$

corresponding to the above choice of basis for $M(H)$). Let $z = -x/y$ be the usual coordinate on the affine chart $\{y \neq 0\}$. We then get a section

$$\mathbf{z}_{\mathrm{LT}} := \pi^*_{\mathrm{GM},0} z \in \mathcal{O}_Y(LT_i)$$

for any i such that $LT_i \to Y$ via the Rapoport-Zink uniformization. We call $\mathbf{z}_{\mathrm{LT}}$ the Lubin-Tate period. We then get a section

$$d\mathbf{z}_{\mathrm{LT}} \in \Omega^1_Y(LT_i).$$

3.6 THE RELATIVE HODGE-TATE FILTRATION

By definition, $\mathcal{H}^{0,1}$ is the kernel of the map

$$\mathrm{HT}_{\mathcal{A}} : T_p\mathcal{A} \otimes_{\hat{\mathbb{Z}}_{p,Y}} \hat{\mathcal{O}}_Y \to \hat{\omega} \tag{3.19}$$

and thus tensoring with $\otimes_{\hat{\mathcal{O}}_Y} \hat{\mathcal{O}}\mathbb{B}^+_{\mathrm{dR},Y}$, then $\mathcal{H}^{0,1} \otimes_{\hat{\mathcal{O}}_Y} \hat{\mathcal{O}}\mathbb{B}^+_{\mathrm{dR},Y}$ is the kernel of the map

$$\mathrm{HT}_{\mathcal{A}} : T_p\mathcal{A} \otimes_{\hat{\mathbb{Z}}_{p,Y}} \hat{\mathcal{O}}\mathbb{B}^+_{\mathrm{dR},Y} \to \omega \otimes_{\mathcal{O}_Y} \hat{\mathcal{O}}\mathbb{B}^+_{\mathrm{dR},Y} \tag{3.20}$$

which reduces modulo $\ker(\theta : \hat{\mathcal{O}}\mathbb{B}^+_{\mathrm{dR},Y} \twoheadrightarrow \hat{\mathcal{O}}_Y)$ to (3.19).

We note that $\hat{z}$ is the reciprocal of the *fundamental period* as defined in [13, Definition 1.2], since it is defined by a relation for (3.19) equivalent to the relation (3.12).

Caraiani and Scholze [9] define the *relative Hodge filtration*

$$0 \to \hat{\omega}^{-1} \to \mathcal{H}^1_{\text{ét}}(\mathcal{A}) \otimes_{\hat{\mathbb{Z}}_{p,Y}} \hat{\mathcal{O}}_Y \to \hat{\omega}(-1) \to 0, \tag{3.21}$$

or after twisting by (1) and applying the isomorphism $\mathcal{H}^1_{\text{ét}}(\mathcal{A})(1) \cong T_p\mathcal{A}$ given by (3.4), equivalently as

$$0 \to \hat{\omega}^{-1}(1) \to T_p\mathcal{A} \otimes_{\hat{\mathbb{Z}}_{p,Y}} \hat{\mathcal{O}}_Y \to \hat{\omega} \to 0. \tag{3.22}$$

Taking the specialization of (3.22) at a geometric point $Y = (A) \in Y(\overline{\mathbb{Q}}_p, \mathcal{O}_{\overline{\mathbb{Q}}_p})$, we obtain the usual Hodge-Tate exact sequence

$$0 \to \mathrm{Lie}(A)(1) \xrightarrow{(HT_A)^\vee(1)} T_pA \otimes_{\mathbb{Z}_p} \mathbb{C}_p \xrightarrow{HT_A} \Omega^1_{A/\mathbb{C}_p} \to 0. \qquad (3.23)$$

We have the following result on the integral Hodge-Tate complex.

Theorem 3.13 ([25], Theorem II.1.1). *Suppose $p > 2$. Taking an integral model $A/\mathcal{O}_{\mathbb{C}_p}$, we have a sequence*

$$0 \to \mathrm{Lie}(A)(1) \xrightarrow{(HT_A)^\vee(1)} T_pA \otimes_{\mathbb{Z}_p} \mathcal{O}_{\mathbb{C}_p} \xrightarrow{HT_A} \Omega^1_{A/\mathcal{O}_{\mathbb{C}_p}} \to 0$$

which is in general not *exact, but which is a complex with cohomology groups killed by $p^{1/(p-1)}$.*

Proposition 3.14. *The penultimate arrow in (3.22) is given by HT_A from (3.20), and its kernel is given by the image of*

$$(HT_A)^\vee(1) : \hat{\omega}^{-1}(1) \hookrightarrow T_pA \otimes_{\hat{\mathbb{Z}}_{p,Y}} \hat{\mathcal{O}}_Y. \qquad (3.24)$$

In particular, since $\mathcal{H}^{0,1} \otimes_{\mathcal{O}_Y} \hat{\mathcal{O}}_Y$ is the kernel of (3.20), we have

$$(HT_A)^\vee(1)(\hat{\omega}^{-1}) = \mathcal{H}^{0,1} \qquad (3.25)$$

and

$$(HT_A)^\vee(1)(\hat{\omega}^{-1} \otimes_{\hat{\mathcal{O}}_Y} \hat{\mathcal{O}}\mathbb{B}^+_{\mathrm{dR}Y}(1)) = \mathcal{H}^{0,1} \otimes_{\hat{\mathcal{O}}_Y} \hat{\mathcal{O}}\mathbb{B}^+_{\mathrm{dR},Y}. \qquad (3.26)$$

Proof. By (3.23), the penultimate arrow in (3.21), which is a map locally of finite free $\hat{\mathcal{O}}_Y$-modules, specializes to HT_A on geometric fibers. Hence by Nakayama's lemma, it specializes to HT_A on geometric stalks, and so the map must be HT_A since $Y_{\mathrm{pro\acute{e}t}}$ has enough points. Since the kernel of HT_A in (3.23) is given by the image of $(HT_A)^\vee(1)$, the same stalk-wise argument works to show that the kernel of (3.20) is given by the image of $(HT_A)^\vee(1)$. This immediately implies (3.25). For the final claim, we again can check on stalks at profinite covers of geometric points: since

$$\hat{\mathcal{O}}\mathbb{B}^+_{\mathrm{dR},Y}/(\ker\theta) \cong \hat{\mathcal{O}}_Y,$$

by (3.25) and Nakayama's lemma (since the image of $\ker\theta$ in the stalk is the maximal ideal of the stalk of $\hat{\mathcal{O}}\mathbb{B}^+_{\mathrm{dR},Y}$) we have (3.26). $\qquad \square$

3.7 THE FAKE HASSE INVARIANT

Henceforth, we define the *fake Hasse invariant* to be as follows.

Definition 3.15.
$$\mathfrak{s} := HT_{\mathcal{A}}(\alpha_{\infty,2}) \in \Gamma(\mathcal{Y}, \hat{\omega}), \tag{3.27}$$

which is a nonvanishing section in $\hat{\omega}(\mathcal{Y}_x)$, and so trivializes $\hat{\omega}$ on $\mathcal{Y}_x$. It is equivalently obtained by pulling back the global section x of $\mathcal{O}_{\mathbb{P}^1}(1)$ along π_{HT}: $\hat{\mathcal{Y}} \to \mathbb{P}^1$.

We will briefly address a formal model of $\mathcal{Y}$ corresponding to the compactification X of Y. As per the results of [56], there is a proétale cover $\hat{\mathcal{X}} \to X$ extending $\hat{\mathcal{Y}} \to Y$. Let $\mathfrak{X}$ be the formal model corresponding to $\hat{\mathcal{X}}$ by applying [56, Lemma II.1.1] to the proper adic space $\hat{\mathcal{X}}$. Then there is a line bundle ω^+ on $\mathfrak{X}$ whose generic fiber is ω:

$$\omega^+ \otimes_{\mathcal{O}_{\mathfrak{X}}} \mathcal{O}_X = \omega.$$

In fact, by the same lemma in loc. cit., we have that

$$\mathfrak{s} \in \Gamma(\mathfrak{X}, \omega^+)$$

and in fact trivializes ω^+ on the adic-theoretic closure of $\hat{\mathcal{Y}}_x$ in $\mathfrak{X}$.

3.8   RELATIVE DE RHAM COHOMOLOGY AND THE HODGE–DE RHAM FILTRATION

Definition 3.16. *Recall the Poincaré sequence (of $\pi^{-1}\mathcal{O}_Y$-modules)*

$$0 \to \mathcal{O}_{\mathcal{A}} \xrightarrow{d} \Omega^1_{\mathcal{A}/Y} \to 0 \tag{3.28}$$

where $d: \mathcal{O}_{\mathcal{A}} \to \Omega^1_{\mathcal{A}/Y}$ is the tensorial exterior differential. Taking the first higher direct image under $\pi: \mathcal{A} \to Y$ of this (3.28), we get the usual de Rham bundle

$$\mathcal{H}^1_{\mathrm{dR}}(\mathcal{A}) := R^1\pi_{\mathrm{dR},*}\mathcal{O}_{\mathcal{A}} = R^1\pi_*(0 \to \mathcal{O}_{\mathcal{A}} \to \Omega^1_{\mathcal{A}/Y} \to 0)$$

on $Y_{pro\acute{e}t}$, which is a rank 2 vector bundle equipped with the usual Hodge–de Rham filtration

$$0 \to \omega \to \mathcal{H}^1_{\mathrm{dR}}(\mathcal{A}/Y) \to \omega^{-1} \to 0 \tag{3.29}$$

and Gauss-Manin connection

$$\nabla: \mathcal{H}^1_{\mathrm{dR}}(\mathcal{A}) \to \mathcal{H}^1_{\mathrm{dR}}(\mathcal{A}) \otimes_{\mathcal{O}_Y} \Omega^1_Y \tag{3.30}$$

which satisfies Griffiths transversality with respect to the Hodge–de Rham filtration.

We extend the filtration (3.29) to

$$\mathcal{H}^1_{\mathrm{dR}}(\mathcal{A}) \otimes_{\mathcal{O}_Y} \mathcal{O}\mathbb{B}^+_{\mathrm{dR},Y}$$

by taking the convolution of the (decreasing) Hodge–de Rham filtration with the natural (decreasing) filtration on $\mathcal{O}\mathbb{B}^+_{\mathrm{dR},Y}$ as defined in Section 2.7. That is, we define

$$\mathrm{Fil}^i(\mathcal{H}^1_{\mathrm{dR}}(\mathcal{A}) \otimes_{\mathcal{O}_Y} \mathcal{O}\mathbb{B}^+_{\mathrm{dR},Y}) = \sum_{j \in \mathbb{Z}} \mathrm{Fil}^{-j} \mathcal{H}^1_{\mathrm{dR}}(\mathcal{A}) \otimes_{\mathcal{O}_Y} \mathrm{Fil}^{i+j} \mathcal{O}\mathbb{B}^+_{\mathrm{dR},Y}.$$

Since the natural connection on $\mathcal{O}\mathbb{B}^+_{\mathrm{dR},Y}$ extends $d : \mathcal{O}_Y \to \Omega^1_Y$, we can extend (3.30) to $\mathcal{H}^1_{\mathrm{dR}}(\mathcal{A})$ by the Leibniz rule to get a connection

$$\nabla : \mathcal{H}^1_{\mathrm{dR}}(\mathcal{A}) \otimes_{\mathcal{O}_Y} \mathcal{O}\mathbb{B}^+_{\mathrm{dR},Y} \to \mathcal{H}^1_{\mathrm{dR}}(\mathcal{A}) \otimes_{\mathcal{O}_Y} \mathcal{O}\mathbb{B}^+_{\mathrm{dR},Y} \otimes_{\mathcal{O}_Y} \Omega^1_Y$$

of filtered locally free $\mathcal{O}\mathbb{B}^+_{\mathrm{dR},Y}$-modules with integrable connection satisfying Griffiths transversality with respect to the convolution filtration on $\mathcal{H}^1_{\mathrm{dR}}(\mathcal{A}) \otimes_{\mathcal{O}_Y} \mathcal{O}\mathbb{B}^+_{\mathrm{dR},Y}$. Here, recall that given a locally free $\mathcal{O}$-module with decreasing filtration $\mathcal{M}$ and a connection

$$\nabla : \mathcal{M} \to \mathcal{M} \otimes_{\mathcal{O}} \Omega^1,$$

for ∇ to satisfy Griffiths transversality means that

$$\nabla(\mathrm{Fil}^i \mathcal{M}) \subset \mathrm{Fil}^{i-1} \mathcal{M} \otimes_{\mathcal{O}} \Omega^1.$$

3.9 RELATIVE p-ADIC DE RHAM COMPARISON THEOREM APPLIED TO $\mathcal{A} \to Y$

We now apply the comparison theorem of Section 2.9 to the smooth proper morphism $\mathcal{A} \to Y$ over $\mathrm{Spa}(\mathbb{Q}_p, \mathbb{Z}_p)$. In the notation of the section, in our situation we have

$$\mathcal{E} = \mathcal{O}_{\mathcal{A}}, \quad \mathbb{L} = \hat{\mathbb{Z}}_{p,\mathcal{A}}.$$

Theorem 2.72 says that we have an isomorphism

$$T_p \mathcal{A} \otimes_{\hat{\mathbb{Z}}_{p,Y}} \mathcal{O}\mathbb{B}_{\mathrm{dR},Y} = \mathcal{H}^1_{\text{ét}}(\mathcal{A}) \otimes_{\hat{\mathbb{Z}}_{p,Y}} \mathcal{O}\mathbb{B}_{\mathrm{dR},Y} \cong \mathcal{H}^1_{\mathrm{dR}}(\mathcal{A}) \otimes_{\mathcal{O}_Y} \mathcal{O}\mathbb{B}_{\mathrm{dR},Y} \quad (3.31)$$

of sheaves on $Y_{\mathrm{proét}}$, compatible with filtrations and connections. The compatibility with the connections implies that the *horizontal* sections (i.e., sections f

with $\nabla(f) = 0$ of the Gauss-Manin connection

$$\nabla : \mathcal{H}_{dR}^1(\mathcal{A}) \otimes_{\mathcal{O}_Y} \mathcal{O}\mathbb{B}_{dR,Y} \cong T_p\mathcal{A} \otimes_{\hat{\mathbb{Z}}_{p,Y}} \mathcal{O}\mathbb{B}_{dR,Y} \to T_p\mathcal{A} \otimes_{\hat{\mathbb{Z}}_{p,Y}} \mathcal{O}\mathbb{B}_{dR,Y} \otimes_{\mathcal{O}_Y} \Omega_Y^1$$
$$\cong \mathcal{H}_{dR}^1(\mathcal{A}) \otimes_{\mathcal{O}_Y} \mathcal{O}\mathbb{B}_{dR,Y} \otimes_{\mathcal{O}_Y} \Omega_Y^1 \tag{3.32}$$

are given by

$$(T_p\mathcal{A} \otimes_{\hat{\mathbb{Z}}_{p,Y}} \mathcal{O}\mathbb{B}_{dR,Y})^{\nabla=0} = T_p\mathcal{A} \otimes_{\hat{\mathbb{Z}}_{p,Y}} \mathbb{B}_{dR,Y}.$$

Definition 3.17. *Recall that we have that the* $\mathbb{B}_{dR,Y}^+$*-local systems*

$$\mathbb{M} = (\mathcal{H}_{\acute{e}t}^1(\mathcal{A}) \otimes_{\hat{\mathbb{Z}}_{p,Y}} \mathcal{O}\mathbb{B}_{dR,Y}^+)^{\nabla=0} = \mathcal{H}_{\acute{e}t}^1(\mathcal{A}) \otimes_{\hat{\mathbb{Z}}_{p,Y}} \mathbb{B}_{dR,Y}^+$$

and

$$\mathbb{M}_0 = (\mathcal{H}_{dR}^1(\mathcal{A}) \otimes_{\mathcal{O}_Y} \mathcal{O}\mathbb{B}_{dR,Y}^+)^{\nabla=0}.$$

From Theorem 2.61, we have

$$\mathbb{M}_0 \otimes_{\mathbb{B}_{dR,Y}^+} \mathcal{O}\mathbb{B}_{dR,Y}^+ = \mathcal{H}_{dR}^1(\mathcal{A}) \otimes_{\mathcal{O}_Y} \mathcal{O}\mathbb{B}_{dR,Y}^+. \tag{3.33}$$

By (2.14), they satisfy

$$\mathbb{M}_0 \subset \mathbb{M}. \tag{3.34}$$

Thus, after tensoring with $\otimes_{\mathbb{B}_{dR,Y}^+}^+ \mathcal{O}\mathbb{B}_{dR,Y}^+$ *and invoking Theorem 2.61, we have the following natural inclusion of* $\mathcal{O}\mathbb{B}_{dR,Y}^+$*-modules which we denote by* ι_{dR}*:*

$$\mathcal{H}_{dR}^1(\mathcal{A}) \otimes_{\mathcal{O}_Y} \mathcal{O}\mathbb{B}_{dR,Y}^+ = \mathbb{M}_0 \otimes_{\mathbb{B}_{dR,Y}^+} \mathcal{O}\mathbb{B}_{dR,Y}^+ \overset{\iota_{dR}}{\subset} \mathbb{M} \otimes_{\mathbb{B}_{dR,Y}^+} \mathcal{O}\mathbb{B}_{dR,Y}^+$$
$$\cong \mathcal{H}_{\acute{e}t}^1(\mathcal{A}) \otimes_{\hat{\mathbb{Z}}_{p,Y}} \mathcal{O}\mathbb{B}_{dR,Y}^+. \tag{3.35}$$

Using the isomorphism

$$\mathcal{H}_{\acute{e}t}^1(\mathcal{A}) \cong T_p\mathcal{A}(-1)$$

furnished by (3.4), this induces an inclusion

$$\mathcal{H}_{dR}^1(\mathcal{A}) \otimes_{\mathcal{O}_Y} \mathcal{O}\mathbb{B}_{dR,Y}^+(1) = \mathbb{M}_0 \otimes_{\mathbb{B}_{dR,Y}^+} \mathcal{O}\mathbb{B}_{dR,Y}^+(1) \overset{\iota_{dR}}{\hookrightarrow} \mathbb{M} \otimes_{\mathbb{B}_{dR,Y}^+} \mathcal{O}\mathbb{B}_{dR,Y}^+(1)$$
$$\cong T_p\mathcal{A} \otimes_{\hat{\mathbb{Z}}_{p,Y}} \mathcal{O}\mathbb{B}_{dR,Y}^+, \tag{3.36}$$

which, upon evaluating on $\mathcal{Y} \to Y$ *and using*

$$\hat{\mathbb{Z}}_{p,\mathcal{Y}}(1) \cong \hat{\mathbb{Z}}_{p,\mathcal{Y}} \cdot t$$

from (4.3), induces

$$\mathcal{H}^1_{\mathrm{dR}}(\mathcal{A}) \otimes_{\mathcal{O}_Y} \mathcal{O}\mathbb{B}^+_{\mathrm{dR},\mathcal{Y}} \cdot t = \mathbb{M}_0 \otimes_{\mathbb{B}^+_{\mathrm{dR},Y}} \mathcal{O}\mathbb{B}^+_{\mathrm{dR},\mathcal{Y}} \cdot t \overset{\iota_{\mathrm{dR}}}{\hookrightarrow} \mathbb{M} \otimes_{\mathbb{B}^+_{\mathrm{dR},Y}} \mathcal{O}\mathbb{B}^+_{\mathrm{dR},\mathcal{Y}} \cdot t$$

$$\cong T_p \mathcal{A} \otimes_{\hat{\mathbb{Z}}_{p,Y}} \mathcal{O}\mathbb{B}^+_{\mathrm{dR},\mathcal{Y}} \xrightarrow{\alpha_\infty^{-1}} \mathcal{O}\mathbb{B}^{+,\oplus 2}_{\mathrm{dR},\mathcal{Y}} \tag{3.37}$$

or equivalently

$$\mathcal{H}^1_{\mathrm{dR}}(\mathcal{A}) \otimes_{\mathcal{O}_Y} \mathcal{O}\mathbb{B}^+_{\mathrm{dR},\mathcal{Y}} = \mathbb{M}_0 \otimes_{\mathbb{B}^+_{\mathrm{dR},Y}} \mathcal{O}\mathbb{B}^+_{\mathrm{dR},\mathcal{Y}} \overset{\iota_{\mathrm{dR}}}{\hookrightarrow} \mathbb{M} \otimes_{\mathbb{B}^+_{\mathrm{dR},Y}} \mathcal{O}\mathbb{B}^+_{\mathrm{dR},\mathcal{Y}}$$

$$\cong T_p \mathcal{A} \otimes_{\hat{\mathbb{Z}}_{p,Y}} \mathcal{O}\mathbb{B}^+_{\mathrm{dR},\mathcal{Y}} \cdot t^{-1} \xrightarrow{\alpha_\infty^{-1}} (\mathcal{O}\mathbb{B}^+_{\mathrm{dR},\mathcal{Y}} \cdot t^{-1})^{+,\oplus 2}. \tag{3.38}$$

Tensoring with $\otimes_{\mathcal{O}\mathbb{B}^+_{\mathrm{dR},Y}} \hat{\mathcal{O}}\mathbb{B}^+_{\mathrm{dR},Y}$, we obtain

$$\mathcal{H}^1_{\mathrm{dR}}(\mathcal{A}) \otimes_{\mathcal{O}_Y} \hat{\mathcal{O}}\mathbb{B}^+_{\mathrm{dR},\mathcal{Y}} = \mathbb{M}_0 \otimes_{\mathbb{B}^+_{\mathrm{dR},Y}} \hat{\mathcal{O}}\mathbb{B}^+_{\mathrm{dR},\mathcal{Y}} \overset{\iota_{\mathrm{dR}}}{\hookrightarrow} \mathbb{M} \otimes_{\mathbb{B}^+_{\mathrm{dR},Y}} \hat{\mathcal{O}}\mathbb{B}^+_{\mathrm{dR},\mathcal{Y}}$$

$$\cong T_p \mathcal{A} \otimes_{\hat{\mathbb{Z}}_{p,Y}} \hat{\mathcal{O}}\mathbb{B}^+_{\mathrm{dR},\mathcal{Y}} \cdot t^{-1} \xrightarrow{\alpha_\infty^{-1}} (\hat{\mathcal{O}}\mathbb{B}^+_{\mathrm{dR},\mathcal{Y}} \cdot t^{-1})^{+,\oplus 2}. \tag{3.39}$$

Chapter Four

The fundamental de Rham periods

In this chapter, we give the construction of the p-adic Maass-Shimura operator. This involves considering various periods and period sheaves arising from the de Rham comparison theorem, and then constructing a suitable horizontal splitting arising from the Hodge filtration.

4.1 A PROÉTALE LOCAL DESCRIPTION OF $\mathcal{O}\mathbb{B}_{\mathrm{dR}}^{(+)}$

Recall that for an affinoid $\mathcal{U} \subset \mathcal{Y}^{\mathrm{aa}}$ we have an étale morphism of adic spaces $\pi_{\mathrm{HT}} : \mathcal{U} \to \Omega \subset \mathbb{P}^1$. Using the p-adic exponential, this adic-locally defines an étale morphism

$$\mathcal{U} \to \mathbb{T} = \mathrm{Spa}(\mathbb{Q}_p\langle T, T^{-1}\rangle, \mathbb{Z}_p\langle T, T^{-1}\rangle),$$

and taking finite étale covers

$$\mathbb{T}^{1/p^m} = \mathrm{Spa}(\mathbb{Q}_p\langle T^{1/p^m}, T^{-1/p^m}\rangle, \mathbb{Z}_p\langle T^{1/p^m}, T^{-1/p^m}\rangle)$$

of $\mathbb{T}$ to extract p^{th}-power roots of the p-adic exponential, we can extend this to étale morphisms $\mathcal{U} \to \mathbb{T}$ defined on larger étale neighborhoods. In this section, we will deal with the more general situation where we have an arbitrary adic space $\mathcal{U}$ over $\mathrm{Spa}(\mathbb{Q}_p, \mathbb{Z}_p)$ with a fixed étale–locally defined étale morphism $\pi : \mathcal{U} \to \mathbb{T}$, and specialize in Chapter 6, Section 6.1, to the situation described above.

Let $K/\mathbb{Q}_p$ denote a perfectoid field, with ring of integers $\mathcal{O}_K$.

$$\tilde{\mathbb{T}} = \varprojlim_m \mathbb{T}^{1/p^m} \in \mathbb{T}_{\mathrm{proét}}. \tag{4.1}$$

Letting $\tilde{\mathbb{T}}_K$ denote the base change to $\mathrm{Spa}(K, \mathcal{O}_K)$, this is an affinoid perfectoid object of $\mathbb{T}_{K,\mathrm{proét}}$. Using our given étale–locally defined étale map $\mathcal{U} \to \mathbb{T}$, let

$$\tilde{\mathcal{U}} = \mathcal{U} \times_{\mathbb{T}} \tilde{\mathbb{T}} \in \mathcal{U}_{\mathrm{proét}}.$$

Let $\tilde{\mathcal{U}}_K \in \mathcal{U}_{K,\mathrm{proét}}$ denote the base change to $\mathrm{Spa}(K, \mathcal{O}_K)$, so that $\tilde{\mathcal{U}}_K$ is an affinoid perfectoid object in $\tilde{\mathcal{U}}_{K,\mathrm{proét}}$. The following proposition and its proof are copied almost verbatim from [57], though now we take into account the corrected definition of $\mathcal{O}\mathbb{B}_{\mathrm{dR}}^{+}$ (see (2.8)).

Proposition 4.1 ([57], Proposition 6.10). *We have a natural isomorphism of sheaves on the localized site* $\mathcal{U}_{pro\acute{e}t}/\tilde{\mathcal{U}}$

$$\mathbb{B}_{\mathrm{dR},\tilde{\mathcal{U}}}^{+}[\![X]\!] \xrightarrow{\sim} \mathcal{O}\mathbb{B}_{\mathrm{dR},\tilde{\mathcal{U}}}^{+}$$

sending $X \mapsto T \otimes 1 - 1 \otimes [T^{\flat}]$. *In particular, we have*

$$\mathbb{B}_{\mathrm{dR},\tilde{\mathcal{U}}}^{+}[\![X]\!][t^{-1}] \xrightarrow{\sim} \mathcal{O}\mathbb{B}_{\mathrm{dR},\tilde{\mathcal{U}}}.$$

Proof. As in loc. cit., the key ingredient of the argument follows.

Claim 4.2. *There is a* unique *morphism*

$$\mathcal{O}_{\tilde{\mathcal{U}}} \to \mathbb{B}_{\mathrm{dR},\tilde{\mathcal{U}}}^{+}[\![X]\!]$$

sending $T \mapsto [T^{\flat}] + X$ *and such that the resulting map*

$$\mathcal{O}_{\tilde{\mathcal{U}}} \to \mathbb{B}_{\mathrm{dR},\tilde{\mathcal{U}}}^{+}[\![X]\!]/\ker\theta = \hat{\mathcal{O}}_{\tilde{\mathcal{U}}}$$

is the natural inclusion.

Once we have shown this claim, we get a natural map

$$(\mathcal{O}_{\mathcal{U}} \otimes_{W(\kappa)} W(\hat{\mathcal{O}}_{\mathcal{U}^{\flat}}^{+}))|_{\tilde{\mathcal{U}}} \to \mathbb{B}_{\mathrm{dR},\mathcal{U}}^{+}|_{\tilde{\mathcal{U}}}[\![X]\!]$$

which induces the inverse of the map in the statement upon passing to the p-adic completion (here we use that $\mathbb{B}_{\mathrm{dR},\mathcal{U}}^{+}|_{\tilde{\mathcal{U}}}$ is p-adically complete), inverting p the $(\ker\theta)$-adic completion.

In order to show the claim, we recall the following lemmas from [57].

Lemma 4.3 ([57], Lemma 6.11). *Let* (R, R^{+}) *be a perfectoid affinoid* $(k, \mathcal{O}_{k})$-*algebra, and let* S *be a finitely generated* $\mathcal{O}_{K}$-*algebra. Then any morphism*

$$f : S \to \mathbb{B}_{\mathrm{dR}}^{+}(R, R^{+})[\![X]\!]$$

such that $\theta(f(S)) \subset R^{+}$ *extends to the* p-*adic completion of* S.

Lemma 4.4 (cf. [57], Lemma 6.12). *Let* $V = \mathrm{Spa}(R, R^{+})$ *be an affinoid adic space of finite type over* $\mathrm{Spa}(W(\kappa)[p^{-1}], W(\kappa))$ *with an étale map* $V \to \mathbb{T}$. *Then* *there exists a finitely generated* $W(\kappa)[T]$-*algebra* R_0^+, *such that* $R_0 = R_0^+[p^{-1}]$ *is étale over*

$$W(\kappa)[p^{-1}][T]$$

and R^+ *is the* p–*adic completion of* R_0^+.

Proof of Claim 4.2. We again follow the argument in loc. cit. Note that we have a map

$$W(\kappa)[p^{-1}][T] \to \mathbb{B}_{\mathrm{dR},\mathcal{U}}^+|_{\tilde{\mathcal{U}}}[\![X]\!] \tag{4.2}$$

sending T to $[T^\flat] + X$. Now take any affinoid perfectoid $V \in \mathcal{U}_{\mathrm{proét}}/\tilde{\mathcal{U}}_K$ writing it as a union of inverse limits of affinoid $V_i \in \mathcal{U}_{\mathrm{proét}}$. Then $\mathcal{O}_\mathcal{U}(V) = \varinjlim_i \mathcal{O}_\mathcal{U}(V_i)$. Applying Lemma 4.4 to $V_i = \mathrm{Spa}(R_i, R_i^+) \to \mathcal{U} \to \mathbb{T}$, we have algebras R_{i0}^+ whose generic fibers R_{i0} are étale over $W(\kappa)[p^{-1}][T]$. By Hensel's lemma, the map (4.2) *uniquely* lifts to a map $R_{i0} \to \mathbb{B}_{\mathrm{dR},\mathcal{U}}^+(V)[\![X]\!]$ and in particular we get a unique lifting of R_{i0}^+. This map in turn extends by Lemma 4.3 to the p-adic completion $R_i^+ = \mathcal{O}_\mathcal{U}^+(V_i)$, and hence we get a unique lifting $\mathcal{O}_\mathcal{U}(V_i) \to \mathbb{B}_{\mathrm{dR},\mathcal{U}}^+[\![X]\!]$. Taking the direct limit, we get a unique lifting $\mathcal{O}_\mathcal{U}(V) \to \mathbb{B}_{\mathrm{dR},\mathcal{U}}^+[\![X]\!]$. Since V was arbitrary and the affinoid perfectoids form a basis of the proétale site, and invoking the fact that $\mathcal{O}_\mathcal{U}$ is a sheaf on $\mathcal{U}_{\mathrm{proét}}$ (and so satisfies the gluing axiom), we have proven the claim. $\square$

4.2   THE FUNDAMENTAL DE RHAM PERIODS

Let

$$t := \log[\epsilon] \in \mathrm{Fil}^1 \mathbb{B}_{\mathrm{dR}}^+(\mathcal{Y}) \tag{4.3}$$

and

$$\epsilon := (\langle \alpha_{\infty,1}, \alpha_{\infty,2} \rangle_0, \langle \alpha_{\infty,1}, \alpha_{\infty,2} \rangle_1, \ldots) \in \hat{\mathbb{Z}}_p(1)(\mathcal{Y})$$

where $\langle \cdot, \cdot \rangle$ is the Weil pairing (3.3), and $\langle \cdot, \cdot \rangle_n$ denotes its image under the projection

$$\hat{\mathbb{Z}}_{p,\mathcal{Y}}(1) = \varprojlim_n \mu_{p^n} \to \mu_{p^n}.$$

Hence t is a generator (i.e., a nonvanishing global section) of $\mathrm{Fil}^1 \mathbb{B}_{\mathrm{dR}}^+(\mathcal{Y})$ over $\mathbb{B}_{\mathrm{dR}}^+(\mathcal{Y})$.

Note also that t gives a natural isomorphism

$$\hat{\mathbb{Z}}_{p,\mathcal{Y}}(1) \cong \hat{\mathbb{Z}}_{p,\mathcal{Y}} \cdot t \quad : \quad \gamma \mapsto \log[\gamma] \tag{4.4}$$

of $\hat{\mathbb{Z}}_{p,\mathcal{Y}}$-modules on $\mathcal{Y}_{\text{proét}}$, and thus by the isomorphism (3.4) induced by the Weil pairing, we have a natural isomorphism

$$\mathcal{H}^1_{\text{ét}}(\mathcal{A})|_{\mathcal{Y}} \cong T_p\mathcal{A}(-1)|_{\mathcal{Y}} \cong T_p\mathcal{A}|_{\mathcal{Y}} \cdot t^{-1} \underset{\alpha_\infty^{-1}}{\xrightarrow{\sim}} \left(\hat{\mathbb{Z}}_{p,\mathcal{Y}} \cdot t^{-1}\right)^{\oplus 2}. \tag{4.5}$$

Using (4.4) and (3.38), we write the global section of ω given by the fake Hasse invariant

$$\mathfrak{s} := HT_{\mathcal{A}}(\alpha_2) \in \hat{\omega}(\mathcal{Y}) \subset \mathcal{H}^1_{\text{ét}}(\mathcal{A}) \otimes_{\hat{\mathbb{Z}}_{p,\mathcal{Y}}} \hat{\mathcal{O}}\mathbb{B}^+_{\text{dR},\mathcal{Y}}(\mathcal{Y}) \xrightarrow{\sim} (\hat{\mathcal{O}}\mathbb{B}^+_{\text{dR},\mathcal{Y}} \cdot t^{-1})^{\oplus 2}$$

as

$$\mathfrak{s} = \mathbf{x}_{\text{dR}}\alpha_{\infty,1}/t + \mathbf{y}_{\text{dR}}\alpha_{\infty,2}/t \tag{4.6}$$

where $\mathbf{x}_{\text{dR}}, \mathbf{y}_{\text{dR}} \in \hat{\mathcal{O}}\mathbb{B}^+_{\text{dR},Y}(\mathcal{Y})$.

Definition 4.5. *We call* $\mathbf{x}_{\text{dR}}, \mathbf{y}_{\text{dR}} \in \hat{\mathcal{O}}\mathbb{B}^+_{\text{dR},Y}(\mathcal{Y})$ *the fundamental de Rham periods.*

Remark 4.6. By the proof of Proposition 2.2.5 of [9], we see that the map

$$\omega \otimes_{\mathcal{O}_Y} \hat{\mathcal{O}}\mathbb{B}^+_{\text{dR},\mathcal{Y}} \subset \mathcal{H}^1_{\text{dR}}(\mathcal{A}) \otimes_{\mathcal{O}_Y} \hat{\mathcal{O}}\mathbb{B}^+_{\text{dR},\mathcal{Y}} \xrightarrow{\theta \circ \iota_{\text{dR}}} \mathcal{H}^1_{\text{ét}}(\mathcal{A}) \otimes_{\hat{\mathbb{Z}}_{p,Y}}$$

$$\hat{\mathcal{O}}_{\mathcal{Y}} \xrightarrow{HT_{\mathcal{A}}} \omega(-1) \otimes_{\mathcal{O}_Y} \hat{\mathcal{O}}_{\mathcal{Y}}$$

obtained from reducing the right-hand side of (3.35) modulo $\ker\theta$ is just the 0 map. Hence

$$\mathbf{x}_{\text{dR}}, \mathbf{y}_{\text{dR}} \in \ker(\theta : \hat{\mathcal{O}}\mathbb{B}^+_{\text{dR},Y} \twoheadrightarrow \hat{\mathcal{O}}_Y)(\mathcal{Y}).$$

4.3 $GL_2(\mathbb{Q}_p)$-TRANSFORMATION PROPERTIES OF THE FUNDAMENTAL DE RHAM PERIODS

For any $\gamma \in GL_2(\mathbb{Q}_p)$, let

$$p^n\gamma = \begin{pmatrix} a & b \\ c & d \end{pmatrix}$$

where $n \in \mathbb{Z}$ is such that $p^n\gamma \in M_2(\mathbb{Z}_p)$ but $p^{n-1} \notin M_2(\mathbb{Z}_p)$. Retain the same notation as in Section 3.3, so that we have

$$\alpha_{\infty,1} \cdot \begin{pmatrix} a & b \\ c & d \end{pmatrix} = a(\check{\phi}_*)^{-1}(\alpha_{\infty,1}) + c(\check{\phi}_*)^{-1}(\alpha_{\infty,2})$$

$$\alpha_{\infty,2} \cdot \begin{pmatrix} a & b \\ c & d \end{pmatrix} = b(\check{\phi}_*)^{-1}(\alpha_{\infty,1}) + d(\check{\phi}_*)^{-1}(\alpha_{\infty,2}).$$

Note that when $\gamma \in GL_2(\mathbb{Z}_p)$, then $(\check{\phi}_*)^{-1}$ is just the identity. From the calculations

$$
\begin{aligned}
\gamma^* \mathfrak{s} = \begin{pmatrix} a & b \\ c & d \end{pmatrix}^* HT_A(\alpha_{\infty,2}) &= HT_A(b(\check{\phi}_*)^{-1}(\alpha_{\infty,1}) + d(\check{\phi}_*)^{-1}(\alpha_{\infty,2})) \\
&= (b\hat{z}^{-1} + d)HT_A((\check{\phi}_*)^{-1}(\alpha_{\infty,2})) \\
&= (b\hat{z}^{-1} + d)(\check{\phi}^{-1})^* \mathfrak{s}
\end{aligned}
\tag{4.7}
$$

and

$$
\begin{aligned}
\gamma^* t = \begin{pmatrix} a & b \\ c & d \end{pmatrix}^* &\log[(\langle \alpha_{\infty,1}, \alpha_{\infty,2}, \rangle_0, \ldots)] \\
&= \log[(\langle a(\check{\phi}_*)^{-1}(\alpha_{\infty,1}) + c(\check{\phi}_*)^{-1}(\alpha_{\infty,2}), b(\check{\phi}_*)^{-1}(\alpha_{\infty,1}) \\
&\quad + d(\check{\phi}_*)^{-1}(\alpha_{\infty,2}))_0, \ldots)] \\
&= (ad - bc) \log[(\langle (\check{\phi}_*)^{-1}(\alpha_{\infty,1}), (\check{\phi}_*)^{-1}(\alpha_{\infty,2}))_0, \ldots)] \\
&= (ad - bc) \log[(\langle \alpha_{\infty,1}, \alpha_{\infty,2} \rangle_0, \ldots)] \\
&= (ad - bc)t,
\end{aligned}
\tag{4.8}
$$

where the penultimate equality follows from

$$
\langle (\check{\phi}_*)^{-1}(\alpha_{\infty,1}), (\check{\phi}_*)^{-1}(\alpha_{\infty,2}) \rangle = \langle \alpha_{\infty,1}, \alpha_{\infty,2} \rangle,
$$

which in turn follows from the functoriality of the Weil pairing: Given an isogeny $\phi: A \to A'$ with dual isogeny $\check{\phi}: \check{A}' \to \check{A}$, and the Weil pairings $\langle \cdot, \cdot \rangle$ and $\langle \cdot, \cdot \rangle'$ associated with A and A', we have

$$
\langle \check{\phi}_*(x), y \rangle = \langle x, \phi_*(y) \rangle'.
$$

Now by the construction of our ϕ (using our definition of $p^n \gamma$), we see that either $(\check{\phi}_*)^{-1}(\alpha_{\infty,1}) = \phi_*(\alpha_{\infty,1})$ or $(\check{\phi}_*)^{-1}(\alpha_{\infty,2}) = \phi_*(\alpha_{\infty,2})$, and hence

$$
\begin{aligned}
\langle (\check{\phi}_*)^{-1}(\alpha_{\infty,i}), (\check{\phi}_*)^{-1}(\alpha_{\infty,3-i}) \rangle &= \langle \phi_*(\alpha_{\infty,i}), (\check{\phi}_*)^{-1}(\alpha_{\infty,3-i}) \rangle \\
&= \langle \alpha_{\infty,i}, \check{\phi}_*(\check{\phi}_*)^{-1}(\alpha_{\infty,3-i}) \rangle \\
&= \langle \alpha_{\infty,i}, \alpha_{\infty,3-i} \rangle
\end{aligned}
$$

for $i = 1$ or 2.

We have the following proposition.

Proposition 4.7. *Let* $\gamma \in GL_2(\mathbb{Q}_p)$ *and*

$$
p^n \gamma = \begin{pmatrix} a & b \\ c & d \end{pmatrix}
$$

where $n \in \mathbb{Z}$ is such that $p^n \gamma \in M_2(\mathbb{Z}_p)$ but $p^{n-1} \notin M_2(\mathbb{Z}_p)$. Then $\mathbf{x}_{\mathrm{dR}}/t$ and $\mathbf{y}_{\mathrm{dR}}/t$ satisfy the transformation laws

$$
\begin{aligned}
\gamma^*(\mathbf{x}_{\mathrm{dR}}/t) &= (d\mathbf{x}_{\mathrm{dR}} - b\mathbf{y}_{\mathrm{dR}})(b\hat{\mathbf{z}}^{-1} + d)(ad - bc)^{-1} \\
&= (b\mathbf{z}_{\mathrm{dR}}^{-1} + d)(b\hat{\mathbf{z}}^{-1} + d)(ad - bc)^{-1}(\mathbf{x}_{\mathrm{dR}}/t),
\end{aligned}
\tag{4.9}
$$

and

$$
\begin{aligned}
\gamma^*(\mathbf{y}_{\mathrm{dR}}/t) &= (-c\mathbf{x}_{\mathrm{dR}} + a\mathbf{y}_{\mathrm{dR}})(b\hat{\mathbf{z}}^{-1} + d)(ad - bc)^{-1} \\
&= (c\mathbf{z}_{\mathrm{dR}} + a)(b\hat{\mathbf{z}}^{-1} + d)(ad - bc)^{-1}(\mathbf{y}_{\mathrm{dR}}/t),
\end{aligned}
\tag{4.10}
$$

where $\mathbf{z}_{\mathrm{dR}}$ is defined in Definition 4.32.

Proof. Let

$$
X := \gamma^* \mathbf{x}_{\mathrm{dR}}, \qquad\qquad Y := \gamma^* \mathbf{y}_{\mathrm{dR}}.
$$

Then by definition,

$$
\iota_{\mathrm{dR}}(\mathfrak{s}) = \iota_{\mathrm{dR}} \circ HT_{\mathcal{A}}(\alpha_{\infty,2}) = \mathbf{x}_{\mathrm{dR}} \alpha_{\infty,1} t^{-1} + \mathbf{y}_{\mathrm{dR}} \alpha_{\infty,2} t^{-1}
$$

and so

$$
\begin{aligned}
\gamma^* \iota_{\mathrm{dR}}(\mathfrak{s}) &= HT_{\mathcal{A}}\left(b\frac{(\check{\phi}_*)^{-1}(\alpha_{\infty,1})}{t} + d\frac{(\check{\phi}_*)^{-1}(\alpha_{\infty,2})}{t} \right) \\
&= (b\hat{\mathbf{z}}^{-1} + d)\left(\mathbf{x}_{\mathrm{dR}} \frac{(\check{\phi}_*)^{-1}(\alpha_{\infty,1})}{t} + \mathbf{y}_{\mathrm{dR}} \frac{(\check{\phi}_*)^{-1}(\alpha_{\infty,2})}{t} \right).
\end{aligned}
$$

From (4.7), we have

$$
\begin{aligned}
(b\hat{\mathbf{z}}^{-1} + d)&\left(\mathbf{x}_{\mathrm{dR}} \frac{(\check{\phi}_*)^{-1}(\alpha_{\infty,1})}{t} + \mathbf{y}_{\mathrm{dR}} \frac{(\check{\phi}_*)^{-1}(\alpha_{\infty,2})}{t} \right) = \gamma^* \mathfrak{s} \\
&= X\left(a\frac{(\check{\phi}_*)^{-1}(\alpha_{\infty,1})}{t} + c\frac{(\check{\phi}_*)^{-1}(\alpha_{\infty,2})}{t} \right)(ad - bc)^{-1} \\
&\quad + Y\left(b\frac{(\check{\phi}_*)^{-1}(\alpha_{\infty,1})}{t} + d\frac{(\check{\phi}_*)^{-1}(\alpha_{\infty,2})}{t} \right)(ad - bc)^{-1} \\
&= (aX + bY)(ad - bc)^{-1}\frac{(\check{\phi}_*)^{-1}(\alpha_{\infty,1})}{t} \\
&\quad + (cX + dY)(ad - bc)^{-1}\frac{(\check{\phi}_*)^{-1}(\alpha_{\infty,2})}{t}
\end{aligned}
$$

which, by equating the coefficients of the basis

$$
\frac{(\check{\phi}_*)^{-1}(\alpha_{\infty,1})}{t}, \qquad \frac{(\check{\phi}_*)^{-1}(\alpha_{\infty,2})}{t},
$$

implies that

$$X = -(d\mathbf{x}_{\mathrm{dR}} - b\mathbf{y}_{\mathrm{dR}})(b\hat{\mathbf{z}}^{-1} + d), \qquad Y = -(-c\mathbf{x}_{\mathrm{dR}} + a\mathbf{y}_{\mathrm{dR}})(b\hat{\mathbf{z}}^{-1} + d)$$

as desired. $\qquad\qquad\qquad\qquad\qquad\qquad\qquad\qquad\qquad\qquad\qquad\qquad\quad\Box$

4.4 THE p-ADIC LEGENDRE RELATION

In this section, we examine the functoriality of the comparison inclusion ι_{dR} more closely. Let $\mu_{p^n,Y} = \nu^* \mu_{p,Y_{\text{ét}}}$ where $\mu_{p^n,Y_{\text{ét}}}$ is the usual constant sheaf associated with μ_{p^n} on $Y_{\text{ét}}$, and we recall $\nu : Y_{\text{proét}} \to Y_{\text{ét}}$ is the natural projection of sites. Denote

$$\omega_{\mu_{p^n}} := \pi_* \Omega^1_{\mu_{p^n}/Y}.$$

Let $\mu_{p^\infty,Y} := \varinjlim_n \mu_{p^n,Y}$ and let

$$\omega_{\mu_{p^\infty}} := \pi_* \Omega^1_{\mu_{p^n}/Y} = \varprojlim_n \omega_{\mu_{p^n}}, \quad \hat{\omega}_{\mu_{p^\infty}} = \omega_{\mu_{p^\infty}} \otimes_{\mathcal{O}_Y} \hat{\mathcal{O}}_Y.$$

The canonical invariant differential

$$\frac{dT}{T} \in \omega_{\mu_{p^\infty}}(Y)$$

is a nonvanishing global section, which hence gives a canonical trivialization

$$\omega_{\mu_{p^\infty}} \cong \mathcal{O}_Y. \tag{4.11}$$

Note that $\omega_{\mu_{p^\infty}}$ has the structure of a filtered module with integrable connection, taking the trivial filtration

$$\mathrm{Fil}^i \omega_{\mu_{p^\infty}} = \begin{cases} \omega_{\mu_{p^\infty}} & i \le 0, \\ 0 & i > 0 \end{cases}$$

and the trivial connection

$$\nabla' : \omega_{\mu_{p^\infty}} \to \omega_{\mu_{p^\infty}} \otimes_{\mathcal{O}_Y} \Omega^1_Y, \quad \nabla'\left(\frac{dT}{T}\right) = 0,$$

which, under the trivialization (4.11), is just $d : \mathcal{O}_Y \to \Omega^1_Y$. Then it is clear that for $(\mathcal{E}, \nabla) = (\omega_{\mu_{p^\infty}}, \nabla')$, we have

$$\begin{aligned} \mathbb{M}_0 &= (\omega_{\mu_{p^\infty}} \otimes_{\mathcal{O}_Y} \mathcal{O}\mathbb{B}^+_{\mathrm{dR},Y})^{\nabla' \otimes \nabla = 0} = (\mathcal{O}\mathbb{B}^+_{\mathrm{dR},Y})^{\nabla = 0} = \mathbb{B}^+_{\mathrm{dR},Y} = \mathrm{Fil}^0 \mathbb{B}_{\mathrm{dR},Y} \\ &= \mathrm{Fil}^0 (\mathcal{O}\mathbb{B}_{\mathrm{dR},Y})^{\nabla = 0} = \mathrm{Fil}^0 (\omega_{\mu_{p^\infty}} \otimes_{\mathcal{O}_Y} \mathcal{O}\mathbb{B}_{\mathrm{dR},Y})^{\nabla' \otimes \nabla = 0} = \mathbb{M} \end{aligned} \tag{4.12}$$

in the notation of Theorem 2.65. In other words, (3.34) is an equality for $\omega_{\mu_{p^\infty}}$.
Let

$$\omega_{\mathcal{A}[p^n]} = \pi_* \Omega^1_{\mathcal{A}[p^n]/Y}$$

so that

$$\varprojlim_n \omega_{\mathcal{A}[p^n]} = \pi_* \Omega^1_{\mathcal{A}[p^n]/Y} = \omega.$$

For any proétale $\mathcal{U} \to Y$ and any $\alpha = (\alpha_1, \alpha_2, \ldots) \in T_p \mathcal{A}(\mathcal{U})$, from (3.19) (or (3.3)) we have maps

$$\alpha_n^* : \omega_{\mu_{p^n}}|_{\mathcal{U}} \to \omega_{\mathcal{A}[p^n]}|_{\mathcal{U}}$$

and

$$\alpha^* : \hat{\omega}_{\mu_{p^\infty}} \to \hat{\omega}, \quad \frac{dT}{T} \mapsto HT_{\mathcal{A}}(\alpha) = \alpha^* \left(\frac{dT}{T} \right).$$

Now note that we have a map of completed filtered $\mathcal{O}_Y$-modules with integrable connections (in the sense of Definition 2.73)

$$\alpha^* : \hat{\omega}_{\mu_{p^\infty}} \to \mathcal{H}^1_{\mathrm{dR}}(\mathcal{A}) \otimes_{\mathcal{O}_Y} \hat{\mathcal{O}}_Y, \tag{4.13}$$

which factors through

$$\alpha^* : \hat{\omega}_{\mu_{p^\infty}} \to \hat{\omega} = \mathrm{Fil}^1(\mathcal{H}^1_{\mathrm{dR}}(\mathcal{A}) \otimes_{\mathcal{O}_Y} \hat{\mathcal{O}}_Y)$$

where $\mathrm{Fil}^i(\mathcal{H}^1_{\mathrm{dR}}(\mathcal{A}) \otimes_{\mathcal{O}_Y} \hat{\mathcal{O}}_Y)$ is the filtration inherited from the Hodge filtration

$$\mathrm{Fil}^i(\mathcal{H}^1_{\mathrm{dR}}(\mathcal{A}) \otimes_{\mathcal{O}_Y} \hat{\mathcal{O}}_Y) = \begin{cases} \mathcal{H}^1_{\mathrm{dR}}(\mathcal{A}) \otimes_{\mathcal{O}_Y} \hat{\mathcal{O}}_Y & i < 1 \\ \hat{\omega} & i = 1 \\ 0 & i > 1. \end{cases}$$

Hence applying Corollary 2.74 to (4.12) $\otimes_{\mathbb{B}^+_{\mathrm{dR},Y}} \hat{\mathcal{O}}\mathbb{B}^+_{\mathrm{dR},\mathcal{Y}}$, and by (3.34) tensored with $\otimes_{\mathbb{B}^+_{\mathrm{dR},Y}} \hat{\mathcal{O}}\mathbb{B}^+_{\mathrm{dR},\mathcal{Y}}$, we have the following commutative diagram.

$$\begin{array}{ccc} \hat{\omega}_{\mu_{p^\infty}}|_{\mathcal{Y}} & \overset{i}{=\!=\!=\!=} & \hat{\mathcal{O}}\mathbb{B}^+_{\mathrm{dR},\mathcal{Y}} \\ \downarrow{\alpha^*} & & \downarrow{\alpha^*} \\ \hat{\omega} \otimes_{\hat{\mathcal{O}}_Y} \hat{\mathcal{O}}\mathbb{B}^+_{\mathrm{dR},\mathcal{Y}} & \overset{\iota_{\mathrm{dR}}}{\hookrightarrow} & T_p \mathcal{A} \otimes_{\mathbb{Z}_p,Y} \hat{\mathcal{O}}\mathbb{B}^+_{\mathrm{dR},\mathcal{Y}} \cdot t^{-1} \end{array} \tag{4.14}$$

where

$$i := (4.12) \otimes_{\mathbb{B}^+_{\mathrm{dR},Y}} \hat{\mathcal{O}}\mathbb{B}^+_{\mathrm{dR},\mathcal{Y}}.$$

Note that

$$i\left(\frac{dT}{T}\right) = 1,$$

$$\hat{\mathcal{O}}\mathbb{B}^+_{\mathrm{dR},Y}(\mathcal{Y}) \ni 1 \xmapsto{\alpha^*} \alpha \in T_p\mathcal{A} \otimes_{\hat{\mathbb{Z}}_{p,Y}} \hat{\mathcal{O}}\mathbb{B}^+_{\mathrm{dR},Y}(\mathcal{Y}) \subset T_p\mathcal{A} \otimes_{\hat{\mathbb{Z}}_{p,Y}} \hat{\mathcal{O}}\mathbb{B}^+_{\mathrm{dR},Y}(\mathcal{Y}) \cdot t^{-1}.$$

$$(4.15)$$

Proposition 4.8. *For any proétale $\mathcal{U} \to Y$ and any $\alpha, \beta \in T_p\mathcal{A}(\mathcal{U})$, we have*

$$\left\langle \iota_{\mathrm{dR}}\left(\alpha^*\frac{dT}{T}\right), \beta \right\rangle = \langle \alpha, \beta \rangle.$$

Proof. From the commutativity of (4.14) we have

$$\iota_{\mathrm{dR}} \circ \alpha^* = \alpha^* \circ i$$

and from the observations (4.15), we have

$$\left\langle \iota_{\mathrm{dR}}\left(\alpha^*\frac{dT}{T}\right), \beta \right\rangle = \left\langle \alpha^*i\left(\frac{dT}{T}\right), \beta \right\rangle = \langle \alpha^*1, \beta \rangle = \langle \alpha, \beta \rangle.$$

□

Theorem 4.9. *The composition*

$$HT_\mathcal{A} \circ \iota_{\mathrm{dR}} : \hat{\omega} \otimes_{\hat{\mathcal{O}}_Y} \hat{\mathcal{O}}\mathbb{B}^+_{\mathrm{dR},\mathcal{Y}} \cdot t \to \hat{\omega} \otimes_{\hat{\mathcal{O}}_Y} \hat{\mathcal{O}}\mathbb{B}^+_{\mathrm{dR},\mathcal{Y}}$$

is the natural inclusion (induced by $\hat{\mathcal{O}}\mathbb{B}^+_{\mathrm{dR},Y} \cdot t \subset \hat{\mathcal{O}}\mathbb{B}^+_{\mathrm{dR},Y}$).

Proof. Let

$$w \in \omega \otimes_{\mathcal{O}_Y} \hat{\mathcal{O}}\mathbb{B}^+_{\mathrm{dR},Y} \cdot t$$

and

$$\iota_{\mathrm{dR}}(w) = x\alpha_{\infty,1} + y\alpha_{\infty,2}.$$

Then

$$HT_A(\iota_{\mathrm{dR}}(w)) = x\alpha^*_{\infty,1}\frac{dT}{T} + y\alpha^*_{\infty,2}\frac{dT}{T}.$$

Now by Proposition 4.8, we have

$$\langle HT_A(\iota_{\mathrm{dR}}(w)), \gamma \rangle = \left\langle x\alpha^*_{\infty,1}\frac{dT}{T} + y\alpha^*_{\infty,2}\frac{dT}{T}, \gamma \right\rangle = \langle x\alpha_{\infty,1} + y\alpha_{\infty,2}, \gamma \rangle$$

$$(4.16)$$

$$= \langle \iota_{\mathrm{dR}}(w), \gamma \rangle.$$

Hence by the nondegeneracy of the Weil pairing $\langle \cdot, \cdot \rangle$, we have $HT_A(\iota_{dR}(w)) = w$, as desired. $\qquad\square$

We have the following "p-adic Legendre" relation of periods in $\hat{\mathcal{O}}\mathbb{B}_{dR}^+(\mathcal{Y}_x)$ associated with the fake Hasse invariant $\mathfrak{s}$.

Corollary 4.10. *We have the following identity of sections of $\hat{\mathcal{O}}\mathbb{B}_{dR,Y}^+(\mathcal{Y}_x)$:*

$$\mathbf{x}_{dR}/\hat{\mathbf{z}} + \mathbf{y}_{dR} = t \tag{4.17}$$

where $\hat{\mathbf{z}}$ is the Hodge-Tate period and $\mathbf{x}_{dR}, \mathbf{y}_{dR}, t \in \hat{\mathcal{O}}\mathbb{B}_{dR}^+(\mathcal{Y})$ are defined in Section 4.2.

Proof. Recall that by definition,

$$\iota_{dR}(t \cdot \mathfrak{s}) = \mathbf{x}_{dR}\alpha_{\infty,1} + \mathbf{y}_{dR}\alpha_{\infty,2}.$$

Applying Theorem 4.9 to $\mathfrak{s} \cdot t \in \omega(\mathcal{Y}_x) \cdot t = \omega \cdot t(\mathcal{Y}_x)$ (the last equality following since $t \in \mathbb{B}_{dR}^+(\mathcal{Y})$ is a global section), we see that

$$t \cdot \mathfrak{s} = HT_A(\iota_{dR}(t \cdot \mathfrak{s})) = \mathbf{x}_{dR}/\hat{\mathbf{z}} \cdot \mathfrak{s} + \mathbf{y}_{dR} \cdot \mathfrak{s} = (\mathbf{x}_{dR}/\hat{\mathbf{z}} + \mathbf{y}_{dR}) \cdot \mathfrak{s}$$

and the desired identity follows. $\qquad\square$

Definition 4.11. *Henceforth, let*

$$\omega_{dR} := \omega \otimes_{\mathcal{O}_Y} \mathcal{O}\mathbb{B}_{dR,Y}^+, \qquad \omega_{dR}^{-1} := \omega \otimes_{\mathcal{O}_Y} \mathcal{O}\mathbb{B}_{dR,Y}^+$$

and

$$\hat{\omega}_{dR} := \omega \otimes_{\mathcal{O}_Y} \hat{\mathcal{O}}\mathbb{B}_{dR,Y}^+, \qquad \hat{\omega}_{dR}^{-1} := \omega \otimes_{\mathcal{O}_Y} \hat{\mathcal{O}}\mathbb{B}_{dR,Y}^+.$$

We note the following corollary of the proof of Theorem 4.9.

Proposition 4.12. *We have*

$$\mathbf{x}_{dR}, \mathbf{y}_{dR} \in t \cdot \hat{\mathcal{O}}\mathbb{B}_{dR,Y}^+(\mathcal{Y}). \tag{4.18}$$

In particular,

$$\iota_{dR}(\hat{\omega}_{dR}) \subset T_p A \otimes_{\hat{\mathbb{Z}}_{p,Y}} \hat{\mathcal{O}}\mathbb{B}_{dR,\mathcal{Y}}^+. \tag{4.19}$$

Proof. For $y \in \mathcal{Y} \setminus \mathcal{Y}_x$, we have $\mathfrak{s}(y) = 0$ (since by definition $\mathcal{Y} \setminus \mathcal{Y}_x = \{\hat{\mathbf{z}} = 0\} = \{\mathfrak{s} = 0\}$), and in particular $t|\mathfrak{s}(y)$. Hence we are reduced to proving (4.18) on $\mathcal{Y}_x$. Let

$$\mathcal{O}_\Delta := \mathcal{O}\mathbb{B}_{dR,\mathcal{Y}}^+/(t), \quad \hat{\mathcal{O}}_\Delta := \hat{\mathcal{O}}\mathbb{B}_{dR,\mathcal{Y}}^+/(t).$$

Note that reducing the p-adic Legendre relation modulo (t) results in

$$\frac{x_{\mathrm{dR}}}{\hat{\mathbf{z}}} + y_{\mathrm{dR}} = 0 \in \mathcal{O}_\Delta(\mathcal{Y}_x) \tag{4.20}$$

where $x_{\mathrm{dR}} = \mathbf{x}_{\mathrm{dR}} \pmod{t}, y_{\mathrm{dR}} = \mathbf{y}_{\mathrm{dR}} \pmod{t} \in \hat{\mathcal{O}}_\Delta(\mathcal{Y})$. (Note that this is well-defined as a sheaf, since $t \in \mathcal{O}\mathbb{B}_{\mathrm{dR},Y}^+(\mathcal{Y})$ is a global section.)

The assertion (4.18) is equivalent to $x_{\mathrm{dR}} = 0$ and $y_{\mathrm{dR}} = 0$, and if suffices to check this at all proétale stalks at all geometric points $y \in \mathcal{Y}_x(\mathbb{C}_p, \mathcal{O}_{\mathbb{C}_p})$. Let a subscript of y denote the image of a section in the stalk $\mathcal{O}_{\Delta, \mathcal{Y}_x, y}$.

For the sake of contradiction, assume that either $x_{\mathrm{dR},y} \neq 0$ or $y_{\mathrm{dR},y} \neq 0$. From (4.20), we have either

$$\hat{\mathbf{z}}_y = -\frac{y_{\mathrm{dR},y}}{x_{\mathrm{dR},y}} \quad \text{or} \quad \frac{1}{\hat{\mathbf{z}}_y} = -\frac{x_{\mathrm{dR},y}}{y_{\mathrm{dR},y}}. \tag{4.21}$$

However, since ω is locally trivialized on Y, then there exists a neighborhood U of $\lambda(y)$ with $w \in \omega(U)$ a nonvanishing global section. Let $\mathcal{U} := \lambda^{-1}(U)$ and $\mathcal{U}_x := \mathcal{U} \cap \mathcal{Y}_x$. Hence, there exists some $u \in \hat{\mathcal{O}}_Y(\mathcal{U}_x)^\times$ with

$$w|_{\mathcal{U}_x} = u \cdot \mathbf{s}|_{\mathcal{U}_x}$$

which implies in particular that

$$u \cdot x_{\mathrm{dR}}|_{\mathcal{U}_x}, \quad u \cdot y_{\mathrm{dR}}|_{\mathcal{U}_x} \in \mathcal{O}_\Delta(\mathcal{U}_x),$$

which, combined with (4.21), implies that

$$\hat{\mathbf{z}}_y = -\frac{y_{\mathrm{dR},y}}{x_{\mathrm{dR},y}} - = -\frac{u_y \cdot y_{\mathrm{dR},y}}{u_y \cdot x_{\mathrm{dR},y}} \in \mathcal{O}_{\Delta,\mathcal{Y},y} \cap \hat{\mathcal{O}}_{\mathcal{Y},y} = \mathcal{O}_{\mathcal{Y},y} \quad \text{or}$$

$$\frac{1}{\hat{\mathbf{z}}_y} = -\frac{x_{\mathrm{dR},y}}{y_{\mathrm{dR},y}} - = -\frac{u_y \cdot x_{\mathrm{dR},y}}{u_y \cdot y_{\mathrm{dR},y}} \in \mathcal{O}_{\Delta,\mathcal{Y},y} \cap \hat{\mathcal{O}}_{\mathcal{Y},y} = \mathcal{O}_{\mathcal{Y},y}.$$

However, this implies that for some quasi-compact, quasi-separated neighborhood $\mathcal{V} = \varprojlim_i V_i \subset \mathcal{Y}$ of y, where each V_i is finite étale over Y, that

$$\hat{\mathbf{z}}|_{\mathcal{V}} \quad \text{or} \quad \frac{1}{\hat{\mathbf{z}}|_{\mathcal{V}}} \in \mathcal{O}_{\mathcal{Y}}(\mathcal{V}) = \varinjlim_i \mathcal{O}_Y(V_i)$$

which implies, since Y is an adic space over $\mathrm{Spa}(\mathbb{Q}_p, \mathbb{Z}_p)$, that $\hat{\mathbf{z}}(\mathcal{V}) \subset \overline{\mathbb{Q}}_p$, or equivalently $\hat{\mathbf{z}}(\mathcal{V}) \subset \mathbb{P}^1(\overline{\mathbb{Q}}_p)$. However, we have the following.

Lemma 4.13. *For any proétale $\mathcal{V} \to \mathcal{Y}_x$, we have*

$$\hat{\mathbf{z}}(\mathcal{V}) \not\subset \mathbb{P}^1(\overline{\mathbb{Q}}_p).$$

Proof of Lemma 4.13. We recall that $\pi_{\mathrm{HT}} : \hat{\mathcal{Y}} \to \mathbb{P}^1$ is smooth and smooth maps of adic spaces are open ([30, Proposition 1.7.8]), and so $\pi_{\mathrm{HT}}(\mathcal{V}) \subset \mathbb{P}^1(\overline{\mathbb{Q}}_p)$ is open in $\mathbb{P}^1$. However, this is a contradiction, since $\mathbb{P}^1(\overline{\mathbb{Q}}_p)$ contains no non-empty open sets (since, for example, any point in $\overline{\mathbb{Q}}_p \subset \mathbb{C}_p$ contains a point in $\mathbb{C}_p \setminus \overline{\mathbb{Q}}_p$), and $\pi_{\mathrm{HT}}(y) \in \pi_{\mathrm{HT}}(\mathcal{V})$ so that $\pi_{\mathrm{HT}}(\mathcal{V}) \neq 0$. So we are done. $\qquad\square$

Corollary 4.14. *The map*

$$(HT_{\mathcal{A}})^{\vee}(1) : \hat{\omega}_{\mathrm{dR}}^{-1} \cdot t \hookrightarrow T_p \mathcal{A} \otimes_{\hat{\mathbb{Z}}_{p,Y}} \hat{\mathcal{OB}}_{\mathrm{dR},\mathcal{y}}^{+}$$

is a section for the natural projection (using (4.19))

$$T_p \mathcal{A} \otimes_{\hat{\mathbb{Z}}_{p,Y}} \hat{\mathcal{OB}}_{\mathrm{dR},\mathcal{y}}^{+} \twoheadrightarrow (T_p \mathcal{A} \otimes_{\hat{\mathbb{Z}}_{p,Y}} \hat{\mathcal{OB}}_{\mathrm{dR},\mathcal{y}}^{+})/\iota_{\mathrm{dR}}(\hat{\omega}_{\mathrm{dR}}).$$

Proof. First note that we have

$$(T_p \mathcal{A} \otimes_{\hat{\mathbb{Z}}_{p,Y}} \hat{\mathcal{OB}}_{\mathrm{dR},\mathcal{y}}^{+})/\iota_{\mathrm{dR}}(\hat{\omega}_{\mathrm{dR}})$$

$$\xrightarrow[\sim]{\alpha_{\infty}^{-1}} \hat{\mathcal{OB}}_{\mathrm{dR},\mathcal{y}}^{+,\oplus 2}/\left((\mathbf{x}_{\mathrm{dR}} \cdot \alpha_{\infty,1}/t + \mathbf{y}_{\mathrm{dR}} \cdot \alpha_{\infty,2}/t)\hat{\mathcal{OB}}_{\mathrm{dR},\mathcal{y}}^{+} \right).$$

By the p-adic Legendre relation (4.17), the right-hand side isomorphic to the sub-$\hat{\mathcal{OB}}_{\mathrm{dR},\mathcal{y}}^{+}$-module

$$(\alpha_{\infty,1} - 1/\hat{\mathbf{z}} \cdot \alpha_{\infty,2})\hat{\mathcal{OB}}_{\mathrm{dR},\mathcal{y}}^{+}$$

of $\hat{\mathcal{OB}}_{\mathrm{dR},\mathcal{y}}^{+,\oplus 2} \xrightarrow[\sim]{\alpha_{\infty}} T_p \mathcal{A} \otimes_{\hat{\mathbb{Z}}_{p,Y}} \hat{\mathcal{OB}}_{\mathrm{dR},\mathcal{y}}^{+}$ on $\mathcal{Y}_x$, and is the sub-$\hat{\mathcal{OB}}_{\mathrm{dR},\mathcal{y}}^{+}$-module

$$(\hat{\mathbf{z}} \cdot \alpha_{\infty,1} - \alpha_{\infty,2})\hat{\mathcal{OB}}_{\mathrm{dR},\mathcal{y}}^{+}$$

on $\mathcal{Y}_y$. Hence it is isomorphic to $\mathcal{H}^{0,1} \otimes_{\hat{\mathcal{O}}_Y} \hat{\mathcal{OB}}_{\mathrm{dR},\mathcal{y}}^{+}$, by the definition of $\mathcal{H}^{0,1}$ (see Section 3.4) and the Hodge-Tate period $\hat{\mathbf{z}}$, and now the corollary follows from Proposition 3.14. $\qquad\square$

4.5   RELATION TO COLMEZ'S "p-ADIC PERIOD PAIRING"

Although not necessary for the construction of our p-adic Maass-Shimura operator, let us record the relation between ι_{dR} and the p-adic comparison theorem of Colmez, which is formulated in the case of abelian varieties as a integration pairing between de Rham cohomology and the Tate module in [15].

Let

$$\mathcal{OB}_{\mathrm{dR},\mathcal{y}}^{+}(y) := \mathcal{OB}_{\mathrm{dR},\mathcal{y},y}^{+} \otimes_{\mathcal{O}_{\mathcal{y},y}} \mathcal{O}_{\mathcal{y},y}/\mathfrak{m}_y$$

denote the fiber of $\mathcal{OB}^+_{\mathrm{dR},\mathcal{Y}}$ (as an $\mathcal{O}_\mathcal{Y}$-module) at y, where $\mathfrak{m}_y \subset \mathcal{O}_{\mathcal{Y},y}$ is the maximal ideal. Let $\iota_{\mathrm{dR},y} = \iota_{\mathrm{dR},y} \pmod{\mathfrak{m}_y}$ denote the induced map on fibers. We note that the following compatibility between the relative and p-adic comparison theorems is implicit in the discussion after Theorem 8.4 in [57] (as Theorem 7.11 and 8.4, or rather their proofs, are used in the proof of the relative comparison theorem Theorem 8.8 in loc. cit.):

$$
\begin{array}{ccc}
\mathcal{H}^1_{\mathrm{dR}}(A)|_{\mathcal{Y},y} \otimes_{\mathcal{O}_{\mathcal{Y},y}} \mathcal{OB}^+_{\mathrm{dR},\mathcal{Y},y} & \xrightarrow{\ \iota_{\mathrm{dR},y}\ } & T_p A \otimes_{\mathbb{Z}_p} \mathcal{OB}^+_{\mathrm{dR},\mathcal{Y},y} \cdot t^{-1} \\
\downarrow{\scriptstyle \mathrm{mod}\ \mathfrak{m}_y} & & \downarrow{\scriptstyle \mathrm{mod}\ \mathfrak{m}_y} \\
H^1_{\mathrm{dR}}(A) \otimes_{\overline{\mathbb{Q}}_p} \mathcal{OB}^+_{\mathrm{dR},\mathcal{Y}}(y) & \xrightarrow{\ \iota_{\mathrm{dR}}(y)\ } & T_p A \otimes_{\mathbb{Z}_p} \mathcal{OB}^+_{\mathrm{dR},Y}(y) \cdot t^{-1} \\
\uparrow{\scriptstyle B^+_{\mathrm{dR}} \subset \mathcal{OB}^+_{\mathrm{dR},\mathcal{Y}}(y)} & & \uparrow{\scriptstyle B^+_{\mathrm{dR}} \subset \mathcal{OB}^+_{\mathrm{dR},\mathcal{Y}}(y)} \\
H^1_{\mathrm{dR}}(A) \otimes_{\overline{\mathbb{Q}}_p} B^+_{\mathrm{dR}} & \xleftarrow{\hspace{1cm}}\ \xrightarrow{\ \iota\ } & T_p A \otimes_{\mathbb{Z}_p} B^+_{\mathrm{dR}} \cdot t^{-1}
\end{array}
\tag{4.22}
$$

where $B^+_{\mathrm{dR}} \subset \mathcal{OB}^+_{\mathrm{dR},\mathcal{Y}}(y)$ denotes the natural inclusion, which factors through

$$
B^+_{\mathrm{dR}} \cong \mathbb{B}^+_{\mathrm{dR},\mathcal{Y},y} \subset \mathcal{OB}^+_{\mathrm{dR},\mathcal{Y},y} \xrightarrow{\ \mathrm{mod}\ \mathfrak{m}_y\ } \mathcal{OB}^+_{\mathrm{dR},\mathcal{Y}}(y),
$$

and ι is the absolute comparison inclusion from Theorem 8.4 of loc. cit.

We now identify ι as the p-adic comparison map of Colmez ι_0 below, which we now recall.

Definition 4.15. *Colmez has shown ([15, Introduction, Section 5, Théorème 11, or Théorème II.3.2]) an absolute p-adic de Rham comparison theorem*

$$
H^1_{\mathrm{dR}}(A/\mathbb{C}_p) \otimes_{\mathbb{C}_p} B^+_{\mathrm{dR}} \cdot t \xrightarrow{\ \iota_0\ } T_p A \otimes_{\mathbb{Z}_p} B^+_{\mathrm{dR}}
\tag{4.23}
$$

which is given by the inclusion

$$
H^1_{\mathrm{dR}}(A/\mathbb{C}_p) \otimes_{\mathbb{C}_p} B^+_{\mathrm{dR}} \subset \mathrm{Hom}_{B^+_{\mathrm{dR}}}(T_p A \otimes_{\mathbb{Z}_p} B^+_{\mathrm{dR}}, B^+_{\mathrm{dR}}) \cong H^1_{\acute{e}t}(A, \mathbb{Z}_p) \otimes_{\mathbb{Z}_p} B^+_{\mathrm{dR}}
$$
$$
\cong T_p A \otimes_{\mathbb{Z}_p} B^+_{\mathrm{dR}} \cdot t^{-1}
$$

(where the last isomorphism is given by the Weil pairing (3.4)) induced by the nondegenerate B^+_{dR}-bilinear "p-adic period pairing"

$$
\langle \cdot, \cdot \rangle_p : (H^1_{\mathrm{dR}}(A/\overline{\mathbb{Q}}_p) \otimes_{\overline{\mathbb{Q}}_p} B^+_{\mathrm{dR}}) \times (T_p A \otimes_{\mathbb{Z}_p} B^+_{\mathrm{dR}}) \to B^+_{\mathrm{dR}}.
\tag{4.24}
$$

Here the p-adic period pairing is defined as follows. Fix an abstract embedding

$$
\mathbb{C}_p \hookrightarrow B^+_{\mathrm{dR}}
\tag{4.25}
$$

(such an embedding is highly non-canonical), and then define

$$\langle \omega, \gamma \rangle_p = \lim_{n \to \infty} p^n \int_{\gamma_n} \omega$$

where $\gamma = (\gamma_n) \in T_p A \otimes_{\mathbb{Z}_p} B_{\mathrm{dR}}^+$, *and here by definition*

$$\int_{\gamma_n} \omega := \int_{x_n}^{y_n} \omega$$

where $(x_n), (y_n)$ *are any sequences "bornées" of elements of* $A(B_{\mathrm{dR}}^+)$ *(see [15, Appendice A] for precise definitions) such that*

$$\theta(y_n - x_n) = \gamma_n$$

for all n. *Moreover, this construction is independent of the choice of embedding* (4.25).

Theorem 4.16. *In the notation of (4.22) and (4.23), we have*

$$\iota = \iota_0,$$

or, equivalently,

$$\langle w, \gamma \rangle_p = \langle \iota_{\mathrm{dR}}(w), \gamma \rangle$$

for any $w \in H_{\mathrm{dR}}^1(A/\mathbb{C}_p) \otimes_{\mathbb{C}_p} B_{\mathrm{dR}}^+$.

Proof. Colmez proves the following "pullback formula" for $\langle \cdot, \cdot \rangle_p$ (which essentially follows from the construction of $\langle \cdot, \cdot \rangle_p$, see p. 142 of loc. cit.): for any

$$\gamma = (\gamma_n) \in \mathrm{Lie} A \otimes_{\mathbb{C}_p} B_{\mathrm{dR}}^+ \cdot t \subset T_p A \otimes_{\mathbb{Z}_p} B_{\mathrm{dR}}^+,$$

and any

$$\beta = (\beta_n) \in T_p A \cong \mathrm{Hom}_{\mathbb{Z}_p}(T_p A, \mathbb{Z}_p \cdot t),$$

we have

$$\left\langle \beta^* \frac{dT}{T}, \gamma \right\rangle_p = \lim_{n \to \infty} p^n \int_{\gamma_n} \beta^* \frac{dT}{T} = \lim_{n \to \infty} p^n \int_{\langle \beta_n, \gamma_n \rangle_n} \frac{dT}{T} = \langle \beta, \gamma \rangle$$

where we recall $\langle \cdot, \cdot \rangle_n$ denotes the truncated Weil pairing, and we are viewing the Weil pairing as a pairing $\langle \cdot, \cdot \rangle : T_p A \times T_p A \to \mathbb{Z}_p(1) \cong \mathbb{Z}_p \cdot t$; here the last equality above follows from [15, Proposition B.1.7].

In particular, we have

$$\left\langle \alpha_2^* \frac{dT}{T}, \alpha_1 - \frac{1}{\hat{z}(y)} \alpha_2 \right\rangle_p = \langle \alpha_2, \alpha_1 \rangle = \left\langle \iota_{\mathrm{dR}}(y) \left(\alpha_2^* \frac{dT}{T} \right), \alpha_1 - \frac{1}{\hat{z}(y)} \alpha_2 \right\rangle$$

where the last equality follows from Proposition 4.8. Hence for any $w \in \Omega^1_{A/\mathbb{C}_p} \otimes_{\mathbb{C}_p} B_{\mathrm{dR}}^+$, we have

$$\langle w, \gamma \rangle_p = \langle \iota_{\mathrm{dR}}(y)(w), \gamma \rangle$$

and so in particular we are reduced to checking that

$$\langle w \pmod{\Omega^1_{A/\mathbb{C}_p} \otimes_{\mathbb{C}_p} B_{\mathrm{dR}}^+}, \gamma \rangle_p = \langle \iota_{\mathrm{dR}}(w) \pmod{\Omega^1_{A/\mathbb{C}_p} \otimes_{\mathbb{C}_p} B_{\mathrm{dR}}^+}, \gamma \rangle$$

for any

$$w \pmod{\Omega^1_{A/\mathbb{C}_p} \otimes_{\mathbb{C}_p} B_{\mathrm{dR}}^+} \in H^1_{\mathrm{dR}}(A/\mathbb{C}_p) \otimes_{\mathbb{C}_p} B_{\mathrm{dR}}^+ / (\Omega^1_{A/\mathbb{C}_p} \otimes_{\mathbb{C}_p} B_{\mathrm{dR}}^+)$$
$$\cong \mathrm{Lie}(A) \otimes_{\mathbb{C}_p} B_{\mathrm{dR}}^+.$$

But now this follows from Corollary 4.14. $\qquad\qquad\square$

4.6 RELATION TO CLASSICAL (SERRE-TATE) THEORY ON THE ORDINARY LOCUS

We take a brief interlude to recall some facts from classical p-adic analysis on the ordinary locus, and its relations to the de Rham periods defined in the previous section. Recall the ordinary locus $Y^{\mathrm{ord}} \subset Y$, with universal ordinary elliptic curve $\mathcal{A}^{\mathrm{ord}} \to Y^{\mathrm{ord}}$; here, $\mathcal{A}^{\mathrm{ord}}$ can be thought of as the restriction $\mathcal{A}|_{Y^{\mathrm{ord}}}$ of the universal elliptic curve $\mathcal{A} \to Y$. In particular, since Y has a formal model $\hat{Y}/\mathbb{Z}_p$ with a moduli interpretation, the universal elliptic curve $\mathcal{A} \to Y$ and hence $\mathcal{A}^{\mathrm{ord}} \to Y^{\mathrm{ord}}$ spread out to formal schemes $\hat{\mathcal{A}} \to \hat{Y}$ and $\hat{\mathcal{A}}^{\mathrm{ord}} \to \hat{Y}^{\mathrm{ord}}$ over $\mathbb{Z}_p$; let $\hat{\mathcal{A}}_0^{\mathrm{ord}}, \hat{Y}_0^{\mathrm{ord}}$ denote the special fibers. Now the adic reduction map $\mathcal{A}^{\mathrm{ord}} \to \hat{\mathcal{A}}_0^{\mathrm{ord}}$ allows us to make the identifications

$$\mathcal{A}^{\mathrm{ord}}[p^n](\overline{\mathbb{F}}_p) = \hat{\mathcal{A}}_0^{\mathrm{ord}}[p^n](\overline{\mathbb{F}}_p)$$

and so

$$(T_p \mathcal{A}^{\mathrm{ord}})^{\mathrm{ét}} = T_p \hat{\mathcal{A}}_0^{\mathrm{ord}}(\overline{\mathbb{F}}_p).$$

Definition 4.17. *The Igusa tower $Y^{\mathrm{Ig}} \to Y^{\mathrm{ord}}$ is a $\mathbb{Z}_p^\times$-torsor which represents the moduli space classifying isomorphism classes of triples*

$$(A, t, \mu_{p^\infty} \hookrightarrow A[p^\infty]),$$

where A is an ordinary elliptic curve, $t \in A[N]$ is a generator, and $\mu_{p^\infty} \hookrightarrow A[p^\infty]$ is an embedding of p-divisible groups. By (3.10), an equivalent way to define Y^{Ig} is as the solution to the moduli problem classifying isomorphism classes of triples

$$(A, t, (T_p A)^{\acute{e}t} \cong \mathbb{Z}_p).$$

One can define (non-canonical) adic étale section $Y^{\mathrm{ord}} \to Y^{\mathrm{Ig}}$ of the $\mathbb{Z}_p^\times$-torsor $Y^{\mathrm{Ig}} \to Y^{\mathrm{ord}}$ defined by choosing an embedding $\mu_{p^\infty} \subset \mathcal{A}^{\mathrm{ord}}[p^\infty]$ (or equivalently $\hat{\mathbb{Z}}_{p,Y}(1) \subset T_p \mathcal{A}^{\mathrm{ord}}$), or in other words by choosing a trivialization

$$\hat{\mathbb{Z}}_{p,Y^{\mathrm{ord}}} \xrightarrow{\sim} (T_p \mathcal{A}^{\mathrm{ord}})^0(-1) \subset T_p \mathcal{A}^{\mathrm{ord}}(-1) \tag{4.26}$$

of the canonical line $(T_p \mathcal{A}^{\mathrm{ord}})^0 \subset T_p \mathcal{A}^{\mathrm{ord}}$; here $(T_p \mathcal{A}^{\mathrm{ord}})^0$ denotes the connected component of the universal ordinary Tate module. From (3.10), we get the isomorphism

$$(T_p \mathcal{A}^{\mathrm{ord}})^{\acute{e}t} \cong \mathrm{Hom}_{\hat{\mathbb{Z}}_{p,Y^{\mathrm{ord}}}}(\mathcal{A}^{\mathrm{ord}}[p^\infty]^0, \mu_{p^\infty}), \tag{4.27}$$

where $(T_p \mathcal{A}^{\mathrm{ord}})^{\acute{e}t} = T_p \hat{\mathcal{A}}_0^{\mathrm{ord}}(\overline{\mathbb{F}}_p)$ is the étale quotient of $T_p \mathcal{A}^{\mathrm{ord}}$ and $\mathcal{A}^{\mathrm{ord}}[p^\infty]^0$ is the connected component of the p-divisible group of the universal ordinary elliptic curve. Hence to have a trivialization as in (4.26) is equivalent to having a trivialization

$$\hat{\mathbb{Z}}_{p,Y^{\mathrm{ord}}} \xrightarrow{\sim} (T_p \mathcal{A}^{\mathrm{ord}})^{\acute{e}t}.$$

Definition 4.18. *Recall that the connected part of the Tate module*

$$T_p \mathcal{A}^{\mathrm{ord},0} \subset T_p \mathcal{A}^{\mathrm{ord}}$$

is called the canonical line. Now we define the affinoid subdomain of the ordinary locus

$$\mathcal{Y}^{\mathrm{Ig}} \subset \mathcal{Y}^{\mathrm{ord}}$$

as the adic space over $\mathrm{Spa}(F, \mathcal{O}_F)$ which classifies isomorphism classes of triples $(A, t, \alpha : \mathbb{Z}_p^{\oplus 2} \xrightarrow{\sim} T_p A)$, where A is ordinary with the additional condition on the p^∞-level structure $\alpha : \mathbb{Z}_p^{\oplus 2} \xrightarrow{\sim} T_p A$ that:

$$\alpha|_{\mathbb{Z}_p \oplus \{0\}} : \mathbb{Z}_p \xrightarrow{\sim} (T_p A)^0 \subset T_p A. \tag{4.28}$$

In other words $\hat{\mathbf{z}}(A, t, \alpha) = \infty$, and

$$\alpha|_{\{0\} \oplus \mathbb{Z}_p} : \mathbb{Z}_p \xrightarrow{\sim} (T_p A)^{\acute{e}t} \subset T_p A. \tag{4.29}$$

Equivalently, we have

$$\mathcal{Y}^{\mathrm{Ig}} = \{\mathbf{z} = \infty\} = \{(A, t, \alpha) \in \mathcal{Y}^{\mathrm{ord}}, \alpha|_{\mathbb{Z}_p \oplus \{0\}} : \mathbb{Z}_p \xrightarrow{\sim} (T_p A)^0\}$$

so we can view $\mathcal{Y}^{\mathrm{Ig}}$ naturally as an affinoid subdomain of $\mathcal{Y}$.

Definition 4.19. *Henceforth, let*

$$B := \left\{ \begin{pmatrix} a & b \\ 0 & d \end{pmatrix} \in GL_2(\mathbb{Z}_p) \right\} \subset GL_2(\mathbb{Z}_p)$$

denote the subgroup of upper triangular matrices. It has a normal subgroup

$$C := \left\{ \begin{pmatrix} \mathbb{Z}_p^\times & \mathbb{Z}_p \\ 0 & 1 \end{pmatrix} \right\} \subset B,$$

which we note is isomorphic to $\mathbb{Z}_p \rtimes \mathbb{Z}_p^\times$, with quotient $B/C = \mathbb{Z}_p^\times$.

Proposition 4.20. *We have that*

1. $\mathcal{Y}^{\mathrm{Ig}} \to Y^{\mathrm{ord}}$ is a B-torsor,
2. $\mathcal{Y}^{\mathrm{Ig}} \to Y^{\mathrm{Ig}}$ is a C-torsor,
3. $Y^{\mathrm{Ig}} \to Y^{\mathrm{ord}}$ is a $\mathbb{Z}_p^\times$-torsor.

Proof. For (1), notice that B is the stabilizer of $\{z_{\mathrm{HT}} = \infty\}$ under the $GL_2(\mathbb{Q}_p)$-action. Moduli-theoretically, in the above discussion we noted that $\mathcal{Y}^{\mathrm{Ig}}$ parametrizes triples (A, t, α) with $\alpha_{\infty,1} : \mathbb{Z}_p \xrightarrow{\sim} (T_p A)^0$ and α_2 arbitrary. It is clear that given a fixed ordinary elliptic curve A, B acts simply transitively on the set of such α.

For (2), suppose that we are given (A, t, α_0) with a trivialization $\alpha : T_p A^{\mathrm{ord,\acute{e}t}} \cong \mathbb{Z}_p$. Then by Cartier duality, this induces

$$T_p A^{\mathrm{ord},0} = \mathrm{Hom}(T_p A^{\mathrm{ord,\acute{e}t}}, \mathbb{Z}_p(1)) \xrightarrow{\alpha_0} \mathbb{Z}_p(1). \qquad (4.30)$$

Then choosing a trivialization $i : \mathbb{Z}_p(1) \xrightarrow{\sim} \mathbb{Z}_p$ (which amounts to fixing a compatible sequence of pth-power roots of unity, or equivalently an invariant differential on the formal Lie group μ_{p^∞}), of which there are $\mathbb{Z}_p^\times$ choices, we get a trivialization

$$\alpha_1 := i \circ (4.30) : T_p A^{\mathrm{ord},0} \xrightarrow{\sim} \mathbb{Z}_p.$$

Now we can extend to any basis $\alpha : \mathbb{Z}_p^{\oplus 2} \xrightarrow{\sim} T_p A$, and for (A, t, α) to be a point of $\mathcal{Y}^{\mathrm{Ig}}$ above (A, t, α_0), α must be compatible with α_0 in the sense that

$$\alpha \pmod{\langle \alpha_1 \rangle} : \mathbb{Z}_p \xrightarrow{\sim} T_p A / T_p A^0 = T_p A^{\mathrm{\acute{e}t}}$$

is equal to α_0; in other words,

$$\alpha_2 \equiv \alpha \pmod{\langle \alpha_1 \rangle}.$$

The subgroup of $GL_2(\mathbb{Z}_p)$ which preserves this congruence relation is exactly C, and so C acts simply transitively on the set of points (A, t, α) above (A, t, α_0).

Finally, (3) is clear from the definition of Y^{Ig}, since there are $\mathbb{Z}_p^{\times}$ ways to trivialize $\alpha : \mathbb{Z}_p \xrightarrow{\sim} T_p A^{\text{ét}}$. $\qquad\square$

Fix a residue disc $D \subset Y^{\mathrm{ord}}$, and let $\mathcal{D} := D \times_{Y^{\mathrm{ord}}} \mathcal{Y}^{\mathrm{Ig}}$. Note that $(T_p \mathcal{A}^{\mathrm{ord}})^{\text{ét}}|_D = T_p \hat{\mathcal{A}}_0^{\mathrm{ord}}|_D(\overline{\mathbb{F}}_p)$ is constant, and hence there is a (non-canonical) section

$$\alpha_2 : (T_p \mathcal{A}^{\mathrm{ord}})^{\text{ét}}|_D \to T_p \mathcal{A}^{\mathrm{ord}}|_D.$$

Note that *choosing* a trivialization

$$\alpha_\infty : \hat{\mathbb{Z}}_{p,Y^{\mathrm{ord}}}^{\oplus 2} \xrightarrow{\sim} T_p \mathcal{A}^{\mathrm{ord}}$$

with

$$\alpha_\infty|_{\hat{\mathbb{Z}}_{p,Y^{\mathrm{ord}}} \oplus \{0\}} : \hat{\mathbb{Z}}_{p,Y^{\mathrm{ord}}} \xrightarrow{\sim} (T_p \mathcal{A}^{\mathrm{ord}})^0$$

and

$$\alpha_\infty|_{\{0\} \oplus \hat{\mathbb{Z}}_{p,Y^{\mathrm{ord}}}} = \alpha : \hat{\mathbb{Z}}_{p,Y^{\mathrm{ord}}} \xrightarrow{\sim} (T_p \mathcal{A}^{\mathrm{ord}})^{\text{ét}}$$

defines an adic étale section

$$D \hookrightarrow \mathcal{D}$$

of the B-torsor $\mathcal{D} \to D$. The $\mathbb{Z}_p^{\times,\oplus 2}$-orbit of the image of this section is a (non-canonical) $\mathbb{Z}_p^{\times,\oplus 2}$-torsor of D.

Proposition 4.21. *On $\mathcal{Y}^{\mathrm{Ig}}$, we have*

$$1/\hat{\mathbf{z}} = 0.$$

As a consequence, on $\mathcal{Y}^{\mathrm{Ig}}$, we have

$$\mathbf{y}_{\mathrm{dR}}/t = 1.$$

Proof. The first part follows by definition of $\mathcal{Y}^{\mathrm{Ig}}$. Now by the p-adic Legendre relation (4.17), we have $\mathbf{y}_{\mathrm{dR}} = t$ on $\mathcal{Y}^{\mathrm{Ig}}$. $\qquad\square$

We now recall Serre-Tate coordinates, which are defined naturally on $\mathcal{Y}^{\mathrm{Ig}}$, stated for the proétale site $Y_{\mathrm{proét}}^{\mathrm{ord}}$. (One can check that all the isomorphisms defined below are functorial for the proétale objects, and so induce isomorphisms of sheaves on $Y_{\mathrm{proét}}^{\mathrm{ord}}$.) For our previously fixed ordinary residue disc $D \subset Y^{\mathrm{ord}}$ and let $\mathcal{D} = \mathcal{Y}^{\mathrm{Ig}} \times_{Y^{\mathrm{ord}}} D$, which is a proétale cover of D and hence an object of $Y_{\mathrm{proét}}$.

Theorem 4.22 (Taking inverse limits over artinian W-algebras of Corollary 4.1.5, Theorem 4.3.1 (quat) [34]). *We have the following natural isomorphisms of sheaves on $Y^{\mathrm{ord}}_{pro\acute{e}t}$:*

$$HT_{\mathcal{A}} : (T_p\mathcal{A}^{\mathrm{ord}})^{\acute{e}t} \otimes_{\hat{\mathbb{Z}}_{p,D}} \mathcal{O}_D \xrightarrow{\sim} \omega|_D$$

$$(HT_{\mathcal{A}})^{\vee,-1} : (T_p\mathcal{A}^{\mathrm{ord}})^0(-1) \otimes_{\hat{\mathbb{Z}}_{p,D}} \mathcal{O}_D$$
$$= \mathrm{Hom}_{\mathbb{Z}_p}((T_p\mathcal{A}^{\mathrm{ord}})^{\acute{e}t}, \hat{\mathbb{Z}}_{p,D}) \otimes_{\hat{\mathbb{Z}}_{p,D}} \mathcal{O}_D \xrightarrow{\sim} \omega^{-1}|_D.$$

Moreover, there is a natural isomorphism of formal schemes over $\mathcal{O}_F$ (viewed as functors of $\mathcal{O}_F$-algebras)

$$\hat{Y}^{\mathrm{ord}}(\cdot) \cong \mathrm{Hom}_{\mathbb{Z}_p}(T_p\hat{\mathcal{A}}_0(\overline{\mathbb{F}}_p) \otimes_{\mathbb{Z}_p} T_p\hat{\mathcal{A}}_0(\overline{\mathbb{F}}_p), \hat{\mathbb{G}}_m(\cdot)) \qquad (4.31)$$

which on the adic generic fibers induces a natural isomorphism of adic spaces over $\mathrm{Spa}(F, \mathcal{O}_F)$ (viewed as functors of $(F, \mathcal{O}_F)$-algebras)

$$D(\cdot) \cong \mathrm{Hom}_{\mathbb{Z}_p}((T_p\mathcal{A}^{\mathrm{ord}})^{\acute{e}t} \otimes_{\mathbb{Z}_p} (T_p\mathcal{A}^{\mathrm{ord}})^{\acute{e}t}|_D, \hat{\mathbb{G}}_m(\cdot)) \otimes_{\mathbb{Z}_p} \mathbb{Q}_p. \qquad (4.32)$$

This defines coordinates on the proétale B-cover $\mathcal{D} \to D$ in the following way: let $\alpha_{\infty,1}, \alpha_{\infty,2}$ be the basis of $(T_p\mathcal{A}^{\mathrm{ord}})|_{\mathcal{D}}$ defined in (4.28) and (4.29). Then the trivialization

$$\alpha_{\infty}|_{\{0\} \oplus \ddot{\mathbb{Z}}_{p,\mathcal{D}}} : \hat{\mathbb{Z}}_{p,\mathcal{D}} \xrightarrow{\sim} (T_p\mathcal{A}^{\mathrm{ord}})^{\acute{e}t}|_{\mathcal{D}}$$

induces, via (4.32), a canonical isomorphism of adic spaces

$$\mathcal{D}(\cdot) \cong \mathrm{Hom}_{\mathbb{Z}_p}((T_p\mathcal{A}^{\mathrm{ord}})^{\acute{e}t}|_{\mathcal{D}} \otimes_{\mathbb{Z}_p} (T_p\mathcal{A}^{\mathrm{ord}})^{\acute{e}t}|_{\mathcal{D}}, \hat{\mathbb{G}}_m(\cdot)) \otimes_{\mathbb{Z}_p} \mathbb{Q}_p$$
$$\xrightarrow{\alpha_{\infty}|_{\{0\} \oplus \ddot{\mathbb{Z}}_{p,\mathcal{D}}}^{-1}} \mathrm{Hom}_{\mathbb{Z}_p}(\hat{\mathbb{Z}}_{p,\mathcal{D}}, \hat{\mathbb{G}}_m(\cdot)) \otimes_{\mathbb{Z}_p} \mathbb{Q}_p. \qquad (4.33)$$

Definition 4.23. *Given a residue disc $D \subset Y^{\mathrm{ord}}$, let $\mathcal{D} = \mathcal{Y}^{\mathrm{Ig}} \times_{Y^{\mathrm{ord}}} D$. The pullback of the natural coordinate T on*

$$\mathbb{T} = \mathrm{Spa}(\mathbb{Q}_p\langle T^{\pm 1} \rangle, \mathbb{Z}_p\langle T^{\pm 1} \rangle)$$

is called the Serre-Tate coordinate on $\mathcal{D}$.

Definition 4.24. *Let*
$$\hat{\mathcal{O}}_{F,\mathcal{D}} = \hat{\mathbb{Z}}_{p,\mathcal{D}} \otimes_{\mathbb{Z}_p} \mathcal{O}_F.$$

We call the map of $\mathcal{O}_F$-modules

$$T - \exp : \mathcal{O}_{\mathcal{D}} \xrightarrow{\sim} \hat{\mathcal{O}}_{F,\mathcal{D}}[\![T-1]\!] \otimes_{\mathbb{Z}_p} \mathbb{Q}_p, \tag{4.34}$$

induced by (4.33) the Serre-Tate expansion map on $\mathcal{D}$. Now let $y \in \mathcal{D}$ be a geometric point, and let $D(\lambda(y))$ denote the residue disc in Y^{ord} centered around $\lambda(y)$ (recalling that $\lambda : \mathcal{Y}^{\mathrm{Ig}} \to Y^{\mathrm{ord}}$ is the natural projection), and let $D(y) = D(\lambda(y)) \times_{Y^{\mathrm{ord}}} \mathcal{Y}^{\mathrm{Ig}}$. Using the fact that dT is a generator of $\Omega^1_{\hat{\mathbb{G}}_m}$, one can show that

$$T - \exp(f) = \sum_{n=0}^{\infty} \frac{1}{n!} \left(\frac{d}{dT} \right)^n \big|_{T=1} f(y)(T-1)^n \tag{4.35}$$

where for a section s of $\mathcal{O}_{\mathcal{D}}$, $s(y)$ denotes the image of s in the residue field at y. Henceforth, we adopt the shorthand

$$f(T) := T - \exp(f).$$

Let

$$\omega_{Y^{\mathrm{ord}}} := \pi_* \Omega^1_{\mathcal{A}^{\mathrm{ord}}/Y^{\mathrm{ord}}}, \quad \hat{\omega}_{Y^{\mathrm{ord}}} = \omega_{Y^{\mathrm{ord}}} \otimes_{\mathcal{O}_{Y^{\mathrm{ord}}}} \hat{\mathcal{O}}_{Y^{\mathrm{ord}}}$$

denote the restrictions of the Hodge and p-adically completed Hodge bundles ω and $\hat{\omega}$. Let $\alpha_\infty : \hat{\mathbb{Z}}_{p,\mathcal{Y}^{\mathrm{Ig}}}^{\oplus 2} \xrightarrow{\sim} T_p \mathcal{A}^{\mathrm{ord}}$ denote the (restriction) of the universal level structure to $\mathcal{Y}^{\mathrm{Ig}}$.

Definition 4.25. *We have a natural section*

$$HT_{\mathcal{A}}(\alpha_{\infty,2}) = \omega_{\mathrm{can}}^{\mathrm{Katz}} = \mathfrak{s}|_{\mathcal{Y}^{\mathrm{Ig}}} \in \hat{\omega}_{Y^{\mathrm{ord}}}(\mathcal{Y}^{\mathrm{Ig}})$$

which is called Katz's canonical differential. Let $D \subset Y^{\mathrm{ord}}$ be a residue disc, and let

$$\mathcal{D} = \mathcal{Y}^{\mathrm{Ig}} \times_{Y^{\mathrm{ord}}} \mathcal{Y}^{\mathrm{Ig}}$$

and recall we denote by T the Serre-Tate coordinate on $\mathcal{D}$. By [34, Main Theorem 4.4.1], we have

$$\sigma(\omega_{\mathrm{can}}^{\mathrm{Katz},\otimes 2})|_{\mathcal{D}} = d \log T.$$

Proposition 4.26. *In fact, we have*

$$\omega_{\mathrm{can}}^{\mathrm{Katz}} \in \omega_{Y^{\mathrm{ord}}}(\mathcal{Y}^{\mathrm{Ig}}).$$

Moreover, $\omega_{\mathrm{can}}^{\mathrm{Katz}}$ descends to a section in $\omega_{Y^{\mathrm{ord}}}(Y^{\mathrm{Ig}})$.

Proof. We have $\omega_{\mathrm{can}}^{\mathrm{Katz}} \in \hat{\omega}(\mathcal{Y}^{\mathrm{Ig}})$. Now let w be any local generator $w \in \omega(U)$, for an open $U \subset Y^{\mathrm{ord}}$ (using the fact that ω is locally free of rank 1). Recall our projection $\lambda : \mathcal{Y} \to Y$, and let $\mathcal{U} = \lambda^{-1}(U) \cap \mathcal{Y}^{\mathrm{Ig}}$. Then we have

$$u := \omega_{\mathrm{can}}^{\mathrm{Katz}}|_{\mathcal{U}} / w|_{\mathcal{U}} \in \hat{\mathcal{O}}_{\mathcal{Y}^{\mathrm{Ig}}}(\mathcal{U})^{\times}.$$

Moreover,

$$\iota_{\mathrm{dR}}(w)|_{\mathcal{U}} = x_0 \cdot \alpha_{\infty,1}/t + y_0 \cdot \alpha_{\infty,2}/t$$

so that $x_0, y_0 \in \mathcal{OB}_{\mathrm{dR},Y}^+(\mathcal{U})$. Applying Theorem 4.9 to w and invoking Proposition 4.21, we see that

$$y_0 = x_0/\hat{\mathbf{z}} + y_0 = t.$$

In particular, we have $y_0 = t = \mathbf{y}_{\mathrm{dR}}|_{\mathcal{U}}$, and so since $\alpha_{\infty,1}, \alpha_{\infty,2}$ is a basis of

$$T_p \mathcal{A}^{\mathrm{ord}} \otimes_{\hat{\mathbb{Z}}_{p,Y^{\mathrm{ord}}}} \hat{\mathcal{OB}}_{\mathrm{dR},\mathcal{Y}^{\mathrm{Ig}}}^{\mathrm{I}} \xrightarrow[\hat{\alpha}_{?}]{\alpha_{\hat{?}}^{-1}} \hat{\mathcal{OB}}_{\mathrm{dR},\mathcal{Y}^{\mathrm{Ig}}}^+,$$

we must have $u = 1$. Hence

$$\omega_{\mathrm{can}}^{\mathrm{Katz}}|_{\mathcal{U}} = w|_{\mathcal{U}} \in \omega|_{\mathcal{Y}^{\mathrm{Ig}}}(\mathcal{Y}^{\mathrm{Ig}}). \tag{4.36}$$

Note that w and U were arbitrary. Hence we can choose $\{(w,U)\}$ such that $\{\mathcal{U}\}$ is an open covering of $\mathcal{Y}^{\mathrm{Ig}}$, and by (4.36), we see that the $w|_{\mathcal{U}}$ glue to a section in $\omega|_{\mathcal{Y}^{\mathrm{Ig}}}(\mathcal{Y}^{\mathrm{Ig}})$ which is in fact equal to $\omega_{\mathrm{can}}^{\mathrm{Katz}}$. So we are done with the first statement.

For the second statement, note that for any $\begin{pmatrix} a & b \\ 0 & d \end{pmatrix} \in B$, by the first part we have

$$\begin{pmatrix} a & b \\ 0 & d \end{pmatrix}^* \omega_{\mathrm{can}}^{\mathrm{Katz}} = \begin{pmatrix} a & b \\ 0 & d \end{pmatrix}^* HT(\alpha_{\infty,2}) = HT(b\alpha_1 + d\alpha_2)$$

$$= (b/\hat{\mathbf{z}} + d)\omega_{\mathrm{can}}^{\mathrm{Katz}} = d\omega_{\mathrm{can}}^{\mathrm{Katz}} \tag{4.37}$$

where the last equality follows by Proposition 4.21. Hence $\omega_{\mathrm{can}}^{\mathrm{Katz}}$ is invariant by C (see Definition 4.18), and so since $\mathcal{Y}^{\mathrm{Ig}} \to Y^{\mathrm{Ig}}$ is a C-cover by Proposition 4.20, and the proétale structure sheaf is defined as the pullback $\nu^* \mathcal{O}_{Y_{\text{ét}}^{\mathrm{ord}}}$ of the étale structure sheaf, then $\omega_{\mathrm{can}}^{\mathrm{Katz}}$ descends to Y^{Ig} by [13, Lemma 2.24], which we recall below.   □

Lemma 4.27 ([13], Lemma 2.24). *Suppose Y is an adic space, let G be a profinite group and suppose that $Y_\infty \in Y_{\text{proét}}$ is a Galois G-cover of Y. Let $U \in Y_{\text{proét}}$ be quasi-compact and quasi-separated and set $U_\infty := U \times_Y Y_\infty$, so that $U_\infty \to U$ is a Galois G-cover. Let $\mathcal{F}$ be a sheaf on $Y_{\text{proét}}$ which is the pullback of a sheaf from $Y_{\text{ét}}$. Then $\mathcal{F}(U) = \mathcal{F}(U_\infty)^G$ (where the left-hand side denotes G-invariants).*

We will also repeatedly invoke the above lemma on proétale descent in later sections.

Proposition 4.28. *The relative Hodge-Tate filtration (3.22) restricted to $\mathcal{Y}^{\mathrm{Ig}}$*

$$0 \to \hat{\omega}_{\mathcal{Y}^{\mathrm{Ig}}}^{-1}(1) \to T_p\mathcal{A}^{\mathrm{ord}} \otimes_{\hat{\mathbb{Z}}_{p,Y^{\mathrm{ord}}}} \hat{\mathcal{O}}_{\mathcal{Y}^{\mathrm{Ig}}} \to \hat{\omega}_{\mathcal{Y}^{\mathrm{Ig}}} \to 0$$

has a splitting, given by

$$T_p\mathcal{A}^{\mathrm{ord}} \otimes_{\hat{\mathbb{Z}}_{p,Y^{\mathrm{ord}}}} \hat{\mathcal{O}}_{\mathcal{Y}^{\mathrm{Ig}}} \to \hat{\omega}_{\mathcal{Y}^{\mathrm{Ig}}}, \quad x\alpha_{\infty,1} + y\alpha_{\infty,2} \mapsto x\mathfrak{s}^{-1}$$

where $s^{-1} \in \hat{\omega}_{\mathcal{Y}^{\mathrm{Ig}}}$ is the Poincaré dual of $\mathfrak{s}|_{\mathcal{Y}^{\mathrm{Ig}}} = \omega_{\mathrm{can}}^{\mathrm{Katz}}$. (That is, $\langle \omega_{\mathrm{can}}^{\mathrm{Katz}}, s^{-1} \rangle_{\mathrm{Poin}} = 1$.)

Proof. The basis $\alpha_{\infty,1}, \alpha_{\infty,2}$ on $\mathcal{Y}^{\mathrm{Ig}} = \{\hat{\mathbf{z}} = \infty\}$ gives a splitting

$$T_p\mathcal{A}^{\mathrm{ord}} \otimes_{\hat{\mathbb{Z}}_{p,Y^{\mathrm{ord}}}} \hat{\mathcal{O}}_{\mathcal{Y}^{\mathrm{Ig}}} \to \hat{\omega}_{\mathcal{Y}^{\mathrm{Ig}}}^{-1}(1), \quad \alpha_{\infty,1} \mapsto s^{-1}(1), \alpha_{\infty,2} \mapsto 0.$$

Note

$$s^{-1}(1) \mapsto \alpha_{\infty,1} - \frac{1}{\hat{\mathbf{z}}}\alpha_{\infty,2}$$

under the Hodge-Tate filtration

$$\hat{\omega}_{\mathcal{Y}^{\mathrm{Ig}}}^{-1}(1) \hookrightarrow T_p\mathcal{A}^{\mathrm{ord}} \otimes_{\hat{\mathbb{Z}}_{p,Y^{\mathrm{ord}}}} \hat{\mathcal{O}}_{\mathcal{Y}^{\mathrm{Ig}}}.$$

This follows because locally $s^{-1}(1) = u\alpha_{\infty,1}/t + u/\hat{\mathbf{z}}\alpha_2$ for some local section u of $\hat{\mathcal{O}}_{\mathcal{Y}^{\mathrm{Ig}}}^{\times}$, but by Theorem 4.31, we have

$$u = u\langle \alpha_{\infty,2}, \alpha_{\infty,1}/t \rangle = \left\langle \alpha_2^* \frac{dT}{T}, u\alpha_1/t - \frac{u}{\hat{\mathbf{z}}}\alpha_2/t \right\rangle = \langle \omega_{\mathrm{can}}^{\mathrm{Katz}}, s^{-1} \rangle_{\mathrm{Poin}} = 1.$$

Now $s^{-1}(1) \mapsto \alpha_{\infty,1} - \frac{1}{z_{\mathrm{HT}}}\alpha_{\infty,2} = \alpha_{\infty,2}$, because $z_{\mathrm{HT}} = \infty$ on $\mathcal{Y}^{\mathrm{Ig}}$, and we are done. $\square$

We end this section by clarifying the relationship between $\mathcal{Y}^{\mathrm{Ig}}$ and Katz's theory, by showing how a "universal unit root splitting" arises on $\mathcal{Y}^{\mathrm{Ig}}$, which is in essence a splitting of the Hodge filtration on $\mathcal{H}_{\mathrm{dR}}^1(\mathcal{A}^{\mathrm{ord}})$ arising from the eigen-decomposition of a Frobenius endomorphism operator arising from the *canonical lifting of Frobenius* $Y^{\mathrm{ord}} \to Y^{\mathrm{ord}}$. Let us recall the construction of the canonical lifting of Frobenius on Y^{ord} as well as some facts about it.

Let $\mathcal{C}_{\mathrm{can}} \subset \mathcal{A}^{\mathrm{ord}}$ denote the canonical subgroup, defined over $\mathcal{O}_F$. Base changing $\mathcal{A}^{\mathrm{ord}}$ to $\mathrm{Spa}(F, \mathcal{O}_F)$, by properties of the canonical subgroup, the isogeny

$$\pi : \mathcal{A}^{\mathrm{ord}} \to \mathcal{A}^{\mathrm{ord}}/\mathcal{C}_{\mathrm{can}}$$

over $\mathrm{Spa}(F, \mathcal{O}_F)$ is a lifting of the relative Frobenius morphism on the special fiber

$$\mathcal{A}_0^{\mathrm{ord}} \to \mathcal{A}_0^{\mathrm{ord},(p)}$$

where

$$\mathcal{A}_0^{\mathrm{ord},(p)} = \mathcal{A}_0^{\mathrm{ord}} \times_{\mathrm{Spec}(\overline{\mathbb{F}}_p), \mathrm{Frob}} \mathrm{Spec}(\overline{\mathbb{F}}_p)$$

and $\mathrm{Frob} : \mathrm{Spec}(\overline{\mathbb{F}}_p) \to \mathrm{Spec}(\overline{\mathbb{F}}_p)$ is the absolute Frobenius (with $\mathrm{Frob}^* f = f^p$). Then let $\phi : Y^{\mathrm{ord}} \to Y^{\mathrm{ord}}$ be the classifying morphism such that we have the (unique) isomorphism

$$\mathcal{A}^{\mathrm{ord}}/\mathcal{C}_{\mathrm{can}} \cong \mathcal{A}^{\mathrm{ord}} \times_{Y^{\mathrm{ord}},\phi} Y^{\mathrm{ord}} =: \mathcal{A}^{\mathrm{ord},(\phi)}.$$

The map $\phi : Y^{\mathrm{ord}} \to Y^{\mathrm{ord}}$, with induced map $\phi^* : \mathcal{O}_{Y^{\mathrm{ord}}} \to \mathcal{O}_{Y^{\mathrm{ord}}}$, is often known as the *canonical lifting of Frobenius* or *canonical Frobenius endomorphism*. Let

$$\phi^{-1} : \mathcal{A}^{\mathrm{ord},(\phi)} \to \mathcal{A}^{\mathrm{ord}}$$

denote the natural projection, so that we get maps

$$\phi^{-1} : \mathcal{H}_{\mathrm{dR}}^1(\mathcal{A}) \to \mathcal{H}_{\mathrm{dR}}^1(\mathcal{A}) \otimes_{\mathcal{O}_{Y^{\mathrm{ord}}},\phi^*} \mathcal{O}_{Y^{\mathrm{ord}}}, \quad \omega|_{Y^{\mathrm{ord}}}$$
$$\to \pi_* \Omega_{\mathcal{A}^{\mathrm{ord},(\phi)}/Y^{\mathrm{ord}}}, \quad \omega^{-1}|_{Y^{\mathrm{ord}}} \to R^1\pi_* \mathcal{O}_{\mathcal{A}^{\mathrm{ord},(\phi)}}.$$

Then

$$F(\phi) : \mathcal{H}_{\mathrm{dR}}^1(\mathcal{A}^{\mathrm{ord}}) \to \mathcal{H}_{\mathrm{dR}}^1(\mathcal{A}^{\mathrm{ord}}), \quad \omega|_{Y^{\mathrm{ord}}} \to \omega|_{Y^{\mathrm{ord}}}, \quad \omega^{-1}|_{Y^{\mathrm{ord}}} \to \omega^{-1}|_{Y^{\mathrm{ord}}},$$

where

$$F(\phi) = \pi^* \circ \phi^{-1},$$

are ϕ-linear endomorphisms. Recall the ordinary Hodge filtration

$$0 \to \omega|_{Y^{\mathrm{ord}}} \to \mathcal{H}_{\mathrm{dR}}^1(\mathcal{A}^{\mathrm{ord}}) \to \omega^{-1}|_{Y^{\mathrm{ord}}} \to 0. \tag{4.38}$$

As shown in [36, Lemma (A2.1)], we have

$$F(\phi) = p\phi^* \text{ on } \omega|_{Y^{\mathrm{ord}}}, \quad F(\phi) = \phi^* \text{ on } \omega^{-1}|_{Y^{\mathrm{ord}}}.$$

The *unit root splitting* of (4.38) is given by an explicit map

$$\omega^{-1}|_{Y^{\mathrm{ord}}} \to \mathcal{H}^1_{\mathrm{dR}}(\mathcal{A})^{F(\phi)=\phi}$$

which is constructed in Section A2.3 of loc. cit. In Theorem 4.29, we explain how the geometry of $\mathcal{Y}^{\mathrm{Ig}}$ can be used to recover the unit root splitting.

In particular, we explain how the universal level structure $\alpha : \hat{\mathbb{Z}}^{\oplus 2}_{p,\mathcal{Y}^{\mathrm{Ig}}} \xrightarrow{\sim} T_p\mathcal{A}^{\mathrm{ord}}$ also gives rise to a "universal unit root splitting" of $\mathcal{H}^1_{\mathrm{dR}}(\mathcal{A})|_{\mathcal{Y}^{\mathrm{Ig}}}$ in some sense, after extending coefficients to $\mathcal{O}\mathbb{B}^+_{\mathrm{dR},\mathcal{Y}^{\mathrm{Ig}}}$. More precisely, consider the space generated by $\alpha_{\infty,1}$ in the extended Tate module:

$$\langle \alpha_{\infty,1} \rangle := \alpha_{\infty,1} \cdot \hat{\mathcal{O}}\mathbb{B}^+_{\mathrm{dR},\mathcal{Y}^{\mathrm{Ig}}} \cdot t^{-1} \subset T_p\mathcal{A}^{\mathrm{ord}} \otimes_{\hat{\mathbb{Z}}_{p,Y^{\mathrm{ord}}}} \hat{\mathcal{O}}\mathbb{B}^+_{\mathrm{dR},\mathcal{Y}^{\mathrm{Ig}}} \cdot t^{-1}.$$

Theorem 4.29. *Under the de Rham comparison inclusion*

$$\mathcal{H}^1_{\mathrm{dR}}(\mathcal{A}^{\mathrm{ord}})|_{\mathcal{Y}^{\mathrm{Ig}}} \subset \mathcal{H}^1_{\mathrm{dR}}(\mathcal{A}^{\mathrm{ord}}) \otimes_{\mathcal{O}_{Y^{\mathrm{ord}}}} \mathcal{O}\mathbb{B}^+_{\mathrm{dR},\mathcal{Y}^{\mathrm{Ig}}} \xrightarrow{\iota_{\mathrm{dR}}} T_p\mathcal{A}^{\mathrm{ord}} \otimes_{\hat{\mathbb{Z}}_{p,Y^{\mathrm{ord}}}} \mathcal{O}\mathbb{B}^+_{\mathrm{dR},\mathcal{Y}^{\mathrm{Ig}}} \cdot t^{-1},$$

we have that

$$\mathcal{L} := \langle \alpha_{\infty,1} \rangle \cap \iota_{\mathrm{dR}}(\mathcal{H}^1_{\mathrm{dR}}(\mathcal{A}^{\mathrm{ord}})|_{\mathcal{Y}^{\mathrm{Ig}}})$$

is the image $\iota_{\mathrm{dR}}(\mathcal{U})$ of the unit root subspace $\mathcal{U} \subset \mathcal{H}^1_{\mathrm{dR}}(\mathcal{A}^{\mathrm{ord}})|_{\mathcal{Y}^{\mathrm{Ig}}}$. In particular, this implies that the map

$$HT : T_p\mathcal{A}^{\mathrm{ord}} \otimes_{\hat{\mathbb{Z}}_{p,Y^{\mathrm{ord}}}} \hat{\mathcal{O}}\mathbb{B}^+_{\mathrm{dR},\mathcal{Y}^{\mathrm{Ig}}} \cdot t^{-1} \twoheadrightarrow \hat{\omega}_{Y^{\mathrm{Ig}}} \otimes_{\hat{\mathcal{O}}_{\mathcal{Y}^{\mathrm{Ig}}}} \hat{\mathcal{O}}\mathbb{B}^+_{\mathrm{dR},\mathcal{Y}^{\mathrm{Ig}}},$$

the kernel of which is $\langle \alpha_{\infty,1} \rangle$, restricts to the unit root splitting

$$\mathcal{H}^1_{\mathrm{dR}}(\mathcal{A}^{\mathrm{ord}})|_{\mathcal{Y}^{\mathrm{Ig}}} \twoheadrightarrow \mathcal{H}^1_{\mathrm{dR}}(\mathcal{A}^{\mathrm{ord}})|_{\mathcal{Y}^{\mathrm{Ig}}}/\mathcal{U} \cong \omega_{\mathcal{Y}^{\mathrm{Ig}}}.$$

Before proving Theorem 4.29, let's recall some notation. Recall we defined a ϕ-linear endomorphism

$$F(\phi) : \mathcal{H}^1_{\mathrm{dR}}(\mathcal{A}^{\mathrm{ord}}) \xrightarrow{\phi^{-1}} \mathcal{H}^1_{\mathrm{dR}}(\mathcal{A}^{\mathrm{ord},(\phi)}) \xrightarrow{\pi^*} \mathcal{H}^1_{\mathrm{dR}}(\mathcal{A}^{\mathrm{ord}})$$

where

$$\pi : \mathcal{A}^{\mathrm{ord}} \to \mathcal{A}^{\mathrm{ord}}/C_{\mathrm{can}} \cong \mathcal{A}^{\mathrm{ord}} \times_{Y^{\mathrm{ord}},\phi} Y^{\mathrm{ord}} =: \mathcal{A}^{\mathrm{ord},(\phi)}$$

is the universal quotient by the canonical subgroup. Similarly, we have defined $F(\phi)$ on $T_p\mathcal{A}^{\mathrm{ord}}$.

Proof of Theorem 4.29. First we show $\iota_{\mathrm{dR}}(\mathcal{U}) \subset \mathcal{L}$. On the inverse image $\mathcal{D}$ of any ordinary residue disc D under $\mathcal{Y}^{\mathrm{Ig}} \to Y^{\mathrm{ord}}$, $\mathcal{U}$ has a horizontal basis ([34,

Main Theorem (quat) 4.3.1]), say $\beta \in \mathcal{U}(\mathcal{D})$, i.e., with $\nabla(\beta) = 0$, and with

$$F(\phi)(\beta) = u\beta \ \text{ for some } \ u \in \mathcal{O}_{\mathcal{Y}^{\mathrm{Ig}}}^{+}(\mathcal{D})^{\times}$$

(since β is in the unit root eigenspace of $F(\phi)$). Note that

$$\phi^{M} : \mathcal{D} \to \mathcal{D}$$

for some power M (since any residue disc is defined over a finite extension of $\mathbb{Q}_p$). Let us replace ϕ^{M} with ϕ and u^{M} by u in the ensuing discussion just for simplicity. Note that for any sheaf $\mathcal{F}$ on Y^{ord}, we get an induced map

$$\phi^{*} : \mathcal{F}(\mathcal{D}) \to \mathcal{F}(\mathcal{D}).$$

Write

$$\iota_{\mathrm{dR}}(\beta) = x\alpha_{\infty,1}/t + y\alpha_{\infty,2}/t$$

with $x, y \in \mathcal{OB}_{\mathrm{dR}, \mathcal{Y}^{\mathrm{Ig}}}^{+}(\mathcal{D})$. Since ι_{dR} respects connections, we have

$$(\mathcal{H}_{\mathrm{dR}}^{1}(\mathcal{A}^{\mathrm{ord}})|_{\mathcal{Y}^{\mathrm{Ig}}})^{\nabla=0} \subset (\mathcal{H}_{\mathrm{dR}}^{1}(\mathcal{A}^{\mathrm{ord}})|_{\mathcal{Y}^{\mathrm{Ig}}} \otimes_{\mathcal{O}_{\mathcal{Y}^{\mathrm{Ig}}}} \mathcal{OB}_{\mathrm{dR}, \mathcal{Y}^{\mathrm{Ig}}}^{+})^{\nabla=0}$$
$$\overset{\iota_{\mathrm{dR}}}{\hookrightarrow} (T_{p}\mathcal{A}^{\mathrm{ord}} \otimes_{\hat{\mathbb{Z}}_{p, Y^{\mathrm{ord}}}} \mathcal{OB}_{\mathrm{dR}, \mathcal{Y}^{\mathrm{Ig}}}^{+} \cdot t^{-1})^{\nabla=0} = T_{p}\mathcal{A}^{\mathrm{ord}} \otimes_{\hat{\mathbb{Z}}_{p, Y^{\mathrm{ord}}}} \mathbb{B}_{\mathrm{dR}, \mathcal{Y}^{\mathrm{Ig}}}^{+} \cdot t^{-1}$$

and so in fact

$$x, y \in \mathbb{B}_{\mathrm{dR}, \mathcal{Y}^{\mathrm{Ig}}}^{+}(\mathcal{D}).$$

Now fix any $n \geq 0$. Then

$$\theta_{n}(\iota_{\mathrm{dR}}(\beta)) = \theta_{n}(x)\alpha_{\infty,1}/t + \theta_{n}(y)\alpha_{\infty,2}/t \in T_{p}\mathcal{A}^{\mathrm{ord}} \otimes_{\hat{\mathbb{Z}}_{p, Y^{\mathrm{ord}}}} \mathbb{B}_{\mathrm{dR}, \mathcal{Y}^{\mathrm{Ig}}}^{+}/(t)^{n}(\mathcal{D}) \cdot t^{-1}$$

with $\theta_{n}(x), \theta_{n}(y) \in \mathbb{B}_{\mathrm{dR}, \mathcal{Y}^{\mathrm{Ig}}}^{+}/(t^{n})(\mathcal{D})$ where θ_{n} is the projection

$$\mathbb{B}_{\mathrm{dR}, \mathcal{Y}^{\mathrm{Ig}}}^{+} \twoheadrightarrow \mathbb{B}_{\mathrm{dR}, \mathcal{Y}^{\mathrm{Ig}}}^{+}/(t^{n}) = \mathbb{B}_{\mathrm{dR}, \mathcal{Y}^{\mathrm{Ig}}}^{+}/(\ker \theta)^{n}.$$

From the moduli-theoretic interpretation of the $GL_{2}(\mathbb{Q}_p)$ action on $\mathcal{Y}$, one can show that $F(\phi) : T_{p}\mathcal{A}^{\mathrm{ord}} \to T_{p}\mathcal{A}^{\mathrm{ord}}$ is just given by the action of the matrix $\begin{pmatrix} 1 & 0 \\ 0 & p \end{pmatrix}$. Note that

$$\begin{pmatrix} 1 & 0 \\ 0 & p \end{pmatrix}^{*} \alpha_{\infty,1} = \alpha_{\infty,1}, \quad \begin{pmatrix} 1 & 0 \\ 0 & p \end{pmatrix}^{*} \alpha_{\infty,2} = p\alpha_{\infty,2}.$$

Hence we have

$$F(\phi) := \begin{pmatrix} 1 & 0 \\ 0 & p \end{pmatrix}^* : T_p\mathcal{A}^{\mathrm{ord}} \otimes_{\hat{\mathbb{Z}}_{p,Y\mathrm{ord}}} \mathbb{B}^+_{\mathrm{dR},\mathcal{Y}^{\mathrm{Ig}}}/(t^n) \cdot t^{-1}$$
$$\to T_p\mathcal{A}^{\mathrm{ord}} \otimes_{\hat{\mathbb{Z}}_{p,Y\mathrm{ord}}} \mathbb{B}^+_{\mathrm{dR},\mathcal{Y}^{\mathrm{Ig}}} \cdot t^{-1},$$

which given by

$$a\alpha_{\infty,1}/t + b\alpha_{\infty,2}/t \mapsto \phi^*(a)\alpha_{\infty,1}/t + \phi^*(b)p\alpha_{\infty,2}/t$$

for any sections $a, b \in \mathbb{B}^+_{\mathrm{dR},\mathcal{Y}^{\mathrm{Ig}}}/(t^n)$ (here

$$\phi^* : \mathbb{B}^+_{\mathrm{dR},\mathcal{Y}^{\mathrm{Ig}}}/(t^n) \to \mathbb{B}^+_{\mathrm{dR},\mathcal{Y}^{\mathrm{Ig}}}/(t^n)$$

is the map of sheaves induced by $\phi : Y^{\mathrm{ord}} \to Y^{\mathrm{ord}}$). In particular, since

$$\phi^* : \mathbb{B}^+_{\mathrm{dR},\mathcal{Y}^{\mathrm{Ig}}}/(t^n)(\mathcal{D}) \to \mathbb{B}^+_{\mathrm{dR},\mathcal{Y}^{\mathrm{Ig}}}/(t^n)(\mathcal{D})$$

and $F(\phi)^m(\beta) = u^m\beta$ for any $m \geq 0$, we have

$$u^m\theta(x)\alpha_{\infty,1}/t + u^m\theta_n(y)\alpha_{\infty,2}/t$$
$$= u^m\iota_{\mathrm{dR}}(\beta) = F(\phi)^m\iota_{\mathrm{dR}}(\beta)$$
$$= (\phi^*)^m(\theta_n(x))\alpha_{\infty,1}/t + (\phi^*)^m(\theta_n(y))p^m\alpha_{\infty,2}/t.$$

In particular, since ϕ^* lifts a power of Frobenius, it preserves integral structure

$$\phi^* : \mathcal{O}^+_{Y^{\mathrm{ord}}} \to \mathcal{O}^+_{Y^{\mathrm{ord}}},$$

and so the family of maps $(\phi^*)^n : \mathbb{B}^+_{\mathrm{dR},\mathcal{Y}^{\mathrm{Ig}}}/(t^n) \to \mathbb{B}^+_{\mathrm{dR},\mathcal{Y}^{\mathrm{Ig}}}/(t^n)$ is uniformly p-adically bounded, and so the above equation implies

$$\theta_n(y) = u^{-m}(\phi^*)^m(\theta_n(y))p^m \to 0, \quad \text{as} \quad m \to \infty$$

(since $u \in \mathcal{O}^+_{\mathcal{Y}^{\mathrm{Ig}}}$ is an invertible local section), which implies $\theta_n(y) = 0$. Since $n \geq 0$ was arbitrary, this implies $\theta_n(y) = 0$ for all $n \geq 0$, which implies $y = 0$. Hence

$$\iota_{\mathrm{dR}}(\beta) = x\alpha_{\infty,1}/t + y\alpha_{\infty,2}/t = x\alpha_{\infty,1}/t \in \langle \alpha_{\infty,1} \rangle,$$

and so $\iota_{\mathrm{dR}}(\beta) \in \mathcal{L}$.

So we have shown that $\iota_{\mathrm{dR}}(\mathcal{U}) \subset \mathcal{L}$. This implies that $HT_{\mathcal{A}^{\mathrm{ord}}} \circ \iota_{\mathrm{dR}}$ factors as

$$\mathcal{H}^1_{\mathrm{dR}}(\mathcal{A}^{\mathrm{ord}})|_{\mathcal{Y}^{\mathrm{Ig}}}/\iota_{\mathrm{dR}}^{-1}(\mathcal{L}) \overset{\iota_{\mathrm{dR}}}{\hookrightarrow} T_p\mathcal{A}^{\mathrm{ord}}/\langle\alpha_{\infty,1}\rangle \overset{HT_{\mathcal{A}^{\mathrm{ord}}}}{\underset{\sim}{\to}} \hat{\omega}_{\mathcal{Y}^{\mathrm{Ig}}} \otimes_{\hat{\mathcal{O}}_{\mathcal{Y}^{\mathrm{Ig}}}} \hat{\mathcal{O}}\mathbb{B}^+_{\mathrm{dR},\mathcal{Y}^{\mathrm{Ig}}}.$$

Hence the restriction $HT_{\mathcal{A}^{\mathrm{ord}}} \circ \iota_{\mathrm{dR}}|_{\omega_{\mathcal{Y}^{\mathrm{Ig}}}}$ factors as

$$\omega_{\mathcal{Y}^{\mathrm{Ig}}} \to \mathcal{H}^1_{\mathrm{dR}}(\mathcal{A}^{\mathrm{ord}})|_{\mathcal{Y}^{\mathrm{Ig}}}/\iota_{\mathrm{dR}}^{-1}(\mathcal{L}) \overset{\iota_{\mathrm{dR}}}{\hookrightarrow} T_p\mathcal{A}^{\mathrm{ord}} \otimes_{\hat{\mathbb{Z}}_{p,\mathcal{Y}\mathrm{ord}}} \hat{\mathcal{O}}\mathbb{B}^+_{\mathrm{dR},\mathcal{Y}^{\mathrm{Ig}}}/\langle\alpha_{\infty,1}\rangle$$

$$\overset{HT_{\mathcal{A}^{\mathrm{ord}}}}{\underset{\sim}{\to}} \hat{\omega}_{\mathcal{Y}^{\mathrm{Ig}}} \otimes_{\hat{\mathcal{O}}_{\mathcal{Y}^{\mathrm{Ig}}}} \hat{\mathcal{O}}\mathbb{B}^+_{\mathrm{dR},\mathcal{Y}^{\mathrm{Ig}}}.$$

However, we know from Theorem 4.9 that $HT_{\mathcal{A}} \circ \iota_{\mathrm{dR}}|_{\mathcal{Y}^{\mathrm{Ig}}}$ is the natural inclusion

$$\omega_{\mathcal{Y}^{\mathrm{Ig}}} \hookrightarrow \hat{\omega}_{\mathcal{Y}^{\mathrm{Ig}}} \otimes_{\hat{\mathcal{O}}_{\mathcal{Y}^{\mathrm{Ig}}}} \hat{\mathcal{O}}\mathbb{B}^+_{\mathrm{dR},\mathcal{Y}^{\mathrm{Ig}}},$$

so the above map must be injective, and in particular the first arrow

$$\omega_{\mathcal{Y}^{\mathrm{Ig}}} \to \mathcal{H}^1_{\mathrm{dR}}(\mathcal{A}^{\mathrm{ord}})|_{\mathcal{Y}^{\mathrm{Ig}}}/\iota_{\mathrm{dR}}^{-1}(\mathcal{L})$$

is injective. However, we know that the map

$$\omega_{\mathcal{Y}^{\mathrm{Ig}}} \to \mathcal{H}^1_{\mathrm{dR}}(\mathcal{A}^{\mathrm{ord}})|_{\mathcal{Y}^{\mathrm{Ig}}}/\mathcal{U} \overset{\sim}{\to} \omega_{\mathcal{Y}^{\mathrm{Ig}}}$$

is injective (since the unit root splitting is indeed a splitting of the Hodge filtration). But now since both of the above maps are injective, and $\mathcal{U} \subset \iota_{\mathrm{dR}}^{-1}(\mathcal{L})$ from the previous part, then we must have $\mathcal{U} = \iota_{\mathrm{dR}}^{-1}(\mathcal{L})$, i.e., $\iota_{\mathrm{dR}}(\mathcal{U}) = \mathcal{L}$. We are done. $\square$

4.7 THE KODAIRA-SPENCER ISOMORPHISM

Recall the Kodaira-Spencer map (known to be an *isomorphism* of locally free $\mathcal{O}_Y$-modules in this setting) which is given by

$$\sigma : \omega^{\otimes 2} \overset{\sim}{\to} \Omega^1_Y, \qquad \sigma(\omega_1 \otimes \omega_2) = \langle \omega_1, \nabla(\omega_2) \rangle_{\mathrm{Poin}} \tag{4.39}$$

where

$$\langle \cdot, \cdot \rangle_{\mathrm{Poin}} : \mathcal{H}^1_{\mathrm{dR}}(\mathcal{A}) \times \mathcal{H}^1_{\mathrm{dR}}(\mathcal{A}) \to \mathcal{O}_Y \tag{4.40}$$

is the alternating, nondegenerate "Poincaré pairing," for which ω is an isotropic subspace, and which encodes the (Serre) duality between

$$\omega := R^0\pi_*\Omega^1_{\mathcal{A}/Y} \quad \text{and} \quad \omega^{-1} := R^1\pi_*\mathcal{O}_{\mathcal{A}}.$$

We extend (4.39) $\mathcal{OB}_{\mathrm{dR},Y}^+$-linearly to a map (hence an isomorphism)

$$\sigma : \omega^{\otimes 2} \otimes_{\mathcal{O}_Y} \mathcal{OB}_{\mathrm{dR},Y}^+ \xrightarrow{\sim} \Omega_Y^1 \otimes_{\mathcal{O}_Y} \mathcal{OB}_{\mathrm{dR},Y}^+$$
$$\sigma(\omega_1 \otimes \omega_2 \otimes f) = \langle \omega_1, \nabla(\omega_2 \otimes f) \rangle_{\mathrm{Poin}} = \langle \omega_1, \nabla(\omega_2) \otimes f \rangle_{\mathrm{Poin}}$$

(4.41)

(where the last equality follows because ω is an isotropic space under the Poincaré pairing). We note that we have the following commutative diagram

$$
\begin{array}{ccc}
\omega^{\otimes 2} & \xrightarrow{\quad \sim \quad}_{\sigma} & \Omega_Y^1 \\
\downarrow & & \downarrow \\
\omega^{\otimes 2} \otimes_{\mathcal{O}_Y} \mathcal{OB}_{\mathrm{dR},Y}^+ & \xrightarrow{\sim}_{\sigma} & \Omega_Y^1 \otimes_{\mathcal{O}_Y} \mathcal{OB}_{\mathrm{dR},Y}^+ \\
\downarrow{\scriptstyle\theta} & & \downarrow{\scriptstyle\theta} \\
\omega^{\otimes 2} \otimes_{\mathcal{O}_Y} \hat{\mathcal{O}}_Y & \xrightarrow{\sim}_{\sigma} & \Omega_Y^1 \otimes_{\mathcal{O}_Y} \hat{\mathcal{O}}_Y
\end{array}
$$

(4.42)

Proposition 4.30. *For any section γ of $T_p A \otimes_{\hat{\mathbb{Z}}_{p,Y}} \hat{\mathcal{O}}_Y^+$ and w^{-1} of $\hat{\omega}^{-1}$, we have*

$$\langle \gamma, (HT_A)^\vee(w^{-1}) \rangle(-1) = \langle HT_A(\gamma), w^{-1} \rangle_{\mathrm{Poin}}.$$

Proof. By [60, Proposition 4.11], we have that the exact sequence

$$0 \to \mathrm{Lie}(A/\mathbb{C}_p)(1) \xrightarrow{(HT_A)^\vee(1)} T_p A \otimes_{\mathbb{Z}_p} \mathbb{C}_p \xrightarrow{HT_A} \Omega^1_{A/\mathbb{C}_p} \to 0$$

is self-dual (using the principal polarization $A \cong \check{A}$) under the Weil pairing, which implies

$$\langle a, (HT_A)^\vee(1)(b) \rangle(-1) = \langle HT_A(a), b \rangle_{\mathrm{Poin}}$$

for any $a \in T_p A \otimes_{\mathbb{Z}_p} \mathbb{C}_p$ and any $b \in \mathrm{Lie}A(1)$. Hence the fibers of (3.21) are self-dual under the Weil pairing, meaning

$$\langle \gamma(y), (HT_A(y))^\vee(w^{-1}(y)) \rangle(-1) = \langle HT_A(y)(\gamma(y)), w^{-1}(y) \rangle_{\mathrm{Poin}}$$

for any $y \in \mathcal{Y}(\mathbb{C}_p, \mathcal{O}_{\mathbb{C}_p})$ (where "(y)" denotes image in the fiber at y), and hence by Nakayama's lemma we have

$$\langle \gamma_y, (HT_{A,y})^\vee(w_y^{-1}) \rangle(-1) = \langle HT_{A,y}(\gamma_y), w_y^{-1} \rangle_{\mathrm{Poin}}$$

on the stalk at y (where here subscript "y" denotes image in the stalk at y). Since y was an arbitrary geometric point, and since $Y_{\mathrm{proét}}$ has enough points, we are done. $\square$

Theorem 4.31. *We have the following commutative diagram*

$$
\begin{array}{ccccc}
\mathcal{H}^1_{\mathrm{dR}}(\mathcal{A}) \otimes_{\mathcal{O}_Y} \hat{\mathcal{O}}\mathbb{B}^+_{\mathrm{dR},\mathcal{Y}} & \times & \mathcal{H}^1_{\mathrm{dR}}(\mathcal{A}) \otimes_{\mathcal{O}_Y} \hat{\mathcal{O}}\mathbb{B}^+_{\mathrm{dR},\mathcal{Y}} & \xrightarrow{\langle\cdot,\cdot\rangle_{\mathrm{Poin}}} & \hat{\mathcal{O}}\mathbb{B}^+_{\mathrm{dR},\mathcal{Y}} \\
\downarrow{\iota_{\mathrm{dR}}} & & \downarrow{\iota_{\mathrm{dR}}} & & \downarrow \\
T_p\mathcal{A} \otimes_{\hat{\mathbb{Z}}_{p,Y}} \hat{\mathcal{O}}\mathbb{B}^+_{\mathrm{dR},\mathcal{Y}} \cdot t^{-1} & \times & T_p\mathcal{A} \otimes_{\hat{\mathbb{Z}}_{p,Y}} \hat{\mathcal{O}}\mathbb{B}^+_{\mathrm{dR},\mathcal{Y}} \cdot t^{-1} & \xrightarrow{\langle\cdot,\cdot\rangle\cdot t^{-1}} & \hat{\mathcal{O}}\mathbb{B}^+_{\mathrm{dR},\mathcal{Y}} \cdot t^{-2}.
\end{array}
$$

$$\tag{4.43}$$

In other words,

$$
\langle w_1, w_2 \rangle_{\mathrm{Poin}} = \langle \iota_{\mathrm{dR}}(w_1), \iota_{\mathrm{dR}}(w_2) \rangle \cdot t^{-1}
$$

for any sections w_1, w_2 of $\mathcal{H}^1_{\mathrm{dR}}(\mathcal{A}) \otimes_{\mathcal{O}_Y} \hat{\mathcal{O}}\mathbb{B}^+_{\mathrm{dR},\mathcal{Y}}$, and in particular

$$
\langle w_1, w_2 \rangle_{\mathrm{Poin}} = \langle \iota_{\mathrm{dR}}(w_1), (HT_{\mathcal{A}})^\vee(w_2) \rangle \cdot t^{-1}
$$

for any section w_1 of $\hat{\omega}_{\mathrm{dR}}$ and any section w_2 of $\hat{\omega}^{-1}_{\mathrm{dR}}$.

Proof. Since both $\langle\cdot,\cdot\rangle_{\mathrm{Poin}}$ and $\langle\cdot,\cdot\rangle$ are *alternating* on spaces of rank 2, and

$$
\hat{\omega}_{\mathrm{dR},\mathcal{Y}} \subset \mathcal{H}^1_{\mathrm{dR}}(\mathcal{A}) \otimes_{\mathcal{O}_Y} \hat{\mathcal{O}}\mathbb{B}^+_{\mathrm{dR},\mathcal{Y}}
$$

is an isotropic subspace, then by Corollary 4.14 it suffices to check that

$$
\langle\cdot,\cdot\rangle'_{\mathrm{Poin}} = \langle \iota_{\mathrm{dR}}(\cdot), (HT_{\mathcal{A}})^\vee(\cdot) \rangle \cdot t^{-1}
$$

where here $\langle\cdot,\cdot\rangle'_{\mathrm{Poin}}$ denotes the pairing

$$
\langle\cdot,\cdot\rangle'_{\mathrm{Poin}} : \hat{\omega}_{\mathrm{dR},\mathcal{Y}} \times \hat{\omega}^{-1}_{\mathrm{dR},\mathcal{Y}} \to \hat{\mathcal{O}}\mathbb{B}^+_{\mathrm{dR},\mathcal{Y}}
$$

induced via the isotropicity of $\hat{\omega}_{\mathrm{dR}}$ by the actual Poincaré pairing $\langle\cdot,\cdot\rangle_{\mathrm{Poin}}$. By Theorem 4.9 and Proposition 4.30 we have that for any section w of $\hat{\omega}_{\mathrm{dR}}$ and any section w' of $\hat{\omega}^{-1}_{\mathrm{dR}}$,

$$
\begin{aligned}
\langle w, w' \rangle'_{\mathrm{Poin}} &\overset{\text{Theorem 4.9}}{=} \langle HT_{\mathcal{A}}(\iota_{\mathrm{dR}}(w)), w' \rangle'_{\mathrm{Poin}} \\
&\overset{\text{Theorem 4.30}}{=} \langle \iota_{\mathrm{dR}}(w), (HT_{\mathcal{A}})^\vee(w') \rangle(-1) \\
&= \langle \iota_{\mathrm{dR}}(w), (HT_{\mathcal{A}})^\vee(w') \rangle \cdot t^{-1}
\end{aligned}
$$

and so we are done. $\square$

4.8   THE FUNDAMENTAL DE RHAM PERIOD $\mathbf{z}_{\mathrm{dR}}$

Definition 4.32. *Note that* $\mathbf{y}_{\mathrm{dR}}/t \in \hat{\mathcal{O}}\mathbb{B}^+_{\mathrm{dR},Y}(\mathcal{Y})$ *by (4.18), and moreover*

$$\frac{\mathbf{y}_{\mathrm{dR}}}{t} \in \hat{\mathcal{O}}\mathbb{B}^+_{\mathrm{dR}}(\{\theta(\mathbf{y}_{\mathrm{dR}}/y) \neq 0\})^\times$$

since a section f of $\mathcal{O}\mathbb{B}^+_{\mathrm{dR}}$ is invertible if and only if its image $\theta(f)$ in $\hat{\mathcal{O}}_Y$ is invertible. (This follows, for example, because $\ker \theta$ is the maximal ideal of any stalk of $\hat{\mathcal{O}}\mathbb{B}^+_{\mathrm{dR},Y}$.)

We define the fundamental de Rham period *as the ratio*

$$\mathbf{z}_{\mathrm{dR}} := -\frac{\mathbf{x}_{\mathrm{dR}}}{\mathbf{y}_{\mathrm{dR}}} \in \hat{\mathcal{O}}\mathbb{B}_{\mathrm{dR}}(\{\theta(\mathbf{y}_{\mathrm{dR}}/t) \neq 0\}).$$

We will see shortly that in fact $\mathbf{z}_{\mathrm{dR}} \in \mathcal{O}\mathbb{B}^+_{\mathrm{dR},Y}(\mathcal{Y}_x)$ (Theorem 4.36).

The identities (4.9) and (4.10) imply the following.

Proposition 4.33.

$$\begin{pmatrix} a & b \\ c & d \end{pmatrix}^* \mathbf{z}_{\mathrm{dR}} = \frac{d\mathbf{z}_{\mathrm{dR}} + b}{c\mathbf{z}_{\mathrm{dR}} + a} \tag{4.44}$$

for any $\begin{pmatrix} a & b \\ c & d \end{pmatrix} \in GL_2(\mathbb{Q}_p).$

Proposition 4.34. *Both*

$$\mathbf{x}_{\mathrm{dR}}, \mathbf{y}_{\mathrm{dR}} \in t \cdot \hat{\mathcal{O}}\mathbb{B}^+_{\mathrm{dR}}(\mathcal{Y}_x),$$

and hence

$$\theta(\mathbf{x}_{\mathrm{dR}}/t), \theta(\mathbf{y}_{\mathrm{dR}}/t) \in \hat{\mathcal{O}}(\mathcal{Y}_x).$$

Defining the following affinoid subdomains

$$\mathcal{Y}_{x,\mathrm{dR}} := \{\mathbf{z} \neq 0, \theta(\mathbf{x}_{\mathrm{dR}}/t) \neq 0\} \qquad \mathcal{Y}_{y,\mathrm{dR}} := \{\mathbf{z} \neq 0, \theta(\mathbf{y}_{\mathrm{dR}}/t) \neq 0\}$$

of $\mathcal{Y}_x = \{\mathbf{z} \neq 0\}$, we in fact have

$$\mathcal{Y}_{y,\mathrm{dR}} = \mathcal{Y}_x$$

as well as

$$\mathcal{Y}_{x,\mathrm{dR}} = \mathcal{Y}_x \setminus \{\mathbf{y}_{\mathrm{dR}}/t = 1\}.$$

Hence,

$$\mathbf{x}_{\mathrm{dR}}/t \in \hat{\mathcal{O}}\mathbb{B}_{\mathrm{dR}}^{+}(\mathcal{Y}_{x,\mathrm{dR}})^{\times} \quad and \quad \mathbf{y}_{\mathrm{dR}}/t \in \hat{\mathcal{O}}\mathbb{B}_{\mathrm{dR}}^{+}(\mathcal{Y}_{x})^{\times}.$$

In particular,

$$\mathbf{z}_{\mathrm{dR}} \in \hat{\mathcal{O}}\mathbb{B}_{\mathrm{dR}}^{+}(\mathcal{Y}_{x})$$

and

$$\mathbf{z}_{\mathrm{dR}} \in \hat{\mathcal{O}}\mathbb{B}_{\mathrm{dR}}^{+}(\mathcal{Y}_{x,\mathrm{dR}})^{\times}.$$

Proof. The first assertion follows immediately from (4.18).

Now we characterize the subdomains $\mathcal{Y}_{x,\mathrm{dR}}, \mathcal{Y}_{y,\mathrm{dR}}$. Since the line bundles w and Ω_Y are locally trivialized on Y, let $U \subset Y$ be any neighborhood where w a generator of $w|_U$ and dz is a generator of $\Omega_Y^1|U$, and let $\mathcal{U} = \lambda^{-1}(U)$ where $\lambda : \mathcal{Y} \to Y$ is the natural projection. Then $w|_{\mathcal{U}}$ is a generator of $\hat{w}|_{\mathcal{U}}$ and so there is a $u \in \hat{\mathcal{O}}_Y(\mathcal{U})^{\times}$ such that

$$u \cdot w|_{\mathcal{U}} = \mathfrak{s}|_{\mathcal{U}},$$

and so

$$u \cdot \mathbf{x}_{\mathrm{dR}}/t, u \cdot \mathbf{y}_{\mathrm{dR}}/t \in \mathcal{O}\mathbb{B}_{\mathrm{dR},Y}^{+}(\mathcal{U}).$$

Recall the Kodaira-Spencer isomorphism σ as defined in (4.39) and its extension (4.41) to $\mathcal{O}\mathbb{B}_{\mathrm{dR},\mathcal{Y}}^{+}$-coefficients. By the definition of σ and by Theorem 4.31, on $\mathcal{U}_x = \mathcal{U} \cap \mathcal{Y}_x$ we have (suppressing the notation "$|_{\mathcal{U}_x}$" for the pullbacks of our sections to $\mathcal{U}_x \to \mathcal{Y}_x$ for brevity)

$$\sigma((u\mathfrak{s})^{\otimes 2})$$

$$= \left\langle u\mathbf{x}_{\mathrm{dR}}\alpha_{\infty,1}t^{-1} + u\mathbf{y}_{\mathrm{dR}}\alpha_{\infty,2}t^{-1}, \frac{d}{dz}\mathbf{x}_{\mathrm{dR}}\alpha_{\infty,1}t^{-1} + \frac{d}{dz}\mathbf{y}_{\mathrm{dR}}\alpha_{\infty,2}t^{-1} \right\rangle_{\mathrm{Poin}} \otimes dz$$

$$= \left\langle u\mathbf{x}_{\mathrm{dR}}\alpha_{\infty,1}t^{-1} + u\mathbf{y}_{\mathrm{dR}}\alpha_{\infty,2}t^{-1}, \frac{d}{dz}(u\mathbf{x}_{\mathrm{dR}})\alpha_{\infty,1}t^{-1} + \frac{d}{dz}(u\mathbf{y}_{\mathrm{dR}})\alpha_{\infty,2}t^{-1} \right\rangle \cdot t^{-1}$$

$$\otimes dz$$

$$= \left(u\mathbf{x}_{\mathrm{dR}}\frac{d}{dz}(u\mathbf{y}_{\mathrm{dR}}) - u\mathbf{y}_{\mathrm{dR}}\frac{d}{dz}(u\mathbf{x}_{\mathrm{dR}}) \right) \langle \alpha_{\infty,1}t^{-1}, \alpha_{\infty,2}t^{-1} \rangle \cdot t^{-1}$$

$$= \left(u\frac{\mathbf{x}_{\mathrm{dR}}}{t}\frac{d}{dz}\left(u\frac{\mathbf{y}_{\mathrm{dR}}}{t}\right) - u\frac{\mathbf{y}_{\mathrm{dR}}}{t}\frac{d}{dz}\left(u\frac{\mathbf{x}_{\mathrm{dR}}}{t}\right) \right) \otimes dz = - \left(u\frac{\mathbf{y}_{\mathrm{dR}}}{t} \right)^2 \frac{d}{dz}\mathbf{z}_{\mathrm{dR}} \otimes dz.$$

$$(4.45)$$

Here the third-to-last equality holds since

$$\langle \alpha_{\infty,1}t^{-1}, \alpha_{\infty,2}t^{-1} \rangle \cdot t^{-1} = t^{-2}$$

by our definition of t (4.3).

Since $(u\mathfrak{s})^{\otimes 2}|_{\mathcal{U}_x}$ is a generator of $\omega|_{\mathcal{U}_x}^{\otimes 2}$ over $\mathcal{O}_{\mathcal{U}_x}$, and $dz|_{\mathcal{U}_x}$ is a generator of $\Omega^1_{\mathcal{Y}_x}$ over $\mathcal{O}_{\mathcal{Y}_x}$, the commutativity of the diagram (4.42) implies that

$$-\left(u\frac{\mathbf{y}_{\mathrm{dR}}}{t}\right)^2\frac{d}{dz}\mathbf{z}_{\mathrm{dR}} = -\frac{1}{\mathbf{z}^2}\left(\frac{\mathbf{y}_{\mathrm{dR}}}{t}\right)^2 + \frac{1}{\mathbf{z}^2}\frac{\mathbf{y}_{\mathrm{dR}}}{t} + \mathbf{z}\frac{d}{dz}\left(\frac{\mathbf{y}_{\mathrm{dR}}}{t}\right) \in \mathcal{O}(\mathcal{U}_x)^\times. \quad (4.46)$$

Hence, from (4.46) above and the fact that $\mathcal{O}_Y \subset \mathcal{OB}^+_{\mathrm{dR},Y} \xrightarrow{\theta} \hat{\mathcal{O}}_Y$ is the natural inclusion (2.9), we have

$$-\theta\left(u\frac{\mathbf{y}_{\mathrm{dR}}}{t}\right)^2\theta\left(\frac{d}{dz}\mathbf{z}_{\mathrm{dR}}\right) = -\left(u\frac{\mathbf{y}_{\mathrm{dR}}}{t}\right)^2\frac{d}{dz}\mathbf{z}_{\mathrm{dR}} \in \mathcal{O}(\mathcal{U}_x)^\times.$$

Hence $\theta\left(\frac{\mathbf{y}_{\mathrm{dR}}}{t}\right) \neq 0$ on $\mathcal{U}_x$. Now the $\mathcal{U}_x$ for varying U cover $\mathcal{Y}_x$, so we have the second claim of the proposition.

For the third claim, by (4.17) on $\mathcal{Y}_x$ we have

$$\mathbf{z}_{\mathrm{dR}} = \hat{\mathbf{z}}(1 - t/\mathbf{y}_{\mathrm{dR}}).$$

Since $\hat{\mathbf{z}} \neq 0$, $\mathbf{z}_{\mathrm{dR}} = 0 \iff \mathbf{y}_{\mathrm{dR}}/t = 1$.

For the final statement, one notes that $\mathbf{x}_{\mathrm{dR}}/t, \mathbf{y}_{\mathrm{dR}}/t$ are invertible in $\mathcal{OB}^+_{\mathrm{dR}}$ if and only if $\theta(\mathbf{x}_{\mathrm{dR}}/t), \theta(\mathbf{y}_{\mathrm{dR}}/t)$ are invertible in $\hat{\mathcal{O}}$. $\square$

4.9   THE CANONICAL DIFFERENTIAL

Definition 4.35. *We define the* canonical differential *as*

$$\omega_{\mathrm{can}} := \frac{t}{\mathbf{y}_{\mathrm{dR}}} \cdot \mathfrak{s} \in \hat{\omega}_{\mathrm{dR}}(\mathcal{Y}_x). \quad (4.47)$$

By (4.7) and (4.10), we have

$$\begin{pmatrix} a & b \\ c & d \end{pmatrix}^* \omega_{\mathrm{can}} = (ad - bc)(c\mathbf{z}_{\mathrm{dR}} + a)^{-1} \cdot \omega_{\mathrm{can}} \quad (4.48)$$

for any $\begin{pmatrix} a & b \\ c & d \end{pmatrix} \in GL_2(\mathbb{Z}_p)$.

We will see shortly that in fact $\omega_{\mathrm{can}} \in \omega_{\mathrm{dR}}(\mathcal{Y}_x)$ (Theorem 4.36).

Note that

$$\iota_{\mathrm{dR}}(\omega_{\mathrm{can}}) = -\mathbf{z}_{\mathrm{dR}} \cdot \alpha_{\infty,1} + \alpha_{\infty,2} \quad (4.49)$$

and (using Theorem 4.31)

$$\sigma(\omega_{\text{can}}^{\otimes 2}) = \langle \omega_{\text{can}}, \nabla(\omega_{\text{can}}) \rangle_{\text{Poin}} = \langle \iota_{\text{dR}}(\omega_{\text{can}}), \nabla(\iota_{\text{dR}}(\omega_{\text{can}})) \rangle \cdot t^{-1}$$
$$= \langle -\mathbf{z}_{\text{dR}} \cdot \alpha_{\infty,1} + \alpha_{\infty,2}, \nabla(-\mathbf{z}_{\text{dR}} \cdot \alpha_{\infty,1} + \alpha_{\infty,2}) \rangle \cdot t^{-1} = d\mathbf{z}_{\text{dR}} \tag{4.50}$$

where here we use the notation $d\mathbf{z}_{\text{dR}} = \nabla(\mathbf{z}_{\text{dR}})$.

Theorem 4.36. *In fact, we have*

$$\mathbf{z}_{\text{dR}} \in \mathcal{OB}_{\text{dR},Y}^+(\mathcal{Y}_x), \quad \mathbf{z}_{\text{dR}} \in \mathcal{OB}_{\text{dR},Y}^+(\mathcal{Y}_{x,\text{dR}})^{\times},$$

and

$$\omega_{\text{can}} \in \omega_{\text{dR}}(\mathcal{Y}_x), \quad d\mathbf{z}_{\text{dR}} \in \Omega_Y^1 \otimes_{\mathcal{O}_Y} \mathcal{OB}_{\text{dR},Y}^+(\mathcal{Y}_x)$$

are generators of $\omega_{\text{dR}}|_{\mathcal{Y}_x}$ and $\Omega_Y^1 \otimes_{\mathcal{O}_Y} \mathcal{OB}_{\text{dR},Y}^+|_{\mathcal{Y}_x}$, respectively.

Proof. Since ω is a locally free sheaf of rank 1 on Y, it is locally trivialized on Y, so taking a local generator $w \in \omega(U)$ on $U \subset Y$ with preimage $\mathcal{U} \subset \mathcal{Y}$ under the projection $\mathcal{Y} \to Y$, we can pull back to $\mathcal{Y}$ and write

$$w|_{\mathcal{U}} = x(w)\alpha_1/t + y(w)\alpha_2/t$$

using the level structure $\alpha = (\alpha_1, \alpha_2)$, where

$$x(w), y(w) \in \mathcal{OB}_{\text{dR},\mathcal{Y}}^+(\mathcal{U})$$

(no hat since w comes from finite level). We have that $y(w)/t$ is invertible on $\mathcal{U}_x := \mathcal{U} \cap \mathcal{Y}_x$ and so

$$\omega_{\text{can}} := w/(y(w)/t) \in \omega_{\text{dR}}(\mathcal{U}_x)$$

is a generator and

$$z(w) := -x(w)/y(w) \in \mathcal{OB}_{\text{dR},\mathcal{Y}}^+(\mathcal{U}_x).$$

Since ω_{can} and $\omega_{\text{can}}|_{\mathcal{U}_x}$ both generate $\omega_{\text{dR}}(\mathcal{U}_x)$, then there is some $u \in \mathcal{OB}_{\text{dR},\mathcal{Y}}^+$
$(\mathcal{U}_x)^{\times}$ such that
$$\omega_{\text{can}}|_{\mathcal{U}_x} = u \cdot \omega_{\text{can}}|_{\mathcal{U}_x}.$$

Then we have

$$(-u \cdot \mathbf{z}_{\text{dR}}|_{\mathcal{U}_x})\alpha_1/t + u\alpha_2 = u \cdot \omega_{\text{can}}|_{\mathcal{U}_x} = \omega_{\text{can}} = -z(w)\alpha_1/t + \alpha_2.$$

Equating the coordinates of α_1, α_2, we then see that $u = 1$ and

$$z_{\mathrm{dR}}|_{\mathcal{U}_x} = z(w) \in \mathcal{OB}^+_{\mathrm{dR}, \mathcal{Y}}(\mathcal{U}_x).$$

Now choosing a cover $\{U\}$ of Y so that $\{\mathcal{U}_x\}$ covers $\mathcal{Y}_x$, we glue using the above equation to see that

$$\omega_{\mathrm{can}} \in \omega_{\mathrm{dR}}(\mathcal{Y}_x) \quad \text{and} \quad \mathbf{z}_{\mathrm{dR}} \in \mathcal{OB}^+_{\mathrm{dR}, Y}(\mathcal{Y}_x).$$

$\square$

Remark 4.37. The point for why $\mathbf{z}_{\mathrm{dR}} \in \mathcal{OB}^+_{\mathrm{dR}, Y}(\mathcal{Y})$, unlike $\mathbf{x}_{\mathrm{dR}}, \mathbf{y}_{\mathrm{dR}} \in \hat{\mathcal{O}}$ $\mathbb{B}^+_{\mathrm{dR}, Y}(\mathcal{Y})$, can be summarized in the following way. The Hodge filtration ω is a subspace

$$\omega \subset \mathcal{H}^1_{\mathrm{dR}}(\mathcal{A}) \subset \mathcal{H}^1_{\mathrm{dR}}(\mathcal{A}) \otimes_{\mathcal{O}_Y} \mathcal{OB}^+_{\mathrm{dR}, Y} \overset{\iota_{\mathrm{dR}}}{\hookrightarrow} T_p \mathcal{A} \otimes_{\hat{\mathbb{Z}}_{p, Y}} \mathcal{OB}^+_{\mathrm{dR}, Y}(-1),$$

and so its position can be measured on $\mathcal{Y}$ using the basis of $T_p \mathcal{A}$ on $\mathcal{Y} \in Y_{\mathrm{pro\acute{e}t}}$ (with no need to go to the associated perfectoid space, i.e., p-adic completion, $\hat{\mathcal{Y}}$); after extending coefficients from ω to ω_{dR}, this position is given by just one coordinate, $\mathbf{z}_{\mathrm{dR}} \in \mathcal{OB}^+_{\mathrm{dR}, Y}(\mathcal{Y}_x)$.

Proposition 4.38. *On $\mathcal{Y}^{\mathrm{Ig}}$ (see Definition 4.18), we have*

$$\omega^{\mathrm{Katz}}_{\mathrm{can}} = \mathfrak{s}|_{\mathcal{Y}^{\mathrm{Ig}}} = \omega_{\mathrm{can}}|_{\mathcal{Y}^{\mathrm{Ig}}}.$$

In other words, $\omega_{\mathrm{can}} \in \omega_\Delta(\mathcal{Y}_x)$ extends Katz's canonical differential $\omega^{\mathrm{Katz}}_{\mathrm{can}}$ (see Definition 4.25) from $\mathcal{Y}^{\mathrm{Ig}} \subset \mathcal{Y}_x$ to all of $\mathcal{Y}_x$.

Proof. This follows from Proposition 4.21 and Definition 4.23. $\square$

We define the supersingular locus $Y^{\mathrm{ss}} := Y \setminus Y^{\mathrm{ord}}$.

Lemma 4.39. *Recall the Lubin-Tate period $\mathbf{z}_{\mathrm{LT}} \in \mathcal{O}_Y(LT_i)$ $(LT_i \to Y^{\mathrm{ss}})$, and consider the derivation*

$$\frac{d}{d\mathbf{z}_{\mathrm{LT}}} : \mathcal{OB}^+_{\mathrm{dR}, \mathcal{Y}^{\mathrm{ss}}} \to \mathcal{OB}^+_{\mathrm{dR}, \mathcal{Y}^{\mathrm{ss}}}.$$

We have

$$\frac{d}{d\mathbf{z}_{\mathrm{LT}}} \mathbf{z}_{\mathrm{dR}} \in \mathcal{OB}^+_{\mathrm{dR}, \mathcal{Y}}(\mathcal{Y}^{\mathrm{ss}})^\times.$$

Proof. By Theorem 4.36, the line

$$\omega_{\mathrm{dR},\mathcal{Y}}(\mathcal{Y}_x) \overset{\iota_{\mathrm{dR}}}{\subset} (\mathcal{H}^1_{\text{ét}}(\mathcal{A}) \otimes_{\hat{\mathbb{Z}}_{p,Y}} \mathcal{OB}^+_{\mathrm{dR},\mathcal{Y}})(\mathcal{Y}_x) \xrightarrow[\alpha_\infty^{-1}]{\sim} (t^{-1} \cdot \mathcal{OB}^+_{\mathrm{dR},\mathcal{Y}})^{\oplus 2}(\mathcal{Y}_x)$$

is generated by

$$\omega_{\mathrm{can}} := \frac{t}{\mathsf{y}_{\mathrm{dR}}} \cdot \mathfrak{s}. \tag{4.51}$$

Then under the Kodaira-Spencer isomorphism (4.39), and using Theorem 4.31 and the equation (4.49), we have that

$$
\begin{aligned}
\sigma(\omega_{\mathrm{can}} \otimes_{\mathcal{OB}^+_{\mathrm{dR},\mathcal{Y}}} \omega_{\mathrm{can}}) &= \langle \omega_{\mathrm{can}}, \nabla(\omega_{\mathrm{can}}) \rangle_{\mathrm{Poin}} = \langle \iota_{\mathrm{dR}}(\omega_{\mathrm{can}}), \nabla(\iota_{\omega_{\mathrm{can}}}) \rangle \cdot t^{-1} \\
&= \langle -\mathsf{z}_{\mathrm{dR}} \alpha_{\infty,1} + \alpha_{\infty,2}, \nabla(-\mathsf{z}_{\mathrm{dR}} \alpha_{\infty,1} + \alpha_{\infty,2}) \rangle \cdot t^{-1} \\
&= \left\langle -\mathsf{z}_{\mathrm{dR}} \alpha_{\infty,1} + \alpha_{\infty,2}, \frac{d}{d\mathsf{z}_{\mathrm{LT}}} \mathsf{z}_{\mathrm{dR}} \alpha_{\infty,1} \right\rangle \cdot t^{-1} \otimes d\mathsf{z}_{\mathrm{LT}} \\
&= \frac{d}{d\mathsf{z}_{\mathrm{LT}}} \mathsf{z}_{\mathrm{dR}} \otimes d\mathsf{z}_{\mathrm{LT}}
\end{aligned}
\tag{4.52}
$$

is a generator of $(\mathcal{OB}^+_{\mathrm{dR},Y} \otimes_{\mathcal{O}_Y} \Omega^1_Y)(\mathcal{Y}^{\mathrm{ss}})$ as an $\mathcal{OB}^+_{\mathrm{dR},\mathcal{Y}}(\mathcal{Y}^{\mathrm{ss}})$-module. Hence

$$\frac{d}{d\mathsf{z}_{\mathrm{LT}}} \mathsf{z}_{\mathrm{dR}} \in \mathcal{OB}^+_{\mathrm{dR}}(\mathcal{Y}^{\mathrm{ss}})^\times$$

which is what we wanted to show.   □

Chapter Five

The p-adic Maass-Shimura operator

5.1 THE "HORIZONTAL" LIFTING OF THE HODGE-TATE FILTRATION

Let $\mathfrak{s}^{-1} \in \omega^{-1}(\mathcal{Y}_x)$ denote the *unique* section such that

$$\langle \mathfrak{s}, \mathfrak{s}^{-1} \rangle_{\mathrm{Poin}} = 1.$$

Write

$$(HT_{\mathcal{A}})^{\vee}(\mathfrak{s}^{-1}) = u(\alpha_{\infty,1} \cdot t^{-1} - \frac{1}{\hat{\mathbf{z}}} \alpha_{\infty,2} \cdot t^{-1})$$

for some $u \in \hat{\mathcal{O}}_Y(\mathcal{Y}_x)^{\times}$. Then by Theorem 4.31, we have

$$1 = \langle \mathfrak{s}, \mathfrak{s}^{-1} \rangle_{\mathrm{Poin}} = \langle \iota_{\mathrm{dR}}(\mathfrak{s}), (HT_{\mathcal{A}})^{\vee}(\mathfrak{s}^{-1}) \rangle \cdot t^{-1}$$

$$= \left\langle \mathbf{x}_{\mathrm{dR}}\alpha_{\infty,1} \cdot t^{-1} + \mathbf{y}_{\mathrm{dR}}\alpha_{\infty,2} \cdot t^{-1}, u(\alpha_{\infty,1} - \frac{1}{\hat{\mathbf{z}}}\alpha_{\infty,2}) \right\rangle \cdot t^{-1} = u$$

and so

$$(HT_{\mathcal{A}})^{\vee}(\mathfrak{s}^{-1}) = \alpha_{\infty,1} \cdot t^{-1} - \frac{1}{\hat{\mathbf{z}}}\alpha_{\infty,2} \cdot t^{-1}$$

and $\mathfrak{s}^{-1}$ generates $\omega_{\mathrm{dR}}^{-1}|_{\mathcal{Y}_x}$. We can then identify the arrows of the Hodge-Tate sequence ((3.21) tensored with $\otimes_{\hat{\mathcal{O}}_Y} \hat{\mathcal{O}}\mathbb{B}_{\mathrm{dR},\mathcal{Y}_x}^+$) explicitly:

$$0 \to \omega_{\mathrm{dR}}^{-1}|_{\mathcal{Y}_x} \cdot t \xrightarrow{(HT_{\mathcal{A}})^{\vee}(1)} T_p\mathcal{A} \otimes_{\hat{\mathbb{Z}}_{p,Y}} \hat{\mathcal{O}}\mathbb{B}_{\mathrm{dR},\mathcal{Y}_x}^+ \xrightarrow[\alpha_{\infty}^{-1}]{\sim} (\hat{\mathcal{O}}\mathbb{B}_{\mathrm{dR},\mathcal{Y}_x}^+)^{\oplus 2} \xrightarrow{HT_{\mathcal{A}}} \omega_{\mathrm{dR}}|_{\mathcal{Y}_x} \to 0$$

$$(5.1)$$

(using Proposition 3.14), where the second arrow is defined by

$$(HT_{\mathcal{A}})^{\vee}(1)(\mathfrak{s}^{-1}t) = \alpha_{\infty,1} - 1/\hat{\mathbf{z}} \cdot \alpha_{\infty,2}$$

and the penultimate arrow is defined by

$$HT_{\mathcal{A}}(\alpha_{\infty,2}) = \mathfrak{s} \quad \text{and} \quad HT_{\mathcal{A}}(\alpha_{\infty,1}) = 1/\hat{\mathbf{z}} \cdot \mathfrak{s}.$$

We have the following relation between the basis $\{\alpha_{\infty,1}, \alpha_{\infty,2}\}$ and the set

$$\{\iota_{\mathrm{dR}}(\omega_{\mathrm{can}}), (HT_A)^\vee(1)(\mathfrak{s}^{-1}t)\}$$

of elements of $T_p A \otimes_{\hat{\mathbb{Z}}_{p,Y}} \hat{\mathcal{O}}\mathbb{B}^+_{\mathrm{dR},\mathcal{Y}_x}$:

$$\begin{pmatrix} \frac{\mathbf{x}_{\mathrm{dR}}}{t} & 1 \\ \frac{\mathbf{y}_{\mathrm{dR}}}{t} & -\frac{1}{\hat{\mathbf{z}}} \end{pmatrix} \begin{pmatrix} \alpha_{\infty,1} \\ \alpha_{\infty,2} \end{pmatrix} = \begin{pmatrix} \iota_{\mathrm{dR}}(\omega_{\mathrm{can}}) \\ (HT_A)^\vee(1)(\mathfrak{s}^{-1}t) \end{pmatrix}.$$

By the p-adic Legendre relation (4.17), the determinant of the matrix on the left-hand side is -1, and hence the set $\{\iota_{\mathrm{dR}}(\omega_{\mathrm{can}}), (HT_A)^\vee(1)(\mathfrak{s}^{-1}t)\}$ is a basis of $T_p A \otimes_{\hat{\mathbb{Z}}_{p,Y}} \hat{\mathcal{O}}\mathbb{B}^+_{\mathrm{dR},\mathcal{Y}_x}$. Thus, since ω_{can} is a generator of $\omega_{\mathrm{dR}}|_{\mathcal{Y}_x}$ by Theorem 4.36, we have the following relative $\hat{\mathcal{O}}\mathbb{B}^+_{\mathrm{dR},\mathcal{Y}_x}$-Hodge-Tate decomposition on $\mathcal{Y}_x$:

$$\omega_{\mathrm{dR}}|_{\mathcal{Y}_x} \oplus \left(\omega_{\mathrm{dR}}^{-1}|_{\mathcal{Y}_x} \cdot t\right) = \omega_{\mathrm{can}} \cdot \hat{\mathcal{O}}\mathbb{B}^+_{\mathrm{dR},\mathcal{Y}_x} \oplus \mathfrak{s}^{-1}t \cdot \hat{\mathcal{O}}\mathbb{B}^+_{\mathrm{dR},\mathcal{Y}_x}$$

$$\xrightarrow[\sim]{\iota_{\mathrm{dR}} \oplus (HT_A)^\vee(1)} \left((-\mathbf{z}_{\mathrm{dR}} \cdot \alpha_{\infty,1} + \cdot \alpha_{\infty,2})\hat{\mathcal{O}}\mathbb{B}^+_{\mathrm{dR},\mathcal{Y}_x}\right) \oplus \left((-\alpha_{\infty,1} - \frac{1}{\hat{\mathbf{z}}} \cdot \alpha_{\infty,2})\hat{\mathcal{O}}\mathbb{B}^+_{\mathrm{dR},\mathcal{Y}_x}\right)$$

$$\overset{(4.17)}{=} \hat{\mathcal{O}}\mathbb{B}^{+,\oplus 2}_{\mathrm{dR},\mathcal{Y}_x} \xrightarrow[\sim]{\alpha_\infty} T_p A \otimes_{\hat{\mathbb{Z}}_{p,Y}} \hat{\mathcal{O}}\mathbb{B}^+_{\mathrm{dR},\mathcal{Y}_x}. \tag{5.2}$$

Indeed, we can view this as a relative Hodge-Tate decomposition, since by definition of $\mathbf{x}_{\mathrm{dR}}$ and $\mathbf{y}_{\mathrm{dR}}$ (Definition 4.5), the inclusion of $\omega_{\mathrm{dR}}|_{\mathcal{Y}_x}$ into the first factor is just given by ι_{dR}, so the above is indeed a splitting of the relative Hodge-Tate (and Hodge) filtrations *after extending coefficients to* $\hat{\mathcal{O}}\mathbb{B}^+_{\mathrm{dR},\mathcal{Y}_x}$.

Definition 5.1. *Henceforth, define*

$$\mathcal{O}_\Delta := \mathcal{O}\mathbb{B}^+_{\mathrm{dR},\mathcal{Y}}/(t), \quad \hat{\mathcal{O}}_\Delta := \hat{\mathcal{O}}\mathbb{B}^+_{\mathrm{dR},\mathcal{Y}}/(t)$$

and

$$\omega_\Delta := \omega \otimes_{\mathcal{O}_Y} \mathcal{O}_\Delta, \quad \hat{\omega}_\Delta := \omega \otimes_{\mathcal{O}_Y} \hat{\mathcal{O}}_\Delta.$$

Note that we are restricting to the localized site $Y_{pro\acute{e}t}/\mathcal{Y}$ in the definition of $\mathcal{O}_\Delta$ and ω_Δ, on which t is a global section (which makes $\mathcal{O}_\Delta$ well-defined and a sheaf). In a (harmless) conflation of notation, we will also let ω_{can} denote

$$\omega_{\mathrm{can}} \pmod{t}$$

where $\omega_{\mathrm{can}} \in \omega_{\mathrm{dR}}(\mathcal{Y}_x)$ from Definition 4.35.

Definition 5.2. *Henceforth, let*

$$x_{\mathrm{dR}} := \mathbf{x}_{\mathrm{dR}}/t \pmod t, \quad y_{\mathrm{dR}} := \mathbf{y}_{\mathrm{dR}}/t \pmod t \in \hat{\mathcal{O}}_\Delta(\mathcal{Y}_x),$$
$$z_{\mathrm{dR}} := \mathbf{z}_{\mathrm{dR}} \pmod t \in \mathcal{O}_\Delta(\mathcal{Y}_x).$$

Proposition 5.3. *The p-adic comparison inclusion ι_{dR}, restricted to*

$$\omega_{\mathrm{dR},\mathcal{Y}_x} \subset \mathcal{H}^1_{\mathrm{dR}}(\mathcal{A}) \otimes_{\mathcal{O}_Y} \mathcal{OB}^+_{\mathrm{dR},\mathcal{Y}_x},$$

i.e., the inclusion

$$\omega_{\mathrm{dR},\mathcal{Y}_x} \overset{\iota_{\mathrm{dR}}}{\hookrightarrow} T_p \mathcal{A} \otimes_{\hat{\mathbb{Z}}_{p,Y}} \mathcal{OB}^+_{\mathrm{dR},\mathcal{Y}_x} \cdot t^{-1},$$

factors through

$$\omega_{\mathrm{dR},\mathcal{Y}_x} \overset{\iota_{\mathrm{dR}}}{\hookrightarrow} T_p \mathcal{A} \otimes_{\hat{\mathbb{Z}}_{p,Y}} \mathcal{OB}^+_{\mathrm{dR},\mathcal{Y}_x},$$

which, reducing modulo t, induces

$$\omega_{\Delta,\mathcal{Y}_x} \overset{\iota_{\mathrm{dR}}}{\hookrightarrow} T_p \mathcal{A} \otimes_{\hat{\mathbb{Z}}_{p,Y}} \mathcal{O}_{\Delta,\mathcal{Y}_x}.$$

Hence we have natural maps

$$\omega|_{\mathcal{Y}_x} \subset \omega_{\mathrm{dR},\mathcal{Y}_x} \overset{\iota_{\mathrm{dR}}}{\subset} T_p \mathcal{A} \otimes_{\hat{\mathbb{Z}}_{p,Y}} \mathcal{O}_{\mathrm{dR},\mathcal{Y}_x} \overset{\theta}{\twoheadrightarrow} T_p \mathcal{A} \otimes_{\hat{\mathbb{Z}}_{p,Y}} \hat{\mathcal{O}}_{\mathcal{Y}_x} \overset{HT_{\mathcal{A}}}{\longrightarrow} \hat{\omega}_{\mathcal{Y}_x} \qquad (5.3)$$

and

$$\omega|_{\mathcal{Y}_x} \subset \omega_{\Delta,\mathcal{Y}_x} \overset{\iota_{\mathrm{dR}}}{\subset} T_p \mathcal{A} \otimes_{\hat{\mathbb{Z}}_{p,Y}} \mathcal{O}_{\Delta,\mathcal{Y}_x} \overset{\theta}{\twoheadrightarrow} T_p \mathcal{A} \otimes_{\hat{\mathbb{Z}}_{p,Y}} \hat{\mathcal{O}}_{\mathcal{Y}_x} \overset{HT_{\mathcal{A}}}{\longrightarrow} \hat{\omega}_{\mathcal{Y}_x}, \qquad (5.4)$$

and in fact this map is the natural inclusion.

Proof. This follows from Theorem 4.36: since $\omega_{\mathrm{can}} \in \omega_{\mathrm{dR}}(\mathcal{Y}_x)$ is a generator,

$$\iota_{\mathrm{dR}}(\omega_{\mathrm{can}}) = -\mathbf{z}_{\mathrm{dR}} \alpha_{\infty,1} + \alpha_{\infty,2},$$

$\mathbf{z}_{\mathrm{dR}} \in \mathcal{OB}^+_{\mathrm{dR},Y}(\mathcal{Y}_x)$, then

$$\iota_{\mathrm{dR}}(\omega|_{\mathcal{Y}_x}) \subset \iota_{\mathrm{dR}}(\omega_{\mathrm{dR},\mathcal{Y}_x}) = \iota_{\mathrm{dR}}(\omega_{\mathrm{can}}) \cdot \mathcal{OB}^+_{\mathrm{dR},\mathcal{Y}_x} \subset T_p \mathcal{A} \otimes_{\hat{\mathbb{Z}}_{p,Y}} \mathcal{OB}^+_{\mathrm{dR},\mathcal{Y}}(\mathcal{Y}_x),$$

from which the first statement follows. The second statement follows immediately from the first. For the third statement, first note that

$$\theta \circ HT_{\mathcal{A}} = HT_{\mathcal{A}} \circ \theta$$

and so by Theorem 4.9, (5.4) is just the natural inclusion. $\square$

We seek to define a p-adic Maass-Shimura operator on ω_Δ. In order to define such an operator over $\mathcal{O}_\Delta$ with satisfactory properties (such as algebraicity and coincidence with complex Maass-Shimura values at CM points), we need to define a "horizontal" lifting of the relative Hodge-Tate filtration, i.e., such that the lifting is generated by an element contained in the horizontal sections of the Gauss-Manin connection ∇. To be more precise, we will define a line

$$\mathcal{L} \subset T_p A \otimes_{\hat{\mathbb{Z}}_{p,Y}} \mathcal{O}_{\Delta,\mathcal{Y}_x}$$

such that

$$\theta(\mathcal{L}) \subset T_p A \otimes_{\hat{\mathbb{Z}}_{p,Y}} \hat{\mathcal{O}}_{\mathcal{Y}_x}$$

coincides with the usual Hodge filtration in (3.22), and that this line is horizontal means that

$$\nabla_w(\mathcal{L}) \subset \mathcal{L}$$

for any nonvanishing section w of $\Omega^1_{\mathcal{Y}} \otimes_{\mathcal{O}_{\mathcal{Y}}} \mathcal{O}_{\Delta,\mathcal{Y}_x}$; here, we denote by ∇_w the map

$$\nabla_w : T_p A \otimes_{\hat{\mathbb{Z}}_{p,Y}} \mathcal{O}_{\Delta,\mathcal{Y}_x} \rightarrow T_p A \otimes_{\hat{\mathbb{Z}}_{p,Y}} \mathcal{O}_{\Delta,\mathcal{Y}_x}$$

locally defined on an open where w does not vanish, and trivializing the usual connection

$$\nabla : T_p A \otimes_{\hat{\mathbb{Z}}_{p,Y}} \mathcal{O}_{\Delta,\mathcal{Y}_x} \rightarrow T_p A \otimes_{\hat{\mathbb{Z}}_{p,Y}} \mathcal{O}_{\Delta,\mathcal{Y}_x} \otimes_{\mathcal{O}_{\mathcal{Y}_x}} \Omega^1_{\mathcal{Y}_x} \overset{w}{\cong} T_p A \otimes_{\hat{\mathbb{Z}}_{p,Y}} \mathcal{O}_{\Delta,\mathcal{Y}_x}.$$

The fact that $\mathcal{L}$ lifts the Hodge filtration will be immediate from its construction (see (5.7)), and we prove the horizontalness of $\mathcal{L}$ in Proposition 5.7; in fact, we show that $\mathcal{L}$ is the unique such line which has both these properties (see Proposition 5.8).

Note that we have the natural map

$$j : \hat{\mathcal{O}}_{\mathcal{Y}} \cong \mathbb{B}^+_{\mathrm{dR},\mathcal{Y}}/(t) \subset \mathcal{O}\mathbb{B}^+_{\mathrm{dR},\mathcal{Y}}/(t) \tag{5.5}$$

which is a section of $\theta : \mathcal{O}\mathbb{B}^+_{\mathrm{dR},\mathcal{Y}} \twoheadrightarrow \hat{\mathcal{O}}_{\mathcal{Y}}$, i.e.,

$$\theta \circ j = \mathrm{id}. \tag{5.6}$$

This embeds $\hat{\mathcal{O}}_{\mathcal{Y}}$ into the horizontal sections of the $\hat{\mathcal{O}}_{\mathcal{Y}}$-linear connection

$$\nabla : \mathcal{O}\mathbb{B}^+_{\mathrm{dR},\mathcal{Y}}/(t) \rightarrow \mathcal{O}\mathbb{B}^+_{\mathrm{dR},\mathcal{Y}}/(t) \otimes_{\mathcal{O}_{\mathcal{Y}}} \Omega^1_{\mathcal{Y}}$$

which is induced by the $\mathbb{B}_{dR,\mathcal{Y}}^+$-linear connection

$$\nabla : \mathcal{O}\mathbb{B}_{dR,\mathcal{Y}}^+ \to \mathcal{O}\mathbb{B}_{dR,\mathcal{Y}}^+ \otimes_{\mathcal{O}_{\mathcal{Y}}} \Omega_{\mathcal{Y}}^1$$

since $t \in \mathbb{B}_{dR}^+(\mathcal{Y})$ is a horizontal section.

Definition 5.4. *Let*

$$\bar{z} = j(\hat{\mathbf{z}})$$

and note that

$$\theta(\bar{z}) = \hat{\mathbf{z}} = \theta(\hat{\mathbf{z}}) \tag{5.7}$$

and in particular

$$\hat{\mathbf{z}} - \bar{z} \in \ker \theta.$$

5.2 THE "HORIZONTAL" RELATIVE HODGE-TATE DECOMPOSITION OVER $\mathcal{O}_\Delta$

Proposition 5.5. *We have*

$$z_{dR} - \bar{z} \in \mathcal{O}_\Delta(\mathcal{Y}_x)^\times. \tag{5.8}$$

Proof. Clearly we have $z_{dR} - \bar{z} \in \mathcal{O}_\Delta(\mathcal{Y}_x)$. Now note that

$$z_{dR} - \bar{z} \equiv z_{dR} - \hat{\mathbf{z}} \quad (\mathrm{mod}\ (\ker \theta))$$

by (5.7), and by 4.17, we have

$$z_{dR} - \hat{\mathbf{z}} = -\frac{\hat{\mathbf{z}}}{y_{dR}} \in \hat{\mathcal{O}}_\Delta(\mathcal{Y}_x)^\times$$

by the definition of $\mathcal{Y}_x$ and Proposition 4.34. Now the proposition follows because

$$\mathcal{O}_\Delta(\mathcal{Y}_x)^\times = \mathcal{O}_\Delta(\mathcal{Y}_x) \cap \hat{\mathcal{O}}_\Delta(\mathcal{Y}_x)^\times. \qquad \square$$

Henceforth, let

$$\mathcal{L} := ((\alpha_{\infty,1} - 1/\bar{z} \cdot \alpha_{\infty,2})\mathcal{O}_\Delta|_{\mathcal{Y}_x}). \tag{5.9}$$

Now we have the following isomorphism of $\mathcal{O}_{\Delta,y_x}$-modules

$$\omega_\Delta|_{\mathcal{Y}_x} \oplus \mathcal{L}$$
$$\xrightarrow[\sim]{\iota_{\mathrm{dR}}} ((-z_{\mathrm{dR}} \cdot \alpha_{\infty,1} + \alpha_{\infty,2})\mathcal{O}_{\Delta,y_x}) \oplus ((\alpha_{\infty,1} - 1/\bar{z} \cdot \alpha_{\infty,2})\mathcal{O}_{\Delta,y_x}) \quad (5.10)$$
$$\overset{(5.8)}{=} \mathcal{O}_{\Delta,y_x}^{\oplus 2} \xrightarrow[\sim]{\alpha_\infty} T_p\mathcal{A} \otimes_{\hat{\mathbb{Z}}_{p,Y}} \mathcal{O}_{\Delta,y_x}$$

where the first factor in the above decomposition is given by

$$\iota_{\mathrm{dR}} : \omega_\Delta \hookrightarrow T_p\mathcal{A} \otimes_{\hat{\mathbb{Z}}_{p,Y}} \mathcal{O}_{\Delta,y_x}.$$

Definition 5.6. *Henceforth, denote the projection to the first factor of the above splitting by*

$$\overline{\mathrm{split}} : T_p\mathcal{A} \otimes_{\hat{\mathbb{Z}}_{p,Y}} \mathcal{O}_\Delta|_{\mathcal{Y}_x} \twoheadrightarrow \omega_{\Delta,y_x}. \quad (5.11)$$

Proposition 5.7. *We have that $\mathcal{L}$ is horizontal for ∇. That is, for any proétale $U \to \mathcal{Y}_x$ and for any nonvanishing*

$$w \in \Omega_{\mathcal{Y}}^1 \otimes_{\mathcal{O}_{\mathcal{Y}}} \mathcal{O}_{\Delta,y_x}(U),$$

we have

$$\nabla_w(\mathcal{L}|_U) \subset \mathcal{L}|_U.$$

In particular, we have

$$\overline{\mathrm{split}} \circ \nabla^j = \nabla^j \circ \overline{\mathrm{split}}$$

for any $j \in \mathbb{Z}_{\geq 0}$.

Proof. Note that the kernel $\mathcal{L}$ of $\overline{\mathrm{split}}$ is generated by the section

$$\alpha_{\infty,1} - 1/\bar{z} \cdot \alpha_{\infty,2} \in T_p\mathcal{A} \otimes_{\hat{\mathbb{Z}}_{p,Y}} \mathcal{O}_{\Delta,y_x}(\mathcal{Y}_x)$$

by definition, and so we can write any section of $\mathcal{L}|_U$ as

$$f \cdot (\alpha_{\infty,1} - 1/\bar{z} \cdot \alpha_{\infty,2}), \quad f \in \mathcal{O}_{\delta,\mathcal{Y}}(U).$$

Then

$$\nabla_w(f \cdot (\alpha_{\infty,1} - 1/\bar{z} \cdot \alpha_{\infty,2}))$$
$$= \nabla_w(f) \cdot (\alpha_{\infty,1} - 1/\bar{z} \cdot \alpha_{\infty,2}) + f \cdot \nabla_w(\alpha_{\infty,1} - 1/\bar{z} \cdot \alpha_{\infty,2})$$
$$= \nabla_w(f) \cdot (\alpha_{\infty,1} - 1/\bar{z} \cdot \alpha_{\infty,2})$$

which is a section of $\mathcal{L}|_U$. $\qquad\square$

Proposition 5.8. $\mathcal{L}$ *is the unique line in* $T_p\mathcal{A} \otimes_{\hat{\mathbb{Z}}_{p,Y}} \mathcal{O}_{\Delta,\mathcal{Y}_x}$ *with*

$$\theta(\mathcal{L}) = \hat{\omega}_{\mathcal{Y}_x}$$

(i.e., $\mathcal{L}$ *lifts the Hodge-Tate filtration as in (3.22)), and which is horizontal in the sense of Proposition 5.7.*

Proof. For any line

$$\mathcal{L}' \subset T_p\mathcal{A} \otimes_{\hat{\mathbb{Z}}_{p,Y}} \mathcal{O}_{\Delta,\mathcal{Y}_x}$$

with

$$\theta(\mathcal{L}') = \hat{\omega}_{\mathcal{Y}_x} = \langle \alpha_{\infty,1} - 1/\hat{z}\alpha_{\infty,2} \rangle \cdot \hat{\mathcal{O}}_{\mathcal{Y}_x},$$

we must have

$$\mathcal{L}' = \langle \alpha_{\infty,1} - (1/\bar{z} + X)\alpha_{\infty,2} \rangle \mathcal{O}_{\Delta,\mathcal{Y}_x}$$

where

$$X \in \ker(\theta : \mathcal{O}_{\Delta,\mathcal{Y}_x} \twoheadrightarrow \hat{\mathcal{O}}_{\mathcal{Y}_x}). \tag{5.12}$$

From the calculation in the proof of Proposition 5.7, we see that $\mathcal{L}$ is horizontal if and only if X is a horizontal section of

$$\nabla : \mathcal{O}_{\Delta,\mathcal{Y}_x} \to \mathcal{O}_{\Delta,\mathcal{Y}_x} \otimes_{\mathcal{O}_{\mathcal{Y}_x}} \Omega^1_{\mathcal{Y}_x}$$

(recall that X being a horizontal section means $\nabla(X) = 0$). However, the sheaf of horizontal sections of ∇ (by its construction) is simply the subsheaf

$$j : \hat{\mathcal{O}}_{\mathcal{Y}_x} \hookrightarrow \mathcal{O}_{\Delta,\mathcal{Y}_x}$$

as defined in (5.5), so if X is horizontal we have $X \in j(\hat{\mathcal{O}}_{\mathcal{Y}_x})$, say $X = j(x)$. But from (5.6), we see that

$$\theta(X) = \theta(j(x)) = x,$$

and so from (5.12) we have

$$0 = \theta(X) = x,$$

which implies

$$X = j(x) = 0. \qquad \square$$

By direct calculation, on $\mathcal{Y}_x$ we have

$$\overline{\text{split}}(\alpha_{\infty,1}) = \frac{1}{\bar{z} - z_{\text{dR}}} \cdot \omega_{\text{can}} \tag{5.13}$$

and

$$\nabla(\omega_{\mathrm{can}}) = \nabla(\iota_{\mathrm{dR}}(\omega_{\mathrm{can}})) = \nabla(-z_{\mathrm{dR}} \cdot \alpha_{\infty,1} + \alpha_{\infty,2})$$

$$= -\alpha_{\infty,1} \cdot dz_{\mathrm{dR}} \xrightarrow{\overline{\mathrm{split}}} \frac{1}{z_{\mathrm{dR}} - \bar{z}} \cdot \omega_{\mathrm{can}} \tag{5.14}$$

where

$$dz_{\mathrm{dR}} := \nabla(z_{\mathrm{dR}}).$$

Proposition 5.9. *Let $D \subset Y^{\mathrm{ord}}$ be a residue disc, and let $\mathcal{D} = \mathcal{Y}^{\mathrm{Ig}} \times_{Y^{\mathrm{ord}}} D$. On the locus $\mathcal{D} \subset \mathcal{Y}_x$, we have that*

$$d\mathbf{z}_{\mathrm{dR}}|_{\mathcal{D}} = d \log T \in \Omega^1(\mathcal{D})$$

where T is the Serre-Tate coordinate on $\mathcal{D}$, as defined in Definition 4.23.

Proof. By (4.45) and Proposition 4.21, on $\mathcal{Y}^{\mathrm{Ig}}$ we have that

$$\sigma(\mathfrak{s}^{\otimes 2}) = \left(\frac{\mathbf{y}_{\mathrm{dR}}}{t}\right)^2 d\mathbf{z}_{\mathrm{dR}} = d\mathbf{z}_{\mathrm{dR}} \tag{5.15}$$

and hence by the commutativity of the diagram (4.42)

$$d\mathbf{z}_{\mathrm{dR}} \in \Omega^1(\mathcal{Y}_x).$$

By the theorem cited in Definition 4.25 and (5.15), we must have

$$d\mathbf{z}_{\mathrm{dR}}|_{\mathcal{D}} = d \log T. \qquad \square$$

From the relative p-adic de Rham comparison theorem (3.31), by taking k^{th} symmetric powers of each side viewed as $\mathcal{OB}_{\mathrm{dR},Y}^+$-modules, one has the comparison

$$\mathrm{Sym}_{\mathcal{O}_Y}^k \mathcal{H}_{\mathrm{dR}}^1(\mathcal{A}) \otimes_{\mathcal{O}_Y} \mathcal{OB}_{\mathrm{dR},Y}^+ \cong \mathrm{Sym}_{\mathcal{OB}_{\mathrm{dR},Y}^+}^k (\mathcal{H}_{\mathrm{dR}}^1(\mathcal{A}) \otimes_{\mathcal{O}_Y} \mathcal{OB}_{\mathrm{dR},Y}^+)$$

$$\overset{\iota_{\mathrm{dR}}}{\cong} \mathrm{Sym}_{\mathcal{OB}_{\mathrm{dR},Y}^+}^k (T_p \mathcal{A} \otimes_{\hat{\mathbb{Z}}_{p,Y}} \mathcal{OB}_{\mathrm{dR},Y}^+) \tag{5.16}$$

compatible with the inherited connections. Here, the filtration on the left-hand side is given by the convolution of the filtration on $\mathrm{Sym}_{\mathcal{O}_Y}^k \mathcal{H}_{\mathrm{dR}}^1(\mathcal{A})$ inherited from the Hodge–de Rham filtration

$$\omega^{\otimes k} \subset \omega^{\otimes k-1} \otimes_{\mathcal{O}_Y} \mathcal{H}^1_{\mathrm{dR}}(\mathcal{A}) \subset \omega^{\otimes k-2} \otimes_{\mathcal{O}_Y} \mathrm{Sym}^2_{\mathcal{O}_Y} \mathcal{H}^1_{\mathrm{dR}}(\mathcal{A}) \subset \ldots \subset \mathrm{Sym}^k_{\mathcal{O}_Y} \mathcal{H}^1_{\mathrm{dR}}(\mathcal{A}) \tag{5.17}$$

with the filtration on $\mathcal{O}\mathbb{B}^+_{\mathrm{dR},Y}$. From (5.16), we have

$$\begin{aligned}
\mathrm{Sym}^k_{\mathcal{O}_Y} \mathcal{H}^1_{\mathrm{dR}}(\mathcal{A}) \otimes_{\mathcal{O}_Y} \mathcal{O}\mathbb{B}^+_{\mathrm{dR},Y} &\cong \mathrm{Sym}^k_{\mathcal{O}\mathbb{B}^+_{\mathrm{dR},Y}} \left(\mathcal{H}^1_{\mathrm{dR}}(\mathcal{A}) \otimes_{\mathcal{O}_Y} \mathcal{O}\mathbb{B}^+_{\mathrm{dR},Y} \right) \\
&\overset{\iota_{\mathrm{dR}}}{\subset} \mathrm{Sym}^k_{\mathcal{O}\mathbb{B}^+_{\mathrm{dR},Y}} \left(T_p\mathcal{A} \otimes_{\hat{\mathbb{Z}}_{p,Y}} \mathcal{O}\mathbb{B}^+_{\mathrm{dR},Y} \right)
\end{aligned} \tag{5.18}$$

compatible with connections. Reducing modulo (t), the splitting (5.11) induces a splitting on symmetric powers.

Definition 5.10. *Henceforth, denote the map on symmetric powers induced by (5.11) also by*

$$\overline{\mathrm{split}} : \mathrm{Sym}^k_{\mathcal{O}_\Delta} \left(T_p\mathcal{A} \otimes_{\hat{\mathbb{Z}}_{p,Y}} \mathcal{O}_\Delta \right) |_{\mathcal{Y}_x} \twoheadrightarrow (\omega_\Delta) |_{\mathcal{Y}_x}^{\otimes_{\mathcal{O}_\Delta} k}. \tag{5.19}$$

5.3　DEFINITION OF THE p-ADIC MAASS-SHIMURA OPERATOR

In this section, we define our p-adic Maass-Shimura operator using the horizontal splitting (5.19).

Definition 5.11. *We define a map*

$$\partial_k : (\omega_\Delta) |_{\mathcal{Y}_x}^{\otimes_{\mathcal{O}_\Delta} k} \to (\omega_\Delta) |_{\mathcal{Y}_x}^{\otimes_{\mathcal{O}_\Delta} k+2}$$

as the following composition:

$$\begin{aligned}
\omega^{\otimes_{\mathcal{O}_{\Delta,\mathcal{Y}_x}} k}_{\Delta,\mathcal{Y}_x} &\overset{\iota_{\mathrm{dR}}}{\subset} \mathrm{Sym}^k_{\mathcal{O}_{\Delta,\mathcal{Y}_x}} \left(T_p\mathcal{A} \otimes_{\hat{\mathbb{Z}}_{p,Y}} \mathcal{O}_{\Delta,\mathcal{Y}_x} \right) \\
&\overset{\nabla}{\longrightarrow} \left(\mathrm{Sym}^k_{\mathcal{O}_{\Delta,\mathcal{Y}_x}} \left(T_p\mathcal{A} \otimes_{\hat{\mathbb{Z}}_{p,Y}} \mathcal{O}_{\Delta,\mathcal{Y}_x} \right) \otimes_{\mathcal{O}_Y} \Omega^1_{\mathcal{Y}_x} \right) \\
&\overset{\sigma^{-1}}{\longrightarrow} \left(\mathrm{Sym}^k_{\mathcal{O}_{\Delta,\mathcal{Y}_x}} \left(T_p\mathcal{A} \otimes_{\hat{\mathbb{Z}}_{p,Y}} \mathcal{O}_{\Delta,\mathcal{Y}_x} \right) \otimes_{\mathcal{O}_Y} \omega|_{\mathcal{Y}_x}^{\otimes 2} \right) \\
&\hookrightarrow \left(\mathrm{Sym}^k_{\mathcal{O}_{\Delta,\mathcal{Y}_x}} \left(T_p\mathcal{A} \otimes_{\hat{\mathbb{Z}}_{p,Y}} \mathcal{O}_\Delta \right) \otimes_{\mathcal{O}_Y} \omega|_{\mathcal{Y}_x}^{\otimes 2} \right) \\
&\overset{\mathrm{split}^\dagger_\Delta}{\longrightarrow} \left((\omega_{\Delta,\mathcal{Y}_x})^{\otimes_{\mathcal{O}_{\Delta,\mathcal{Y}_x}} k} \otimes_{\mathcal{O}_{\mathcal{Y}_x}} \omega|_{\mathcal{Y}_x}^{\otimes 2} \right) \cong \omega^{\otimes_{\mathcal{O}_{\Delta,\mathcal{Y}_x}} k+2}_{\Delta,\mathcal{Y}_x}.
\end{aligned} \tag{5.20}$$

5.4 THE p-ADIC MAASS-SHIMURA OPERATOR IN COORDINATES AND GENERALIZED p-ADIC MODULAR FORMS

Given any proétale $\mathcal{U} \to \mathcal{Y}_x$ and

$$\omega \in (\omega_\Delta)^{\otimes_{\mathcal{O}_\Delta}{}^k}(\mathcal{U}),$$

we write

$$\omega = F \cdot (\omega_{\mathrm{can}})^{\otimes_{\mathcal{O}_\Delta}{}^k} \xmapsto{\iota_{\mathrm{dR}}} F \cdot (-z_{\mathrm{dR}}\alpha_{\infty,1} + \alpha_{\infty,2})^{\otimes_{\mathcal{O}_{\mathbb{B}_{\mathrm{dR}},\mathcal{Y}}}{}^k}$$

where ω_{can} is defined in Definition 4.35. For brevity, write

$$\omega_{\mathrm{can}}^{\otimes k} = \omega_{\mathrm{can}}^{\otimes_{\mathcal{O}_\Delta}{}^k} \quad \text{and} \quad \iota_{\mathrm{dR}}(\omega_{\mathrm{can}})^{\otimes k} = \iota_{\mathrm{dR}}(\omega_{\mathrm{can}})^{\otimes_{\mathcal{O}_\Delta}{}^k}.$$

Then $\partial_k(\omega)$ is given by the composition

$$\omega \mapsto \nabla(\omega) = \nabla(\iota_{\mathrm{dR}}(\omega)) = \nabla\left(\iota_{\mathrm{dR}}(F \cdot \omega_{\mathrm{can}}^{\otimes k})\right) = \frac{d}{dz_{\mathrm{dR}}} F \cdot \iota_{\mathrm{dR}}(\omega_{\mathrm{can}})^{\otimes k} \cdot dz_{\mathrm{dR}}$$

$$+ \sum_{i=1}^k \iota_{\mathrm{dR}}(\omega_{\mathrm{can}})^{\otimes i-1} \cdot \nabla\left(\iota_{\mathrm{dR}}(\omega_{\mathrm{can}})\right) \cdot \iota_{\mathrm{dR}}(\omega_{\mathrm{can}})^{\otimes k-i+1}$$

$$= \frac{d}{dz_{\mathrm{dR}}} F \cdot \iota_{\mathrm{dR}}(\omega_{\mathrm{can}})^{\otimes k} \cdot dz_{\mathrm{dR}}$$

$$+ \sum_{i=1}^k \iota_{\mathrm{dR}}(\omega_{\mathrm{can}})^{\otimes i-1} \cdot (-\alpha_{\infty,1} \cdot dz_{\mathrm{dR}}) \cdot \iota_{\mathrm{dR}}(\omega_{\mathrm{can}})^{\otimes k-i+1}$$

$$= \frac{d}{dz_{\mathrm{dR}}} F \cdot \iota_{\mathrm{dR}}(\omega_{\mathrm{can}})^{\otimes k} \cdot dz_{\mathrm{dR}} - k \cdot \iota_{\mathrm{dR}}(\omega_{\mathrm{can}})^{\otimes k-1} \cdot \alpha_{\infty,1} \cdot dz_{\mathrm{dR}}$$

$$\xmapsto{\sigma^{-1},(4.50) \pmod t} \frac{d}{dz_{\mathrm{dR}}} F \cdot \iota_{\mathrm{dR}}(\omega_{\mathrm{can}})^{\otimes k+2} - k \cdot \iota_{\mathrm{dR}}(\omega_{\mathrm{can}})^{\otimes k+1} \cdot \alpha_{\infty,1}$$

$$\xmapsto{\mathrm{split}} \left(\frac{d}{dz_{\mathrm{dR}}} + \frac{k}{z_{\mathrm{dR}} - \bar{z}}\right) F \cdot \omega_{\mathrm{can}}^{\otimes k+2} \qquad (5.21)$$

where we have used the calculation

$$\nabla(\iota_{\mathrm{dR}}(\omega_{\mathrm{can}})) = \nabla(-z_{\mathrm{dR}}\alpha_{\infty,1} + \alpha_{\infty,2}) = -\alpha_{\infty,1} \cdot dz_{\mathrm{dR}}$$

and the last arrow uses (5.13).

Definition 5.12. *Now we define*

$$\delta_k : \mathcal{O}_\Delta \to \mathcal{O}_\Delta$$

by

$$\partial_k \left(F \cdot \omega_{\mathrm{can}}^{\otimes k} \right) = (\delta_k F) \cdot \omega_{\mathrm{can}}^{\otimes k+2},$$

the above calculation (5.21) shows that

$$\delta_k = \frac{d}{dz_{\mathrm{dR}}} + \frac{k}{z_{\mathrm{dR}} - \bar{z}}. \tag{5.22}$$

Definition 5.13. *Given an open or closed subset* $\mathcal{U} \subset \mathcal{Y}$, $k \in \mathbb{Z}$, *and a subgroup* $\Gamma \subset GL_2(\mathbb{Z}_p)$ *with* $\mathcal{U} \cdot \Gamma = \mathcal{U}$, *we define a p-adic modular form* F *on* $\mathcal{U}$ *for* Γ *of weight* k *to be a* $F \in \mathcal{O}_\Delta|_{\mathcal{U}}(\mathcal{U})$ *such that*

$$\begin{pmatrix} a & b \\ c & d \end{pmatrix}^* F = (ad - bc)^{-k}(cz_{\mathrm{dR}}|_{\mathcal{U}} + a)^k F \tag{5.23}$$

for any $\begin{pmatrix} a & b \\ c & d \end{pmatrix} \in \Gamma$. *We also make an analogous definition for* $k \in \mathbb{Z}/(p-1) \times \mathbb{Z}_p$, *provided that*

$$(ad - bc)^{-k}(cz_{\mathrm{dR}}|_{\mathcal{U}} + a)^k$$

makes sense on $\mathcal{U}$. *(Note that we embed*

$$\mathbb{Z} \subset \mathbb{Z}/(p-1) \times \mathbb{Z}_p$$

diagonally to make all notions of weight compatible.) Let $M_{k,\Delta}(\Gamma)(\mathcal{U})$ *denote the space of p-adic modular forms on* $\mathcal{U}$ *for* Γ *of weight* k.

Recall the natural projection $\lambda : \mathcal{Y} \to Y$.

Proposition 5.14. *Let* $U \subset \lambda(\mathcal{Y}_x)$ *be an open or closed subset, let* $\mathcal{U} = \lambda^{-1}(U)$, *let* $\Gamma \subset GL_2(\mathbb{Z}_p)$ *be a subgroup with* $\mathcal{U} \cdot \Gamma = \mathcal{U}$, *and let* $k \in \mathbb{Z}_{\geq 0}$. *Then we have*

$$\omega^{\otimes k}|_U(U) \xrightarrow{\lambda^*} \omega^{\otimes k}|_{\mathcal{U}}(\mathcal{U}) \to \omega_\Delta^{\otimes k}|_{\mathcal{U}}(\mathcal{U}) \xrightarrow{\sim} M_{k,\Delta}(\Gamma)(\mathcal{U}) \cdot \omega_{\mathrm{can}}^{\otimes k} \xrightarrow[\sim]{\cdot(\omega_{\mathrm{can}}^{\otimes k})^{-1}} M_{k,\Delta}(\Gamma)(\mathcal{U}).$$

Proof. The first arrow is obvious. Since $\omega_{\mathrm{can}}|_{\mathcal{U}}$ trivializes $\omega_\Delta|_{\mathcal{U}}(\mathcal{U})$ (Theorem 4.36), we can write any $\omega \in \omega^{\otimes k}(\mathcal{U})$ as

$$\omega = F \cdot \omega_{\mathrm{can}}^{\otimes k}$$

with $f \in \mathcal{O}_\Delta|_\mathcal{U}(\mathcal{U})$. Then since w is invariant under the action of Γ (as it's defined on a subset of Y), from (4.48) we see that F is a p-adic modular form on $\mathcal{U}$ for Γ of weight k. The rest of the arrows are obvious. □

Definition 5.15. *We let*

$$\partial_k^j = \partial_{k+2j-2} \circ \partial_{k+2j-4} \circ \cdots \circ \partial_{k+2} \circ \partial_k$$

and

$$\delta_k^j = \delta_{k+2j-2} \circ \delta_{k+2j-4} \circ \cdots \circ \delta_{k+2} \circ \delta_k.$$

It is clear by its definition that ∂_k^j and δ_k^j are $\hat{\mathcal{O}}_\mathcal{Y} \cong \mathbb{B}_{\mathrm{dR},\mathcal{Y}}^+/(t)$-linear.

Definition 5.16. *Given an open or closed subset $\mathcal{U} \subset \mathcal{Y}$, $k \in \mathbb{Z}$, and a subgroup $\Gamma \subset GL_2(\mathbb{Z}_p)$ with $\mathcal{U} \cdot \Gamma = \mathcal{U}$, we define a p-adic modular form F on $\mathcal{U}$ for Γ of weight k to be a $F \in \hat{\mathcal{O}}_\mathcal{Y}|_\mathcal{U}(\mathcal{U})$ such that*

$$\begin{pmatrix} a & b \\ c & d \end{pmatrix}^* F = (ad - bc)^{-k}(c z_{\mathrm{dR}}|_\mathcal{U} + a)^k F \tag{5.24}$$

for any

$$\begin{pmatrix} a & b \\ c & d \end{pmatrix} \in \Gamma.$$

Let $M_{k,\Delta}(\Gamma)(\mathcal{U})$ denote the space of p-adic modular forms on $\mathcal{U}$ for Γ of weight k.

Proposition 5.17. *We have*

$$\delta_k^j : M_{k,\Delta}(\Gamma)(\mathcal{U}) \to M_{k+2j,\Delta}(\Gamma)(\mathcal{U})$$

and

$$\delta_k^j = \sum_{i=0}^{j} \binom{k-1+j}{i}\binom{j}{i}\frac{i!}{(z_{\mathrm{dR}} - \bar{z})^i}\left(\frac{d}{dz_{\mathrm{dR}}}\right)^{j-i}. \tag{5.25}$$

Proof. This is a direct calculation, using (3.15), (4.9) and (4.44). More precisely, one uses induction: if $F \in \mathcal{O}_\Delta(\mathcal{U})$ has weight k', then by direct computation one verifies that

$$\left(\frac{d}{dz_{\mathrm{dR}}} + \frac{k'}{z_{\mathrm{dR}} - \bar{z}}\right) F$$

has weight $k' + 2$ in the sense of (5.24). Another "coordinate free" proof is to observe that each step in the construction of the map

$$f \mapsto \partial_{k'} f$$

is Γ-equivariant. $\qquad\qquad\qquad\qquad\qquad\qquad\qquad\qquad\qquad\qquad\qquad\qquad$ $\square$

For the rest of this section, we will make the identification of sheaves on $Y_{\text{proét}}$

$$\omega^{\otimes 2} = \Omega^1_Y$$

using the Kodaira-Spencer isomorphism σ as defined in (4.39).

Proposition 5.18. *We have*

$$\partial^j_k = \overline{\text{split}} \circ \nabla^j \circ \iota_{\text{dR}}.$$

Proof. By definition, we have

$$\partial^j_k = (\overline{\text{split}} \circ \nabla \circ \iota_{\text{dR}})^j.$$

Since the p-adic de Rham comparison ι_{dR} (see (3.37) and (3.38)) is compatible with connections, we have

$$\iota_{\text{dR}} \circ \nabla = \nabla \circ \iota_{\text{dR}}. \qquad\qquad (5.26)$$

By Proposition 5.7, we have

$$\nabla \circ \overline{\text{split}} = \overline{\text{split}} \circ \nabla. \qquad\qquad (5.27)$$

Hence, we have

$$\partial^j_k = (\overline{\text{split}} \circ \nabla \circ \iota_{\text{dR}})^j \stackrel{(5.26)}{=} (\overline{\text{split}} \circ \nabla)^j \circ \iota_{\text{dR}} \stackrel{(5.27)}{=} \overline{\text{split}} \circ \nabla^j \circ \iota_{\text{dR}}$$

which is what we wanted to show. $\qquad\qquad\qquad\qquad\qquad\qquad\qquad\qquad$ $\square$

Definition 5.19. *Henceforth, let*

$$d^j_k = \theta \circ \partial^j_k$$

and

$$
\theta_k^j = \theta \circ \delta_k^j = \theta \left(\sum_{i=0}^{j} \binom{k-1+j}{i} \binom{j}{i} \frac{i!}{(z_{\mathrm{dR}} - \bar{z})^i} \left(\frac{d}{dz_{\mathrm{dR}}} \right)^{j-i} \right)
$$

$$
= \sum_{i=0}^{j} \binom{k-1+j}{i} \binom{j}{i} \frac{i!}{\theta(z_{\mathrm{dR}} - \bar{z})^i} \theta \circ \left(\frac{d}{dz_{\mathrm{dR}}} \right)^{j-i} \tag{5.28}
$$

$$
= \sum_{i=0}^{j} \binom{k-1+j}{i} \binom{j}{i} i! \left(-\frac{\theta(y_{\mathrm{dR}})}{\hat{\mathbf{z}}} \right)^i \theta \circ \left(\frac{d}{dz_{\mathrm{dR}}} \right)^{j-i}
$$

where the first line follows from (5.25) and the last equality follows from (4.17).

Definition 5.20. *Given an open or closed subset $\mathcal{U} \subset \mathcal{Y}$, $k \in \mathbb{Z}$, and a subgroup $\Gamma \subset GL_2(\mathbb{Z}_p)$ with $\mathcal{U} \cdot \Gamma = \mathcal{U}$, we define a p-adic modular form F on $\mathcal{U}$ for Γ of weight k to be a $F \in \hat{\mathcal{O}}_{\mathcal{Y}}|_{\mathcal{U}}(\mathcal{U})$ such that*

$$
\begin{pmatrix} a & b \\ c & d \end{pmatrix}^{*} F = (ad - bc)^{-k} (c z_{\mathrm{dR}}|_{\mathcal{U}} + a)^k F \tag{5.29}
$$

for any

$$
\begin{pmatrix} a & b \\ c & d \end{pmatrix} \in \Gamma.
$$

We also make an analogous definition for $k \in \mathbb{Z}/(p-1) \times \mathbb{Z}_p$, provided that $(ad - bc)^{-k} (c z_{\mathrm{dR}}|_{\mathcal{U}} + a)^k$ makes sense on $\mathcal{U}$. (Note that we embed $\mathbb{Z} \subset \mathbb{Z}/(p-1) \times \mathbb{Z}_p$ diagonally to make all notions of weight compatible.) Let $M_{k,\Delta}(\Gamma)(\mathcal{U})$ denote the space of p-adic modular forms on $\mathcal{U}$ for Γ of weight k.

Proposition 5.21. *We have*

$$
\theta_k^j : \hat{M}_k(\Gamma)(\mathcal{U}) \to \hat{M}_{k+2j}(\Gamma)(\mathcal{U}).
$$

Proof. This follows from applying θ to Proposition 5.17. $\qquad\square$

5.5 THE p-ADIC MAASS-SHIMURA OPERATOR WITH "NEARLY HOLOMORPHIC COEFFICIENTS"

In this section, we define another p-adic Maass-Shimura operator *without the use of the universal cover* $\theta : \hat{\mathcal{O}}_{\Delta} \twoheadrightarrow \hat{\mathcal{O}}_Y$ by extending from $\mathcal{O}_{\Delta}$ to a slightly larger sheaf of coefficients. Defining an operator with $\hat{\mathcal{O}}_{\Delta}$-coefficients has the problem that there is no natural way to define a nontrivial connection on $\hat{\mathcal{O}}_{\Delta}$; defining such a connection would entail defining a connection on $\hat{\mathcal{O}}_Y \subset \hat{\mathcal{O}}_{\Delta}$, and we have already seen that there are no nontrivial differentials on $\hat{\mathcal{O}}_{\mathcal{Y}}(\mathcal{Y}) = \mathcal{O}_{\hat{\mathcal{Y}}}(\hat{\mathcal{Y}})$. The key to overcoming this is to instead work on an intermediate sheaf

$$\mathcal{O}_\Delta \subset \mathcal{O}_\Delta^\dagger \subset \hat{\mathcal{O}}_\Delta$$

which is large enough to construct a splitting of the Hodge filtration and on which a natural connection *extending the one on* $\mathcal{O}_\Delta$ can be defined. It will turn out that the value of the p-adic Maass-Shimura operator defined over $\mathcal{O}_\Delta^\dagger$ at CM points coincides with the value of θ^j at CM points, and hence are also equal to the values of complex Maass-Shimura operators at CM points.

Definition 5.22. *We define our sheaves of "p-adic nearly holomorphic" functions and modular forms, which have a natural connection. Define*

$$\mathcal{O}_{\mathrm{dR}}^\dagger := \mathcal{OB}_{\mathrm{dR},\mathcal{Y}_x}^+ \left[\frac{1}{\mathbf{z}_{\mathrm{dR}} - \hat{\mathbf{z}}}, \hat{\mathbf{z}}, \frac{1}{\hat{\mathbf{z}}} \right] \subset \hat{\mathcal{OB}}_{\mathrm{dR},\mathcal{Y}_x}^+$$

to be the subpresheaf of $\hat{\mathcal{OB}}_{\mathrm{dR},\mathcal{Y}_x}^+$ *obtained from adjoining the elements* $\frac{1}{\mathbf{z}_{\mathrm{dR}} - \hat{\mathbf{z}}}, \hat{\mathbf{z}}, \frac{1}{\hat{\mathbf{z}}}$ $\in \hat{\mathcal{OB}}_{\mathrm{dR},\mathcal{Y}}^+(\mathcal{Y}_x)$ *to* $\mathcal{OB}_{\mathrm{dR},\mathcal{Y}_x}^+$ *viewed inside the ambient presheaf* $\hat{\mathcal{OB}}_{\mathrm{dR},\mathcal{Y}_x}^+$. *Note that since we adjoin specific elements satisfying the sheaf gluing axioms to the sheaf* $\mathcal{OB}_{\mathrm{dR},\mathcal{Y}_x}^+$ *in order to define* $\mathcal{O}_{\mathrm{dR}}^\dagger$, *then in fact* $\mathcal{O}_{\mathrm{dR}}^\dagger$ *is a sheaf. We have the following natural inclusions*

$$\mathcal{OB}_{\mathrm{dR},\mathcal{Y}_x}^+ \subset \mathcal{O}_{\mathrm{dR}}^\dagger \subset \hat{\mathcal{OB}}_{\Delta,\mathcal{Y}_x}^+.$$

By Proposition 5.24, we can extend the connection

$$\nabla : \mathcal{OB}_{\mathrm{dR},\mathcal{Y}_x}^+ \to \mathcal{OB}_{\mathrm{dR},\mathcal{Y}_x}^+ \otimes_{\mathcal{O}_{\mathcal{Y}_x}} \Omega_{\mathcal{Y}_x}^1$$

to

$$\nabla : \mathcal{O}_{\mathrm{dR}}^\dagger \to \mathcal{O}_{\mathrm{dR}}^\dagger \otimes_{\mathcal{O}_{\mathcal{Y}_x}} \Omega_{\mathcal{Y}_x}^1$$

by defining

$$\nabla(\hat{\mathbf{z}}) = 0, \nabla(1/\hat{\mathbf{z}}) = 0 \quad \text{which implies} \quad \nabla\left(\frac{1}{\mathbf{z}_{\mathrm{dR}} - \hat{\mathbf{z}}} \right) = -\frac{1}{(\mathbf{z}_{\mathrm{dR}} - \hat{\mathbf{z}})^2} d\mathbf{z}_{\mathrm{dR}}.$$

$$(5.30)$$

The intuition behind this is to treat $\hat{\mathbf{z}}$ *as the "antiholomorphic" coordinate, whereas* $\mathbf{z}_{\mathrm{dR}}$ *should be thought of as the holomorphic coordinate in* $\mathcal{OB}_{\mathrm{dR},\mathcal{Y}_x}^+ \subset \mathcal{O}_{\mathrm{dR}}^\dagger$. *Then the antiholomorphic variable* $\hat{\mathbf{z}}$ *should be holomorphically horizontal in the sense that*

$$\nabla_{d\mathbf{z}_{\mathrm{dR}}}(\hat{\mathbf{z}}) = 0, \quad \text{or equivalently} \quad \frac{d}{d\mathbf{z}_{\mathrm{dR}}} \hat{\mathbf{z}} = 0$$

and then formally differentiating $\frac{1}{\mathbf{z}_{\mathrm{dR}}-\hat{\mathbf{z}}}$ by $\frac{d}{d\mathbf{z}_{\mathrm{dR}}}$ results in (5.30).

By the p-adic Legendre relation (4.17) (divided by t and reduced modulo (t)), we have

$$\frac{1}{\mathbf{z}_{\mathrm{dR}}-\hat{\mathbf{z}}}=-\frac{\mathbf{y}_{\mathrm{dR}}}{t\hat{\mathbf{z}}}$$

and so another equivalent definition of $\mathcal{O}_{\mathrm{dR}}^{\dagger}$ is

$$\mathcal{O}_{\mathrm{dR}}^{\dagger}:=\mathcal{O}\mathbb{B}_{\mathrm{dR},\mathcal{Y}_x}^{+}\left[\frac{\mathbf{y}_{\mathrm{dR}}}{t\hat{\mathbf{z}}},\hat{\mathbf{z}},\frac{1}{\hat{\mathbf{z}}}\right].$$

We have a distinguished subsheaf

$$\mathcal{O}_{\mathrm{dR}}^{\circ}:=\mathcal{O}\mathbb{B}_{\mathrm{dR},\mathcal{Y}_x}^{+}\left[\frac{1}{\mathbf{z}_{\mathrm{dR}}-\hat{\mathbf{z}}}\right]\subset\mathcal{O}_{\mathrm{dR}}^{\dagger}$$

which inherits a natural connection from $\mathcal{O}_{\mathrm{dR}}^{\dagger}$.

Definition 5.23. We also define

$$\omega_{\mathrm{dR}}^{\dagger}:=\omega\otimes_{\mathcal{O}_Y}\mathcal{O}_{\mathrm{dR}}^{\dagger},\quad\omega_{\mathrm{dR}}^{\circ}:=\omega\otimes_{\mathcal{O}_Y}\mathcal{O}_{\mathrm{dR}}^{\circ}.$$

Define

$$\mathcal{O}_{\Delta}^{\dagger}:=\mathcal{O}_{\Delta,\mathcal{Y}_x}\left[\frac{1}{\mathbf{z}_{\mathrm{dR}}-\hat{\mathbf{z}}},\hat{\mathbf{z}},\frac{1}{\hat{\mathbf{z}}}\right]\subset\hat{\mathcal{O}}_{\Delta,\mathcal{Y}_x}$$

to be the subpresheaf of $\hat{\mathcal{O}}_{\Delta,\mathcal{Y}_x}$ obtained from adjoining the elements $\frac{1}{\mathbf{z}_{\mathrm{dR}}-\hat{\mathbf{z}}},\hat{\mathbf{z}},\frac{1}{\hat{\mathbf{z}}}\in$ $\hat{\mathcal{O}}_{\Delta}(\mathcal{Y}_x)$ to $\mathcal{O}_{\Delta,\mathcal{Y}_x}$ viewed inside the ambient presheaf $\hat{\mathcal{O}}_{\Delta,\mathcal{Y}_x}$. Again, it is evident from construction that $\mathcal{O}_{\Delta}^{\dagger}$ is a sheaf. We have the following inclusions

$$\mathcal{O}_{\Delta,\mathcal{Y}_x}\subset\mathcal{O}_{\Delta}^{\dagger}\subset\hat{\mathcal{O}}_{\Delta,\mathcal{Y}_x}.$$

By Proposition 5.24, we can extend the connection

$$\nabla:\mathcal{O}_{\Delta,\mathcal{Y}_x}\to\mathcal{O}_{\Delta,\mathcal{Y}_x}\otimes_{\mathcal{O}_{\mathcal{Y}_x}}\Omega_{\mathcal{Y}_x}^1$$

to

$$\nabla:\mathcal{O}_{\Delta}^{\dagger}\to\mathcal{O}_{\Delta}^{\dagger}\otimes_{\mathcal{O}_{\mathcal{Y}_x}}\Omega_{\mathcal{Y}_x}^1$$

by defining

$$\nabla(d\hat{\mathbf{z}})=0,\quad\text{which implies}\quad\nabla\left(\frac{1}{\mathbf{z}_{\mathrm{dR}}-\hat{\mathbf{z}}}\right)=-\frac{1}{(\mathbf{z}_{\mathrm{dR}}-\hat{\mathbf{z}})^2}d\mathbf{z}_{\mathrm{dR}}.\qquad(5.31)$$

Again, the intuition behind (5.31) is to treat z_{dR} as the holomorphic coordinate, and $\hat{\mathbf{z}}$ as the antiholomorphic coordinate.

By the p-adic Legendre relation (4.17), we have

$$\frac{1}{z_{\mathrm{dR}} - \hat{\mathbf{z}}} = -\frac{y_{\mathrm{dR}}}{\hat{\mathbf{z}}}$$

and so another equivalent definition of $\mathcal{O}_{\Delta}^{\dagger}$ is

$$\mathcal{O}_{\Delta}^{\dagger} := \mathcal{O}_{\Delta, \mathcal{Y}_x} \left[\frac{y_{\mathrm{dR}}}{\hat{\mathbf{z}}}, \hat{\mathbf{z}}, \frac{1}{\hat{\mathbf{z}}} \right].$$

We have a distinguished subsheaf

$$\mathcal{O}_{\Delta}^{\circ} := \mathcal{O}_{\Delta} \left[\frac{1}{z_{\mathrm{dR}} - \hat{\mathbf{z}}} \right] \subset \mathcal{O}_{\Delta}^{\dagger}$$

which inherits a natural connection from $\mathcal{O}_{\Delta}^{\dagger}$.

Proposition 5.24. *There is a connection*

$$\nabla : \mathcal{O}_{\mathrm{dR}}^{\dagger} \to \mathcal{O}_{\mathrm{dR}}^{\dagger} \otimes_{\mathcal{O}_{\mathcal{Y}_x}} \Omega_{\mathcal{Y}_x}^1$$

given by (5.30) which extends the natural connection $\nabla : \mathcal{O}\mathbb{B}_{\mathrm{dR}, \mathcal{Y}_x}^{+} \to \mathcal{O}\mathbb{B}_{\mathrm{dR}, \mathcal{Y}_x}^{+} \otimes_{\mathcal{O}_{\mathcal{Y}_x}} \Omega_{\mathcal{Y}_x}^1$. There is also a connection

$$\nabla : \mathcal{O}_{\Delta}^{\dagger} \to \mathcal{O}_{\Delta}^{\dagger} \otimes_{\mathcal{O}_{\mathcal{Y}_x}} \Omega_{\mathcal{Y}_x}^1$$

given by (5.31) which extends the natural connection $\nabla : \mathcal{O}_{\Delta} \to \mathcal{O}_{\Delta} \otimes_{\mathcal{O}_{\mathcal{Y}_x}} \Omega_{\mathcal{Y}_x}^1$.

Proof. First, note that we have a map

$$\theta_{\mathrm{partial}} : \hat{\mathcal{O}}\mathbb{B}_{\mathrm{dR}, \mathcal{Y}_x}^{+} := \mathcal{O}\mathbb{B}_{\mathrm{dR}, \mathcal{Y}_x}^{+} \otimes_{\mathcal{O}_{\mathcal{Y}_x}} \hat{\mathcal{O}}_{\mathcal{Y}_x} \xrightarrow{\theta \otimes 1} \hat{\mathcal{O}}_{\mathcal{Y}_x} \otimes_{\mathcal{O}_{\mathcal{Y}_x}} \hat{\mathcal{O}}_{\mathcal{Y}_x}$$

given by applying the natural $\theta : \mathcal{O}\mathbb{B}_{\mathrm{dR}, \mathcal{Y}_x}^{+} \twoheadrightarrow \hat{\mathcal{O}}_{\mathcal{Y}_x}$ to the first factor of the above tensor product. In a slight abuse of notation, we denote $1/\hat{\mathbf{z}} = \theta_{\mathrm{partial}}(1/\hat{\mathbf{z}})$.

We now show the following.

Lemma 5.25. *The element $\frac{1}{\hat{\mathbf{z}}} \in \hat{\mathcal{O}}_{\mathcal{Y}_x} \otimes_{\mathcal{O}_{\mathcal{Y}_x}} \hat{\mathcal{O}}_{\mathcal{Y}_x}(\mathcal{Y}_x)$ is transcendental over*

$$\mathcal{O}_{\mathcal{Y}_x} \otimes 1 \subset \hat{\mathcal{O}}_{\mathcal{Y}_x} \otimes_{\mathcal{O}_{\mathcal{Y}_x}} \hat{\mathcal{O}}_{\mathcal{Y}_x}.$$

Proof of Lemma 5.25. Suppose otherwise, and we have a proétale $U \to \mathcal{Y}_x$ and $F = \sum_{n=0}^{d} \in (\mathcal{O}_{\mathcal{Y}_x} \otimes 1)(U)[X]$ with

$$F(1/\hat{\mathbf{z}}|_U) = 0. \tag{5.32}$$

Write $U = \varprojlim_i U_i$ with U_i étale over Y. Then in particular, the coefficients $a_n \in \mathcal{O}_{\mathcal{Y}_x}(U) = \lim_i \mathcal{O}_Y(U_i)$ and so take values at geometric fibers in $\overline{\mathbb{Q}}_p$. Hence (5.32) implies that $\hat{\mathbf{z}}$ takes values in $\overline{\mathbb{Q}}_p$ on U, which implies $\pi_{\mathrm{HT}}(U) \subset \overline{\mathbb{Q}}_p$, which contradicts Lemma 4.13. $\qquad\qquad\square$

As a corollary, we have:

Lemma 5.26. *The element $\frac{1}{\hat{\mathbf{z}}} \in \hat{\mathcal{O}}\mathbb{B}^+_{\mathrm{dR},\mathcal{Y}_x}(\mathcal{Y}_x)$ is transcendental over $\mathcal{O}\mathbb{B}^+_{\mathrm{dR},\mathcal{Y}_x} \subset \hat{\mathcal{O}}\mathbb{B}^+_{\mathrm{dR},\mathcal{Y}_x}$.*

Proof of Lemma 5.26. Suppose there is a proétale map $U \to \mathcal{Y}_x$ and a polynomial $F(X) = \sum_{n=0}^{d} a_n X^n \in \mathcal{O}\mathbb{B}^+_{\mathrm{dR},\mathcal{Y}_x}(U)[X]$ with

$$F(1/\hat{\mathbf{z}}|_U) = 0. \tag{5.33}$$

We will show that $F = 0$, which we do by contradiction, so assume $F \neq 0$. Recall the filtrations

$$\mathrm{Fil}^i \mathcal{O}\mathbb{B}^+_{\mathrm{dR},\mathcal{Y}_x} = (\ker \theta)^i \mathcal{O}\mathbb{B}^+_{\mathrm{dR},\mathcal{Y}_x}$$

and

$$\mathrm{Fil}^i \mathcal{O}\mathbb{B}^+_{\mathrm{dR},\mathcal{Y}_x} = \sum_{j \in \mathbb{Z}_{\geq 0}} t^{-j} \mathrm{Fil}^{i+j} \mathcal{O}\mathbb{B}^+_{\mathrm{dR},\mathcal{Y}_x},$$

and let i be the minimal degree among the $a_n \in \mathcal{O}\mathbb{B}^+_{\mathrm{dR},\mathcal{Y}_x}(U)$ so that $i < \infty$ since $F \neq 0$.

$$F(X) = \sum_{n=0}^{d} a_n X^n \in \mathrm{Fil}^i \mathcal{O}\mathbb{B}^+_{\mathrm{dR},\mathcal{Y}_x}(U)[X].$$

Then by Proposition 4.1 base changing to a perfectoid field K, locally there exists a proétale (surjective) cover $\tilde{U}_K \to U$ such that

$$\mathcal{O}\mathbb{B}^+_{\mathrm{dR},\tilde{U}_K} \cong \mathbb{B}^+_{\mathrm{dR},\tilde{U}_K}[\![Y]\!]$$

and hence

$$\mathrm{gr}^i \mathcal{O}\mathbb{B}_{\mathrm{dR},\tilde{U}_K} \cong t^i \hat{\mathcal{O}}_{\tilde{U}_K}[\![Y/t]\!]. \tag{5.34}$$

Let $\bar{a}_n$ denote the image of $a_n \in \mathrm{Fil}^i \mathcal{O}\mathbb{B}^+_{\mathrm{dR},\mathcal{Y}_x}(U)$ in $\mathrm{gr}^i \mathcal{O}\mathbb{B}_{\mathrm{dR},\tilde{U}_K}(\tilde{U}_K)$, and let $\bar{F}(1/\hat{\mathbf{z}}|_{\tilde{U}_K}) \in \mathrm{gr}^i \mathcal{O}\mathbb{B}_{\mathrm{dR},\tilde{U}_K}(\tilde{U}_K)[X]$ denote the corresponding image of F. Note that by the minimality of i chosen above, we have

$$\bar{F} \neq 0. \tag{5.35}$$

Then passing (5.33) to $\mathrm{gr}^i \mathcal{OB}_{\mathrm{dR}, \tilde{U}_K}$, we have

$$\bar{F}(1/\hat{\mathbf{z}}|_{\tilde{U}_K}) = \sum_{n=0}^{d} \bar{a}_n (1/\hat{\mathbf{z}}|_{\tilde{U}_K})^n = 0 \in \mathrm{gr}^i \mathcal{OB}_{\mathrm{dR}, \tilde{U}_K}(\tilde{U}_K)[X]$$

$$\overset{(5.34)}{\cong} t^i \hat{\mathcal{O}}_{\tilde{U}_K}(\tilde{U}_K)[\![Y/t]\!][X]. \tag{5.36}$$

Dividing by t^i and applying the map

$$\rho : \hat{\mathcal{O}}_{\tilde{U}_K}(\tilde{U}_K)[\![Y/t]\!] \hat{\mathcal{O}}_{\tilde{U}_K}(\tilde{U}_K) Y/t \mapsto 0,$$

we get

$$\sum_{n=0}^{d} \rho\left(\frac{\bar{a}_n}{t^i}\right) (1/\hat{\mathbf{z}}|_{\tilde{U}_K})^n = 0 \in \hat{\mathcal{O}}_{\tilde{U}_K}(\tilde{U}_K)[X]. \tag{5.37}$$

Note that we can write (5.37) as a finite sum

$$t^i \sum_j \lambda_j f_j (1/\hat{\mathbf{z}}|_{\tilde{U}_K}) = 0 \tag{5.38}$$

where the $\lambda_j \in \hat{\mathcal{O}}_{\tilde{U}_K}(\tilde{U}_K)$ are linearly independent over $\mathcal{O}_{\tilde{U}_K}(\tilde{U}_K)$ and $f_j \in \mathcal{O}_{\tilde{U}_K}[\![X]\!]$. Hence (5.38) implies that we have

$$f_i(1/\hat{\mathbf{z}}|_{\tilde{U}_K}) = 0 \tag{5.39}$$

for all i. Now by Lemma 5.25, we have $f_i = 0$ for all i, which implies $\bar{F} = 0$, a contradiction to (5.35). $\square$

Now by Lemma 5.26, we can define a connection

$$\nabla : \mathcal{OB}_{\mathrm{dR}, \mathcal{Y}_x}^+[\hat{\mathbf{z}}, 1/\hat{\mathbf{z}}] \to \mathcal{OB}_{\mathrm{dR}, \mathcal{Y}_x}^+[\hat{\mathbf{z}}, 1/\hat{\mathbf{z}}] \otimes_{\mathcal{O}_{\mathcal{Y}_x}} \Omega_{\mathcal{Y}_x}^1$$

by letting ∇ be the natural connection on $\mathcal{OB}_{\mathrm{dR}, \mathcal{Y}_x}^+$ and declaring $\nabla(\hat{\mathbf{z}}) = \nabla(1/\hat{\mathbf{z}}) = 0$. This is indeed a connection by the transcendence of $1/\hat{\mathbf{z}}$ over $\mathcal{OB}_{\mathrm{dR}, \mathcal{Y}_x}^+$. Now there is a unique way to extend to

$$\mathcal{O}_{\mathrm{dR}}^\dagger = \mathcal{OB}_{\mathrm{dR}, \mathcal{Y}_x}^+\left[\frac{1}{\mathbf{z}_{\mathrm{dR}} - \hat{\mathbf{z}}}, \hat{\mathbf{z}}, 1/\hat{\mathbf{z}}\right] \to \mathcal{OB}_{\mathrm{dR}, \mathcal{Y}_x}^+\left[\frac{1}{\mathbf{z}_{\mathrm{dR}} - \hat{\mathbf{z}}}, \hat{\mathbf{z}}, 1/\hat{\mathbf{z}}\right] \otimes_{\mathcal{O}_{\mathcal{Y}_x}} \Omega_{\mathcal{Y}_x}^1$$

$$= \mathcal{O}_{\mathrm{dR}}^\dagger \otimes_{\mathcal{O}_{\mathcal{Y}_x}} \Omega_{\mathcal{Y}_x}^1$$

which respects the Leibniz rule and linearity. Namely, we have

$$\nabla(\mathbf{z}_{\mathrm{dR}} - \hat{\mathbf{z}}) = d\mathbf{z}_{\mathrm{dR}}$$

and so necessarily

$$0 = \nabla(1) = \nabla\left(\frac{1}{\mathbf{z}_{\mathrm{dR}} - \hat{\mathbf{z}}}(\mathbf{z}_{\mathrm{dR}} - \hat{\mathbf{z}})\right)$$

$$= \nabla\left(\frac{1}{\mathbf{z}_{\mathrm{dR}} - \hat{\mathbf{z}}}\right)(\mathbf{z}_{\mathrm{dR}} - \hat{\mathbf{z}}) + \frac{1}{\mathbf{z}_{\mathrm{dR}} - \hat{\mathbf{z}}}\nabla(\mathbf{z}_{\mathrm{dR}} - \hat{\mathbf{z}})$$

$$= \nabla\left(\frac{1}{\mathbf{z}_{\mathrm{dR}} - \hat{\mathbf{z}}}\right)(\mathbf{z}_{\mathrm{dR}} - \hat{\mathbf{z}}) + \frac{1}{\mathbf{z}_{\mathrm{dR}} - \hat{\mathbf{z}}}d\mathbf{z}_{\mathrm{dR}}$$

$$\implies \nabla\left(\frac{1}{\mathbf{z}_{\mathrm{dR}} - \hat{\mathbf{z}}}\right) = -\frac{1}{(\mathbf{z}_{\mathrm{dR}} - \hat{\mathbf{z}})^2}d\mathbf{z}_{\mathrm{dR}}.$$

Hence we have the part of the proposition for $\mathcal{O}_{\mathrm{dR}}^\dagger$. Reducing modulo (t) and recalling that t is horizontal for ∇, we have the part of the proposition for $\mathcal{O}_\Delta$. $\qquad\square$

Definition 5.27. *We also define*

$$\omega_\Delta^\dagger := \omega \otimes_{\mathcal{O}_Y} \mathcal{O}_\Delta^\dagger, \quad \omega_\Delta^\diamond := \omega \otimes_{\mathcal{O}_Y} \mathcal{O}_\Delta^\diamond.$$

We note that we have

$$\mathcal{O}_\Delta^\dagger = \mathcal{O}_{\mathrm{dR}}^\dagger/(t), \quad \mathcal{O}_\Delta^\diamond = \mathcal{O}_{\mathrm{dR}}^\diamond/(t), \quad \omega_\Delta^\dagger = \omega_{\mathrm{dR}}^\dagger \otimes_{\mathcal{O}_{\mathrm{dR}}^\dagger} \mathcal{O}_\Delta^\dagger, \quad \omega_\Delta^\diamond = \omega_{\mathrm{dR}}^\diamond \otimes_{\mathcal{O}_{\mathrm{dR}}^\diamond} \mathcal{O}_\Delta^\diamond,$$

compatible with the natural connections (as t is a horizontal section). We will eventually work on the level of $\mathcal{O}_\Delta^\dagger$ for constructing our p-adic L-function, as $\mathcal{O}_\Delta^\dagger$ has better p-adic properties than $\mathcal{O}_{\mathrm{dR}}^\dagger$. For example, we have a natural p-adic topology on $\mathcal{O}_\Delta^\dagger$, i.e., the topology of uniform convergence on polynomials in $1/(\mathbf{z}_{\mathrm{dR}} - \hat{\mathbf{z}}), 1/\hat{\mathbf{z}}$ over $\mathcal{O}_{\Delta, y_x}$. Note that we have a natural map

$$\theta : \mathcal{O}_{\mathrm{dR}}^\dagger \to \hat{\mathcal{O}}_{y_x}$$

inherited from $\theta : \hat{\mathcal{O}}\mathbb{B}_{\mathrm{dR}, y_x}^+ \twoheadrightarrow \hat{\mathcal{O}}_{y_x}$, as well as a natural map

$$\theta : \mathcal{O}_\Delta^\dagger \twoheadrightarrow \hat{\mathcal{O}}_{y_x}$$

inherited from $\theta : \hat{\mathcal{O}}_{\Delta, y_x} \twoheadrightarrow \hat{\mathcal{O}}_{y_x}$. Once again, as in (2.9), we have that the compositions

$$\mathcal{O}_{y_x} \subset \mathcal{O}_{\mathrm{dR}}^\dagger \xrightarrow{\theta} \hat{\mathcal{O}}_{y_x}, \quad \mathcal{O}_{y_x} \subset \mathcal{O}_\Delta^\dagger \xrightarrow{\theta} \hat{\mathcal{O}}_{y_x}$$

are the natural p-adic completion maps.

Finally, note that since the connections on $\mathcal{O}_{\mathrm{dR}}^{\dagger}$ and $\mathcal{O}_{\Delta}^{\dagger}$ extend the connections on $\mathcal{O}\mathbb{B}_{\mathrm{dR},\mathcal{Y}_x}^{+} \subset \mathcal{O}_{\mathrm{dR}}^{\dagger}$ and $\mathcal{O}_{\Delta,\mathcal{Y}_x} \subset \mathcal{O}_{\Delta}^{\dagger}$, then the p-adic comparison (3.36) (tensored with $\otimes_{\mathcal{O}\mathbb{B}_{\mathrm{dR},\mathcal{Y}_x}^{+}} \mathcal{O}_{\mathrm{dR}}^{\dagger}$ and $\otimes_{\mathcal{O}_{\Delta,\mathcal{Y}_x}} \mathcal{O}_{\Delta}^{\dagger}$)

$$\mathcal{H}_{\mathrm{dR}}^1(\mathcal{A}) \otimes_{\mathcal{O}_Y} \mathcal{O}_{\mathrm{dR}}^{\dagger} \overset{\iota_{\mathrm{dR}}}{\leftrightsquigarrow} T_p\mathcal{A} \otimes_{\hat{\mathbb{Z}}_{p,Y}} \mathcal{O}_{\mathrm{dR}}^{\dagger} \cdot t^{-1}$$

and

$$\mathcal{H}_{\mathrm{dR}}^1(\mathcal{A}) \otimes_{\mathcal{O}_Y} \mathcal{O}_{\Delta}^{\dagger} \overset{\iota_{\Delta}}{\leftrightsquigarrow} T_p\mathcal{A} \otimes_{\hat{\mathbb{Z}}_{p,Y}} \mathcal{O}_{\Delta}^{\dagger} \cdot t^{-1} \tag{5.40}$$

are still compatible with connections.

5.6 THE RELATIVE HODGE-TATE DECOMPOSITION OVER $\mathcal{O}_{\Delta}^{\dagger}$

Note that the basis elements $\iota_{\mathrm{dR}}(\omega_{\mathrm{can}}), (HT_{\mathcal{A}})^{\vee}(1)(\mathfrak{s}^{-1}t)$ of $T_p\mathcal{A} \otimes_{\hat{\mathbb{Z}}_{p,Y}} \hat{\mathcal{O}}\mathbb{B}_{\mathrm{dR},\mathcal{Y}_x}^{+}$ in fact live over the smaller nearly holomorphic sheaves $\mathcal{O}_{\mathrm{dR}}^{\dagger}$ above:

$$\iota_{\mathrm{dR}}(\omega_{\mathrm{can}}) = -\mathbf{z}_{\mathrm{dR}}\alpha_{\infty,1} + \alpha_{\infty,2} \in T_p\mathcal{A} \otimes_{\hat{\mathbb{Z}}_{p,Y}} \mathcal{O}_{\mathrm{dR}}^{\dagger}(\mathcal{Y}_x)$$

$$(HT_{\mathcal{A}})^{\vee}(1)(\mathfrak{s}^{-1}t) = \alpha_{\infty,1} - \frac{1}{\hat{\mathbf{z}}}\alpha_{\infty,2} \in T_p\mathcal{A} \otimes_{\hat{\mathbb{Z}}_{p,Y}} \mathcal{O}_{\mathrm{dR}}^{\dagger}(\mathcal{Y}_x),$$

and hence the relative Hodge-Tate decomposition (5.2), defined over $\hat{\mathcal{O}}\mathbb{B}_{\mathrm{dR},\mathcal{Y}_x}^{+}$, in fact lives over the smaller sheaf $\mathcal{O}_{\mathrm{dR}}^{\dagger}$:

$$\omega_{\mathrm{dR}}^{\dagger} \oplus \mathfrak{s}^{-1}t \cdot \mathcal{O}_{\mathrm{dR}}^{\dagger} = \omega_{\mathrm{can}} \cdot \mathcal{O}_{\mathrm{dR}}^{\dagger} \oplus \mathfrak{s}^{-1}t \cdot \mathcal{O}_{\mathrm{dR}}^{\dagger}$$

$$\xrightarrow[\sim]{\iota_{\mathrm{dR}} \oplus (HT_{\mathcal{A}})^{\vee}(1)} \left((-\mathbf{z}_{\mathrm{dR}} \cdot \alpha_{\infty,1} + \cdot\alpha_{\infty,2})\mathcal{O}_{\mathrm{dR}}^{\dagger}\right) \oplus \left((\alpha_{\infty,1} - \frac{1}{\hat{\mathbf{z}}} \cdot \alpha_{\infty,2})\mathcal{O}_{\mathrm{dR}}^{\dagger}\right)$$

$$\overset{(4.17)}{=} \mathcal{O}_{\mathrm{dR}}^{\dagger,\oplus 2} \xrightarrow[\sim]{\alpha_{\infty}} T_p\mathcal{A} \otimes_{\hat{\mathbb{Z}}_{p,Y}} \mathcal{O}_{\mathrm{dR}}^{\dagger}. \tag{5.41}$$

Reducing (5.41) modulo (t), we get a relative $\mathcal{O}_{\Delta}|_{\mathcal{Y}_x}$-Hodge-Tate decomposition on $\mathcal{Y}_x$:

$$\omega_{\Delta}^{\dagger} \oplus \mathfrak{s}^{-1}t \cdot \mathcal{O}_{\Delta}^{\dagger} = \omega_{\mathrm{can}} \cdot \mathcal{O}_{\Delta}^{\dagger} \oplus \mathfrak{s}^{-1}t \cdot \mathcal{O}_{\Delta}^{\dagger}$$

$$\xrightarrow[\sim]{\iota_{\mathrm{dR}} \oplus (HT_{\mathcal{A}})^{\vee}(1)} \left((-\mathbf{z}_{\mathrm{dR}} \cdot \alpha_{\infty,1} + \cdot\alpha_{\infty,2})\mathcal{O}_{\Delta}^{\dagger}\right) \oplus \left((\alpha_{\infty,1} - \frac{1}{\hat{\mathbf{z}}} \cdot \alpha_{\infty,2})\mathcal{O}_{\Delta}^{\dagger}\right) \tag{5.42}$$

$$\overset{(4.17)}{=} \mathcal{O}_{\Delta}^{\dagger,\oplus 2} \xrightarrow[\sim]{\alpha_{\infty}} T_p\mathcal{A} \otimes_{\hat{\mathbb{Z}}_{p,Y}} \mathcal{O}_{\Delta}^{\dagger}.$$

Definition 5.28. *Henceforth, denote the projection to the first factor in* (5.41) *by the above splitting by*

$$\mathrm{split}_{\mathrm{dR}}^{\dagger} : T_p\mathcal{A} \otimes_{\hat{\mathbb{Z}}_{p,Y}} \mathcal{O}_{\mathrm{dR}}^{\dagger} \twoheadrightarrow \omega_{\mathrm{dR}}^{\dagger} \tag{5.43}$$

and denote the projection to the first factor in (5.42) $\otimes_{\mathcal{O}_{\mathcal{Y}_x}} \hat{\mathcal{O}}_{\mathcal{Y}_x}$ *by*

$$\mathrm{split}_{\Delta}^{\dagger} : T_p\mathcal{A} \otimes_{\hat{\mathbb{Z}}_{p,Y}} \mathcal{O}_{\Delta}^{\dagger} \twoheadrightarrow \omega_{\Delta}^{\dagger}. \tag{5.44}$$

By direct calculation, on $\mathcal{Y}_x$ we have

$$\mathrm{split}_{\mathrm{dR}}^{\dagger}(\alpha_{\infty,1}) = \frac{1}{\hat{\mathbf{z}} - \mathbf{z}_{\mathrm{dR}}} \cdot \omega_{\mathrm{can}}, \quad \mathrm{split}_{\Delta}^{\dagger}(\alpha_{\infty,1}) = \frac{1}{\hat{\mathbf{z}} - \mathbf{z}_{\mathrm{dR}}} \cdot \omega_{\mathrm{can}} \tag{5.45}$$

over $\hat{\mathcal{O}}_{\mathrm{dR},\mathcal{Y}_x}$, and

$$\mathrm{split}_{\Delta}^{\dagger}(\alpha_{\infty,1}) = \frac{1}{\hat{\mathbf{z}} - z_{\mathrm{dR}}} \cdot \omega_{\mathrm{can}}, \quad \mathrm{split}_{\Delta}^{\dagger}(\alpha_{\infty,1}) = \frac{1}{\hat{\mathbf{z}} - z_{\mathrm{dR}}} \cdot \omega_{\mathrm{can}} \tag{5.46}$$

over $\hat{\mathcal{O}}_{\Delta,\mathcal{Y}_x}$. Moreover, we have

$$\nabla(\omega_{\mathrm{can}}) = \nabla(\iota_{\mathrm{dR}}(\omega_{\mathrm{can}})) = \nabla(-\mathbf{z}_{\mathrm{dR}} \cdot \alpha_{\infty,1} + \alpha_{\infty,2})$$
$$= -\alpha_{\infty,1} \cdot d\mathbf{z}_{\mathrm{dR}} \xrightarrow{\overline{\mathrm{split}}} \frac{1}{\mathbf{z}_{\mathrm{dR}} - \hat{\mathbf{z}}} \cdot \omega_{\mathrm{can}} \tag{5.47}$$

over $\hat{\mathcal{O}}\mathbb{B}_{\mathrm{dR},\mathcal{Y}_x}^{+}$ where

$$d\mathbf{z}_{\mathrm{dR}} := \nabla(\mathbf{z}_{\mathrm{dR}}),$$

and

$$\nabla(\omega_{\mathrm{can}}) = \nabla(\iota_{\mathrm{dR}}(\omega_{\mathrm{can}})) = \nabla(-z_{\mathrm{dR}} \cdot \alpha_{\infty,1} + \alpha_{\infty,2})$$
$$= -\alpha_{\infty,1} \cdot dz_{\mathrm{dR}} \xrightarrow{\overline{\mathrm{split}}} \frac{1}{z_{\mathrm{dR}} - \hat{\mathbf{z}}} \cdot \omega_{\mathrm{can}} \tag{5.48}$$

over $\hat{\mathcal{O}}_{\Delta,\mathcal{Y}_x}$, where (as before)

$$dz_{\mathrm{dR}} = \nabla(z_{\mathrm{dR}}).$$

Definition 5.29. *Henceforth, denote the map on symmetric powers induced by* (5.11) *also by*

$$\mathrm{split}_{\mathrm{dR}}^{\dagger} : \mathrm{Sym}_{\mathcal{O}_{\mathrm{dR}}^{\dagger}}^{k}\left(T_p\mathcal{A} \otimes_{\hat{\mathbb{Z}}_{p,Y}} \mathcal{O}_{\mathrm{dR}}^{\dagger}\right) \twoheadrightarrow \omega_{\mathrm{dR},\mathcal{Y}_x}^{\otimes_{\mathcal{O}_{\mathrm{dR}}^{\dagger}} k} \tag{5.49}$$

and

$$\mathrm{split}^{\dagger}_{\Delta} : \mathrm{Sym}^{k}_{\mathcal{O}^{\dagger}_{\Delta}}\left(T_p\mathcal{A}\otimes_{\hat{\mathbb{Z}}_{p,Y}}\mathcal{O}^{\dagger}_{\Delta}\right) \twoheadrightarrow \omega_{\Delta,\mathcal{Y}_x}^{\otimes_{\mathcal{O}^{\dagger}_{\Delta}}k}. \tag{5.50}$$

Definition 5.30. *We define a map*

$$\partial^{\dagger}_{k,\mathrm{dR}} : \left(\omega^{\dagger}_{\mathrm{dR}}\right)^{\otimes_{\mathcal{O}^{\dagger}_{\mathrm{dR}}}k} \to \left(\omega^{\dagger}_{\mathrm{dR}}\right)^{\otimes_{\mathcal{O}^{\dagger}_{\mathrm{dR}}}k+2}$$

as the following composition:

$$
\begin{aligned}
\left(\omega^{\dagger}_{\mathrm{dR}}\right)^{\otimes_{\mathcal{O}^{\dagger}_{\mathrm{dR}}}k} &\overset{\iota_{\mathrm{dR}}}{\subset} \mathrm{Sym}^{k}_{\mathcal{O}^{\dagger}_{\mathrm{dR}}}\left(T_p\mathcal{A}\otimes_{\hat{\mathbb{Z}}_{p,Y}}\mathcal{O}^{\dagger}_{\mathrm{dR}}\right)\\
&\overset{\nabla}{\to}\left(\mathrm{Sym}^{k}_{\mathcal{O}^{\dagger}_{\mathrm{dR}}}\left(T_p\mathcal{A}\otimes_{\hat{\mathbb{Z}}_{p,Y}}\mathcal{O}^{\dagger}_{\mathrm{dR}}\right)\otimes_{\mathcal{O}_{\mathcal{Y}}}\Omega^1_{\mathcal{Y}_x}\right)\\
&\overset{\sigma^{-1}}{\longrightarrow}\left(\mathrm{Sym}^{k}_{\mathcal{O}^{\dagger}_{\mathrm{dR}}}\left(T_p\mathcal{A}\otimes_{\hat{\mathbb{Z}}_{p,Y}}\mathcal{O}^{\dagger}_{\mathrm{dR}}\right)\otimes_{\mathcal{O}_{\mathcal{Y}}}\omega|_{\mathcal{Y}_x}^{\otimes 2}\right)\\
&\overset{\mathrm{split}^{\dagger}_{\mathrm{dR}}}{\longrightarrow}\left(\left(\omega^{\dagger}_{\mathrm{dR}}\right)^{\otimes_{\mathcal{O}^{\dagger}_{\mathrm{dR}}}k}\otimes_{\mathcal{O}_{\mathcal{Y}_x}}\omega|_{\mathcal{Y}_x}^{\otimes 2}\right)\cong\left(\omega^{\dagger}_{\mathrm{dR}}\right)^{\otimes_{\mathcal{O}^{\dagger}_{\mathrm{dR}}}k+2}.
\end{aligned}
\tag{5.51}
$$

We define a map

$$\partial^{\dagger}_{k,\Delta} : \left(\omega^{\dagger}_{\Delta}\right)^{\otimes_{\mathcal{O}^{\dagger}_{\Delta}}k} \to \left(\omega^{\dagger}_{\Delta}\right)^{\otimes_{\mathcal{O}^{\dagger}_{\Delta}}k+2}$$

as the following composition:

$$
\begin{aligned}
\left(\omega^{\dagger}_{\Delta}\right)^{\otimes_{\mathcal{O}^{\dagger}_{\Delta}}k} &\overset{\iota_{\mathrm{dR}}}{\subset} \mathrm{Sym}^{k}_{\mathcal{O}^{\dagger}_{\Delta}}\left(T_p\mathcal{A}\otimes_{\hat{\mathbb{Z}}_{p,Y}}\mathcal{O}^{\dagger}_{\Delta}\right)\\
&\overset{\nabla}{\to}\left(\mathrm{Sym}^{k}_{\mathcal{O}^{\dagger}_{\Delta}}\left(T_p\mathcal{A}\otimes_{\hat{\mathbb{Z}}_{p,Y}}\mathcal{O}^{\dagger}_{\Delta}\right)\otimes_{\mathcal{O}_{\mathcal{Y}}}\Omega^1_{\mathcal{Y}_x}\right)\\
&\overset{\sigma^{-1}}{\longrightarrow}\left(\mathrm{Sym}^{k}_{\mathcal{O}^{\dagger}_{\Delta}}\left(T_p\mathcal{A}\otimes_{\hat{\mathbb{Z}}_{p,Y}}\mathcal{O}^{\dagger}_{\Delta}\right)\otimes_{\mathcal{O}_{\mathcal{Y}}}\omega|_{\mathcal{Y}_x}^{\otimes 2}\right)\\
&\overset{\mathrm{split}^{\dagger}_{\Delta}}{\longrightarrow}\left(\left(\omega^{\dagger}_{\Delta}\right)^{\otimes_{\mathcal{O}^{\dagger}_{\Delta}}k}\otimes_{\mathcal{O}_{\mathcal{Y}_x}}\omega|_{\mathcal{Y}_x}^{\otimes 2}\right)\cong\left(\omega^{\dagger}_{\Delta}\right)^{\otimes_{\mathcal{O}^{\dagger}_{\Delta}}k+2}.
\end{aligned}
\tag{5.52}
$$

We define

$$\partial^{\dagger,j}_{k,\mathrm{dR}}=\partial^{\dagger}_{k,\mathrm{dR}+2(j-1)}\circ\cdots\circ\partial^{\dagger}_{k,\mathrm{dR}},\quad \partial^{\dagger,j}_{k,\Delta}=\partial^{\dagger}_{k,\mathrm{dR}+2(j-1)}\circ\cdots\circ\partial^{\dagger}_{k,\Delta},$$

where the right-hand sides of the above equations denote the j-fold compositions. It is easy to see from the definitions (and the fact that t is horizontal for all connections) that we have

$$\partial^{\dagger,j}_{k,\mathrm{dR}}\quad(\mathrm{mod}\ t)=\partial^{\dagger,j}_{k,\Delta}.$$

5.7 THE p-ADIC MAASS-SHIMURA OPERATOR IN COORDINATES AND GENERALIZED p-ADIC NEARLY HOLOMORPHIC MODULAR FORMS

Definition 5.31. *Now we define*

$$\delta_{k,\mathrm{dR}}^{\dagger,j} : \mathcal{O}_{\mathrm{dR}}^{\dagger} \to \mathcal{O}_{\mathrm{dR}}^{\dagger}$$

by

$$\partial_{k,\mathrm{dR}}^{\dagger,j}\left(F \cdot \omega_{\mathrm{can}}^{\otimes k}\right) = \left(\delta_{k,\mathrm{dR}}^{\dagger,j}F\right) \cdot \omega_{\mathrm{can}}^{\otimes k+2j},$$

and

$$\delta_{k,\Delta}^{\dagger,j} : \mathcal{O}_{\Delta}^{\dagger} \to \mathcal{O}_{\Delta}^{\dagger}$$

by

$$\partial_{k,\Delta}^{\dagger,j}\left(F \cdot \omega_{\mathrm{can}}^{\otimes k}\right) = \left(\delta_{k,\Delta}^{\dagger,j}F\right) \cdot \omega_{\mathrm{can}}^{\otimes k+2j}$$

and we let $\delta_{k,\mathrm{dR}}^{\dagger} := \delta_{k,\mathrm{dR}}^{\dagger,1}, \delta_{k,\Delta}^{\dagger} := \delta_{k,\Delta}^{\dagger,1}$. *It is clear from the definitions that*

$$\delta_{k,\mathrm{dR}}^{\dagger,j} = \delta_{k+2j-2,\mathrm{dR}}^{\dagger} \circ \cdots \circ \delta_{k,\mathrm{dR}}^{\dagger,1}, \quad \delta_{k,\Delta}^{\dagger,j} = \delta_{k+2j-2,\Delta}^{\dagger} \circ \cdots \circ \delta_{k,\Delta}^{\dagger,1}.$$

Moreover, we have

$$\delta_{k,\mathrm{dR}}^{\dagger,j} \pmod{t} = \delta_{k,\Delta}^{\dagger,j}.$$

Finally, we have the following key proposition, which again shows the "horizontalness" of the splittings $\mathrm{split}_{\mathrm{dR}}^{\dagger}$ and $\mathrm{split}_{\Delta}^{\dagger}$. Let

$$\mathcal{L}_{\mathrm{dR}}^{\dagger} := \ker\left(\mathrm{split}_{\mathrm{dR}}^{\dagger}\right), \quad \mathcal{L}_{\Delta}^{\dagger} := \ker\left(\mathrm{split}_{\Delta}^{\dagger}\right).$$

Proposition 5.32. *We have that* $\mathcal{L}_{\mathrm{dR}}^{\dagger}$ *is horizontal for* ∇. *That is, for any proétale* $U \to \mathcal{Y}_x$ *and for any nonvanishing*

$$w \in \Omega_{\mathcal{Y}}^1 \otimes_{\mathcal{O}_{\mathcal{Y}}} \mathcal{O}_{\mathrm{dR}}^{\dagger}(U),$$

we have

$$\nabla_w(\mathcal{L}_{\mathrm{dR}}^{\dagger}|_U) \subset \mathcal{L}_{\mathrm{dR}}^{\dagger}|_U.$$

The analogous statement is true for $\mathcal{L}_{\Delta}^{\dagger}$. *In particular, we have*

$$\mathrm{split}_{\mathrm{dR}}^{\dagger} \circ \nabla^j = \nabla^j \circ \mathrm{split}_{\mathrm{dR}}^{\dagger}, \quad \mathrm{split}_{\Delta}^{\dagger} \circ \nabla^j = \nabla^j \circ \mathrm{split}_{\Delta}^{\dagger}$$

for any $j \in \mathbb{Z}_{\geq 0}$.

Proof. It suffices to prove the statement for $\mathcal{L}_{\mathrm{dR}}^{\dagger}$, as the statement for $\mathcal{L}_{\Delta}^{\dagger}$ then follows from reducing everything modulo (t). Note that $\mathcal{L}_{\mathrm{dR}}^{\dagger}$ is generated by the section

$$(HT_{\mathcal{A}})^{\vee}(1)(\mathfrak{s}^{-1}t) = \alpha_{\infty,1} - 1/\hat{\mathbf{z}} \cdot \alpha_{\infty,2} \in T_p\mathcal{A} \otimes_{\hat{\mathbb{Z}}_{p,Y}} \mathcal{O}_{\mathrm{dR}}^{\dagger}(\mathcal{Y}_x)$$

by definition, and so we can write any section of $\mathcal{L}_{\mathrm{dR}}^{\dagger}|_U$ as

$$f \cdot (\alpha_{\infty,1} - 1/\hat{\mathbf{z}} \cdot \alpha_{\infty,2}), \quad f \in \mathcal{O}_{\mathrm{dR}}^{\dagger}(U).$$

Then

$$\nabla_w(f \cdot (\alpha_{\infty,1} - 1/\hat{\mathbf{z}} \cdot \alpha_{\infty,2})) = \nabla_w(f) \cdot (\alpha_{\infty,1} - 1/\hat{\mathbf{z}} \cdot \alpha_{\infty,2})$$
$$+ f \cdot \nabla_w(\alpha_{\infty,1} - 1/\hat{\mathbf{z}} \cdot \alpha_{\infty,2}) = \nabla_w(f) \cdot (\alpha_{\infty,1} - 1/\hat{\mathbf{z}} \cdot \alpha_{\infty,2})$$

which is a section of $\mathcal{L}_{\mathrm{dR}}^{\dagger}|_U$. $\square$

Again, for the rest of this section we will identify $\omega^{\otimes 2} = \Omega_Y^1$ using the Kodaira-Spencer isomorphism σ (4.39). Moreover, for brevity, we sometimes suppress the subscripts of $\otimes$'s appearing in powers, as it will be obvious from context which category of modules each object lives in.

Proposition 5.33. *We have*

$$\partial_{k,\mathrm{dR}}^{\dagger,j} = \mathrm{split}_{\mathrm{dR}}^{\dagger} \circ \nabla^j \circ \iota_{\mathrm{dR}}, \quad and \quad \partial_{k,\Delta}^{\dagger,j} = \mathrm{split}_{\Delta}^{\dagger} \circ \nabla^j \iota_{\mathrm{dR}}.$$

Proof. Again, it suffices to prove the statement for $\partial_{k,\mathrm{dR}}^{\dagger,j}$, as the analogous statement follows from reducing modulo (t). By definition, we have

$$\partial_{k,\mathrm{dR}}^j = (\mathrm{split}_{\mathrm{dR}}^{\dagger} \circ \nabla \circ \iota_{\mathrm{dR}})^j.$$

Since ι_{dR} (see (5.40)) is compatible with connections, we have

$$\iota_{\mathrm{dR}} \circ \nabla = \nabla \circ \iota_{\mathrm{dR}}. \tag{5.53}$$

By Proposition 5.32, we have

$$\nabla \circ \mathrm{split}_{\mathrm{dR}}^{\dagger} = \mathrm{split}_{\mathrm{dR}}^{\dagger} \circ \nabla. \tag{5.54}$$

Hence, we have

$$\partial_{\mathrm{dR}k}^{\dagger,j} = (\mathrm{split}_{\mathrm{dR}}^{\dagger} \circ \nabla \circ \iota_{\mathrm{dR}})^j \overset{(5.26)}{=} (\mathrm{split}_{\mathrm{dR}}^{\dagger} \circ \nabla)^j \circ \iota_{\mathrm{dR}} \overset{(5.27)}{=} \mathrm{split}_{\mathrm{dR}}^{\dagger} \circ \nabla^j \circ \iota_{\mathrm{dR}}$$

which is what we wanted to show. $\square$

Theorem 5.34. *We have the following formulas for $\delta_{k,\mathrm{dR}}^{\dagger,j}$ and $\delta_{k,\Delta}^{\dagger,j}$:*

$$\delta_{k,\mathrm{dR}}^{\dagger,j} = \sum_{i=0}^{j} \frac{\binom{j}{i}\binom{j+k-1}{i} i!}{(\mathbf{z}_{\mathrm{dR}} - \hat{\mathbf{z}})^i} \left(\frac{d}{d\mathbf{z}_{\mathrm{dR}}} \right)^{j-i}, \quad \delta_{k,\mathrm{dR}}^{\dagger,j} = \sum_{i=0}^{j} \frac{\binom{j}{i}\binom{j+k-1}{i} i!}{(z_{\mathrm{dR}} - \hat{\mathbf{z}})^i} \left(\frac{d}{d\mathbf{z}_{\mathrm{dR}}} \right)^{j-i}.$$

Proof. Again it suffices to do this for $\partial_{k,\mathrm{dR}}^{\dagger}$, as then the statement for $\partial_{k,\Delta}^{\dagger}$ follows by reducing modulo (t).

Suppose w is any section of $(\omega_{\mathrm{dR},\mathcal{Y}_x}^{\dagger})^{\otimes k}$. Then it can be written as

$$w = F \cdot \omega_{\mathrm{can}}^{\otimes k}$$

where F is a section of $\mathcal{O}_{\mathrm{dR}}^{\dagger}$. Then through computation we have

$$\partial_{k,\mathrm{dR}}^{\dagger}(w) = \nabla(\iota_{\mathrm{dR}}(w)) = \nabla(F \cdot (-\mathbf{z}_{\mathrm{dR}}\alpha_{\infty,1} + \alpha_{\infty,2})^k)$$

$$= \left(\frac{d}{d\mathbf{z}_{\mathrm{dR}}} F \cdot (-\mathbf{z}_{\mathrm{dR}}\alpha_{\infty,1} + \alpha_{\infty,2})^k - kF \cdot \alpha_{\infty,1} \cdot (-\mathbf{z}_{\mathrm{dR}}\alpha_{\infty,1} + \alpha_{\infty,2})^{k-1} \right)$$

$$\cdot d\mathbf{z}_{\mathrm{dR}} \stackrel{\sigma_v^{-1}\mathrm{split}_{\mathrm{dR}}^{\dagger}}{\mapsto} \left(\frac{d}{d\mathbf{z}_{\mathrm{dR}}} + \frac{k}{\mathbf{z}_{\mathrm{dR}} - \hat{\mathbf{z}}} \right) F \cdot \omega_{\mathrm{can}}^{\otimes k+2}, \tag{5.55}$$

which proves the formula for $j = 1$. Note that k was arbitrary, so the formula for general $j \in \mathbb{Z}_{\geq 0}$ follows by an easy induction argument on j. $\qquad\square$

Our next proposition says that the operators $\partial_{k,\mathrm{dR}}^{\dagger,j}, \partial_{k,\Delta}^{\dagger,j}$ in fact factor over the smaller sheaves $\mathcal{O}_{\mathrm{dR}}^{\diamond} \subset \mathcal{O}_{\mathrm{dR}}^{\dagger}$ and $\mathcal{O}_{\Delta}^{\diamond} \subset \mathcal{O}_{\Delta}^{\dagger}$.

Definition 5.35. *Let*

$$\partial_{k,\mathrm{dR}}^{\diamond,j} := \partial_{k,\mathrm{dR}}^{\dagger,j}\big|_{(\omega_{\mathrm{dR}}^{\diamond})^{\otimes k}}$$

and

$$\partial_{k,\Delta}^{\diamond,j} := \partial_{k,\Delta}^{\dagger,j}\big|_{(\omega_{\Delta}^{\diamond})^{\otimes k}}.$$

Proposition 5.36. *In fact, we have that the following restrictions of $\partial_{k,\mathrm{dR}}^{\dagger,j}$ and $\partial_{k,\Delta}^{\dagger,j}$ preserve smaller sheaves of differentials:*

$$\partial_{k,\mathrm{dR}}^{\diamond,j} := \partial_{k,\mathrm{dR}}^{\dagger,j}\big|_{(\omega_{\mathrm{dR}}^{\diamond})^{\otimes k}} : (\omega_{\mathrm{dR}}^{\diamond})^{\otimes k} \to (\omega_{\mathrm{dR}}^{\diamond})^{\otimes k+2j}$$

and

$$\partial_{k,\Delta}^{\diamond,j} := \partial_{k,\Delta}^{\dagger,j}\big|_{(\omega_{\Delta}^{\diamond})^{\otimes k}} : (\omega_{\Delta}^{\diamond})^{\otimes k} \to (\omega_{\Delta}^{\diamond})^{\otimes k+2j}.$$

Similarly,

$$\delta_{k,\mathrm{dR}}^{\diamond,j} := \delta_{k,\mathrm{dR}}^{\dagger,j}|_{\mathcal{O}_{\mathrm{dR}}^{\diamond}} : \mathcal{O}_{\mathrm{dR}}^{\diamond} \to \mathcal{O}_{\mathrm{dR}}^{\diamond}$$

and

$$\delta_{k,\Delta}^{\diamond,j} := \delta_{k,\Delta}^{\dagger,j}|_{\mathcal{O}_{\Delta}^{\diamond}} : \mathcal{O}_{\Delta}^{\diamond} \to \mathcal{O}_{\Delta}^{\diamond}.$$

Proof. First note that

$$\omega_{\mathrm{can}} \in \omega_{\mathrm{dR}}^{\diamond}(\mathcal{Y}_x)$$

is a generator, and reducing modulo (t) we also have that

$$\omega_{\mathrm{can}} \in \omega_{\Delta}^{\diamond}(\mathcal{Y}_x)$$

is a generator.

Now the proposition follows immediately from the explicit formulas in Theorem 5.34 and observing that $\frac{d}{d\mathbf{z}_{\mathrm{dR}}}$ and $\frac{d}{dz_{\mathrm{dR}}}$ preserve their respective subsheaves:

$$\frac{d}{d\mathbf{z}_{\mathrm{dR}}} : \mathcal{O}_{\mathrm{dR}}^{\diamond} \to \mathcal{O}_{\mathrm{dR}}^{\diamond}, \quad \frac{d}{dz_{\mathrm{dR}}} : \mathcal{O}_{\Delta}^{\diamond} \to \mathcal{O}_{\Delta}^{\diamond}. \qquad \square$$

We give the following definitions of generalized p-adic modular forms on various subloci of $\mathcal{Y}_x$.

Definition 5.37. *Given an open or closed subset $\mathcal{U} \subset \mathcal{Y}$, $k \in \mathbb{Z}$, and a subgroup $\Gamma \subset GL_2(\mathbb{Z}_p)$ with $\mathcal{U} \cdot \Gamma = \mathcal{U}$, we define a de Rham real analytic modular form F on $\mathcal{U}$ for Γ of weight k to be a $F \in \mathcal{O}_{\mathrm{dR}}^{\dagger}|_{\mathcal{U}}(\mathcal{U})$ such that*

$$\begin{pmatrix} a & b \\ c & d \end{pmatrix}^{*} F = (ad - bc)^{-k}(c\mathbf{z}_{\mathrm{dR}}|_{\mathcal{U}} + a)^{k} F \tag{5.56}$$

for any $\begin{pmatrix} a & b \\ c & d \end{pmatrix} \in \Gamma$. We also make an analogous definition for $k \in \mathbb{Z}/(p-1) \times \mathbb{Z}_p$, provided that

$$(ad - bc)^{-k}(c\mathbf{z}_{\mathrm{dR}}|_{\mathcal{U}} + a)^{k}$$

makes sense on $\mathcal{U}$. (Note that we embed

$$\mathbb{Z} \subset \mathbb{Z}/(p-1) \times \mathbb{Z}_p$$

diagonally to make all notions of weight compatible.) Let $M_{k,\mathrm{dR}}^{\dagger}(\Gamma)(\mathcal{U})$ denote the space of de Rham real analytic modular forms on $\mathcal{U}$ for Γ of weight k.

Similarly, we define a de Rham nearly holomorphic modular form F on $\mathcal{U}$ for Γ of weight k to have the same definition as a de Rham real analytic modular *form given above, except replacing all the $\mathcal{O}_{\mathrm{dR}}^\dagger$'s with $\mathcal{O}_{\mathrm{dR}}^\diamond$'s. We denote the space of de Rham nearly holomorphic modular forms on $\mathcal{U}$ for Γ of weight k by $M_{k,\mathrm{dR}}^\diamond(\Gamma)(\mathcal{U})$.*

Given an open or closed subset $\mathcal{U} \subset \mathcal{Y}$, $k \in \mathbb{Z}$, and a subgroup $\Gamma \subset GL_2(\mathbb{Z}_p)$ with $\mathcal{U} \cdot \Gamma = \mathcal{U}$, we define a p-adic real analytic modular form F on $\mathcal{U}$ for Γ of weight k to be a $F \in \mathcal{O}_\Delta^\dagger|_{\mathcal{U}}(\mathcal{U})$ such that

$$\begin{pmatrix} a & b \\ c & d \end{pmatrix}^* F = (ad - bc)^{-k}(cz_{\mathrm{dR}}|_{\mathcal{U}} + a)^k F \qquad (5.57)$$

for any $\begin{pmatrix} a & b \\ c & d \end{pmatrix} \in \Gamma$. We also make an analogous definition for $k \in \mathbb{Z}/(p-1) \times \mathbb{Z}_p$, provided that

$$(ad - bc)^{-k}(cz_{\mathrm{dR}}|_{\mathcal{U}} + a)^k$$

makes sense on $\mathcal{U}$. (Note that we embed

$$\mathbb{Z} \subset \mathbb{Z}/(p-1) \times \mathbb{Z}_p$$

diagonally to make all notions of weight compatible.) Let $M_{k,\Delta}(\Gamma)(\mathcal{U})$ denote the space of p-adic real analytic modular forms on $\mathcal{U}$ for Γ of weight k.

Similarly, we define a p-adic nearly holomorphic modular form F on $\mathcal{U}$ for Γ of weight k to have the same definition as a p-adic real analytic modular form given above, except replacing all the $\mathcal{O}_\Delta^\dagger$'s with $\mathcal{O}_\Delta^\diamond$'s. We denote the space of p-adic nearly holomorphic modular forms on $\mathcal{U}$ for Γ of weight k by $M_{k,\Delta}^\diamond(\Gamma)(\mathcal{U})$.

Recall the natural projection $\lambda : \mathcal{Y} \to Y$.

Proposition 5.38. *Let $U \subset \lambda(\mathcal{Y}_x)$ be an open or closed subset, let $\mathcal{U} = \lambda^{-1}(U)$, let $\Gamma \subset GL_2(\mathbb{Z}_p)$ be any subgroup with $\mathcal{U} \cdot \Gamma = \mathcal{U}$, and let $k \in \mathbb{Z}_{\geq 0}$. Then we have under the natural maps*

$$\omega^{\otimes k}|_U(U) \xrightarrow{\lambda^*} \omega^{\otimes k}|_{\mathcal{U}}(\mathcal{U}) \to \left(\omega_{\mathrm{dR}}^\dagger\right)^{\otimes k}|_{\mathcal{U}}(\mathcal{U}) \xrightarrow{\sim} M_{k,\mathrm{dR}}^\dagger(\Gamma)(\mathcal{U}) \cdot \omega_{\mathrm{can}}^{\otimes k}$$

$$\xrightarrow[\sim]{\cdot(\omega_{\mathrm{can}}^{\otimes k})^{-1}} M_{k,\mathrm{dR}}^\dagger(\Gamma)(\mathcal{U})$$

and

$$\omega^{\otimes k}|_U(U) \xrightarrow{\lambda^*} \omega^{\otimes k}|_{\mathcal{U}}(\mathcal{U}) \to \left(\omega_\Delta^\dagger\right)^{\otimes k}|_{\mathcal{U}}(\mathcal{U}) \xrightarrow{\sim} M_{k,\Delta}^\dagger(\Gamma)(\mathcal{U}) \cdot \omega_{\mathrm{can}}^{\otimes k}$$

$$\xrightarrow[\sim]{\cdot(\omega_{\mathrm{can}}^{\otimes k})^{-1}} M_{k,\Delta}^\dagger(\Gamma)(\mathcal{U}).$$

Similarly on the "◇ level," we have

$$\omega^{\otimes k}|_U(U) \xrightarrow{\lambda^*} \omega^{\otimes k}|_{\mathcal{U}}(\mathcal{U}) \to (\omega_\Delta^\circ)^{\otimes k}|_{\mathcal{U}}(\mathcal{U}) \xrightarrow{\sim} M_{k,\mathrm{dR}}^\circ(\Gamma)(\mathcal{U}) \cdot \omega_{\mathrm{can}}^{\otimes k}$$

$$\xrightarrow[\sim]{\cdot(\omega_{\mathrm{can}}^{\otimes k})^{-1}} M_{k,\mathrm{dR}}^\circ(\Gamma)(\mathcal{U})$$

and

$$\omega^{\otimes k}|_U(U) \xrightarrow{\lambda^*} \omega^{\otimes k}|_{\mathcal{U}}(\mathcal{U}) \to (\omega_\Delta^\circ)^{\otimes k}|_{\mathcal{U}}(\mathcal{U}) \xrightarrow{\sim} M_{k,\Delta}^\circ(\Gamma)(\mathcal{U}) \cdot \omega_{\mathrm{can}}^{\otimes k}$$

$$\xrightarrow[\sim]{\cdot(\omega_{\mathrm{can}}^{\otimes k})^{-1}} M_{k,\Delta}^\circ(\Gamma)(\mathcal{U}).$$

Proof. The first arrow is obvious. Since $\omega_{\mathrm{can}}|_{\mathcal{U}}$ trivializes $\omega_{\mathrm{dR}}|_{\mathcal{U}}(\mathcal{U}) \subset \omega_{\mathrm{dR}}^\dagger|_{\mathcal{U}}(\mathcal{U})$ (Theorem 4.36), we can write any $\omega \in \omega^{\otimes k}|_{\mathcal{U}}(\mathcal{U})$ as

$$\omega = F \cdot \omega_{\mathrm{can}}^{\otimes k}$$

with $f \in \mathcal{O}_\Delta(\mathcal{U})$. Then since w is invariant under the action of Γ (as it's defined on a subset of Y), from (4.48) we see that F is a p-adic modular form on $\mathcal{U}$ for $GL_2(\mathbb{Z}_p)$ of weight k. The rest of the arrows are obvious. The assertions for Δ follow from reducing the statements for dR modulo (t).

The results over $\mathcal{O}_{\mathrm{dR}}^\circ$ and $\mathcal{O}_\Delta^\circ$ now follow immediately from Proposition 5.36. □

Proposition 5.39. *We have*

$$\delta_{k,\mathrm{dR}}^{\dagger,j} : M_{k,\mathrm{dR}}^\dagger(\Gamma)(\mathcal{U}) \to M_{k+2j,\mathrm{dR}}^\dagger(\Gamma)(\mathcal{U}), \quad \delta_{k,\mathrm{dR}}^{\circ,j} : M_{k,\mathrm{dR}}^\circ(\Gamma)(\mathcal{U}) \to M_{k+2j,\mathrm{dR}}^\circ(\Gamma)(\mathcal{U})$$

and

$$\delta_{k,\Delta}^{\dagger,j} : M_{k,\Delta}^\dagger(\Gamma)(\mathcal{U}) \to M_{k+2j,\Delta}^\dagger(\Gamma)(\mathcal{U}), \quad \delta_{k,\Delta}^{\circ,j} : M_{k,\Delta}^\circ(\Gamma)(\mathcal{U}) \to M_{k+2j,\Delta}^\circ(\Gamma)(\mathcal{U}).$$

Proof. On the "† level," this is a direct calculation, using (3.15), (4.9), (4.44) and Theorem 5.34. More precisely, one uses induction: if $F \in \mathcal{O}_{\mathrm{dR}}^\dagger(\mathcal{U})$ has weight k', then by direct computation one verifies that

$$\left(\frac{d}{d\mathbf{z}_{\mathrm{dR}}} + \frac{k'}{\mathbf{z}_{\mathrm{dR}} - \hat{\mathbf{z}}} \right) F$$

has weight $k' + 2$ in the sense of (5.56), and similarly for $F \in \mathcal{O}_\Delta^\dagger(\mathcal{U})$. Another "coordinate free" proof is to observe that each step in the construction of the maps

$$f \mapsto \partial_{k',\mathrm{dR}}^{\dagger}, \quad f \mapsto \partial_{k',\Delta}^{\dagger}$$

are Γ-equivariant.

The results on the "$\diamond$ level" now follow from Proposition 5.36. $\square$

Remark 5.40. We will often suppress the subscript of "k" in $\partial_{k,\mathrm{dR}}^{\dagger,j}, \partial_{k,\Delta}^{\dagger,j}$, etc. when the weight of the source k is clear. Observe also that we have the chain of inclusions

$$M_{k,\mathrm{dR}}^{\diamond}(\Gamma)(\mathcal{U}) \subset M_{k,\mathrm{dR}}^{\dagger}(\Gamma)(\mathcal{U})$$

and

$$M_{k,\Delta}(\Gamma)(\mathcal{U}) \subset M_{k,\Delta}^{\diamond}(\Gamma)(\mathcal{U}) \subset M_{k,\Delta}^{\dagger}(\Gamma)(\mathcal{U}).$$

The inclusions $M_{k,\mathrm{dR}}^{\diamond}(\Gamma)(\mathcal{U}) \subset M_{k,\mathrm{dR}}^{\dagger}(\Gamma)(\mathcal{U})$ and $M_{k,\Delta}^{\diamond}(\Gamma)(\mathcal{U}) \subset M_{k,\Delta}^{\dagger}(\Gamma)(\mathcal{U})$ are compatible with differential operators, by Proposition 5.36.

5.8   RELATION OF d_k^j AND $(d_k^{\dagger})^i$ TO THE ORDINARY ATKIN-SERRE OPERATOR $d_{k,\mathrm{AS}}^j$ AND KATZ'S p-ADIC MODULAR FORMS

Recall the Atkin-Serre operator

$$d_{\mathrm{AS},k} : \omega|_{\mathcal{Y}^{\mathrm{Ig}}}^{\otimes k} \to \omega|_{\mathcal{Y}^{\mathrm{Ig}}}^{\otimes k+2}$$

which acts, with respect to the Serre-Tate coordinate T on the covering $\mathcal{D} = \mathcal{Y}^{\mathrm{Ig}} \times_{Y^{\mathrm{ord}}} D$ of the ordinary residue disc D, as

$$d_{\mathrm{AS},k}(F \cdot \omega_{\mathrm{can}}^{\mathrm{Katz},\otimes k}) = (\theta_{\mathrm{AS}}F) \cdot \omega_{\mathrm{can}}^{\mathrm{Katz},\otimes k+2}$$

where

$$\theta_{\mathrm{AS}} = \frac{Td}{dT}.$$

Note that the operator θ_{AS}^j, defined in terms of the Serre-Tate coordinate T, does not depend on the weight k, unlike θ_k^j. Let

$$d_{\mathrm{AS},k}^j = d_{\mathrm{AS},k+2j-2} \circ \cdots \circ d_k$$

denote the j-fold composition, and similarly let

$$\theta_{\mathrm{AS}}^j = \theta_{\mathrm{AS},k+2j-2} \circ \cdots \circ \theta_{\mathrm{AS},k}$$

denote the j-fold composition.

Theorem 5.41. *On* $\mathcal{Y}^{\mathrm{Ig}}$, *we have*

$$d_k^j|_{\omega^{\otimes k}}|_{\mathcal{Y}^{\mathrm{Ig}}} = \partial_{k,\Delta}^j|_{\omega^{\otimes k}}|_{\mathcal{Y}^{\mathrm{Ig}}} = \partial_{k,\mathrm{dR}}^j|_{\omega^{\otimes k}}|_{\mathcal{Y}^{\mathrm{Ig}}} = d_{\mathrm{AS},k}^j$$

and

$$\theta_k^j|_{\mathcal{O}_{\mathcal{Y}^{\mathrm{Ig}}}} = \delta_{k,\Delta}^j|_{\mathcal{O}_{\mathcal{Y}^{\mathrm{Ig}}}} = \delta_{k,\mathrm{dR}}^j|_{\mathcal{O}_{\mathcal{Y}^{\mathrm{Ig}}}} = \theta_{\mathrm{AS}}^j.$$

Proof. Throughout the proof, suppress the notations "$|_{\mathcal{Y}^{\mathrm{Ig}}}$" and "$|_{\mathcal{O}_{\mathcal{Y}^{\mathrm{Ig}}}}$" for brevity. From the p-adic Legendre relation (4.17), we have on $\mathcal{Y}^{\mathrm{Ig}} = \{\hat{\mathbf{z}} = \infty\}$:

$$\theta(1/(z_{\mathrm{dR}} - \bar{z})) = -\theta(y_{\mathrm{dR}})/\hat{\mathbf{z}} = 0, 1/(\mathbf{z}_{\mathrm{dR}} - \hat{\mathbf{z}}) = -\mathbf{y}_{\mathrm{dR}}/(t\hat{\mathbf{z}}) = 0, 1/(z_{\mathrm{dR}} - \hat{\mathbf{z}})$$
$$= -y_{\mathrm{dR}}/\hat{\mathbf{z}} = 0.$$

Hence from (5.25) and Theorem 5.34, we have

$$\theta_k^j = \theta \circ \delta_k^j = \theta\left(\frac{d}{dz_{\mathrm{dR}}}\right)^j, \quad \delta_{k,\mathrm{dR}}^{\dagger,j} = \left(\frac{d}{d\mathbf{z}_{\mathrm{dR}}}\right)^j, \quad \delta_{k,\Delta}^{\dagger,j} = \left(\frac{d}{dz_{\mathrm{dR}}}\right)^j. \tag{5.58}$$

By Proposition 5.9, we have

$$\frac{d}{d\mathbf{z}_{\mathrm{dR}}} = \frac{d}{dz_{\mathrm{dR}}} = \frac{d}{d\log T} = \frac{Td}{dT} \tag{5.59}$$

and so combining (5.58) and (5.59), we have

$$d_k^j = \theta(d_{\mathrm{AS},k}^j) = d_{\mathrm{AS},k}^j, \quad \partial_{k,\mathrm{dR}}^{\dagger,j} = \partial_{k,\Delta}^{\dagger,j}$$

where the last equality in the first equation follows from the fact that

$$\mathcal{O}_Y \subset \mathcal{O}\mathbb{B}_{\mathrm{dR},Y}^+ \xrightarrow{\theta} \hat{\mathcal{O}}_Y$$

is the natural inclusion. $\square$

Now recall Katz's notion of p-adic modular forms ([35]) as functions on Y^{Ig} which transform under $\mathrm{Gal}(Y^{\mathrm{Ig}}/Y^{\mathrm{ord}}) \cong \mathbb{Z}_p^\times$ by a certain weight character.

Definition 5.42. *Let* $M_k^{\mathrm{Katz}} \subset \mathcal{O}_{Y^{\mathrm{ord}}}(Y^{\mathrm{Ig}})$ *denote the subring of elements such that*

$$d^* f = d^{-k} f$$

for any $d \in \mathrm{Gal}(Y^{\mathrm{Ig}}/Y^{\mathrm{ord}}) \cong \mathbb{Z}_p^\times$. *This space is often called the* space of Katz p-adic modular forms of weight k.

Proposition 5.43. *The natural action of B on the flag of spaces*

$$M_{k,\Delta}(B)(\mathcal{Y}^{\mathrm{Ig}}) \subset M_{k,\Delta}^{\diamond}(B)(\mathcal{Y}^{\mathrm{Ig}}) \subset M_{k,\Delta}^{\dagger}(B)(\mathcal{Y}^{\mathrm{Ig}})$$

descends to $\mathbb{Z}_p^{\times} \cong B/C \cong \mathrm{Gal}(Y^{\mathrm{Ig}}/Y^{\mathrm{ord}})$. In particular, we have a natural inclusion

$$M_k^{\mathrm{Katz}} \hookrightarrow M_{k,\Delta}(B)(\mathcal{Y}^{\mathrm{Ig}}) \subset M_{k,\Delta}^{\diamond}(B)(\mathcal{Y}^{\mathrm{Ig}}) \subset M_{k,\Delta}^{\dagger}(B)(\mathcal{Y}^{\mathrm{Ig}})$$

which is compatible with the action of $\mathbb{Z}_p^{\times} \cong B/C \cong \mathrm{Gal}(Y^{\mathrm{Ig}}/Y^{\mathrm{ord}})$.

Proof. From Proposition 4.26, we have

$$\omega_{\mathrm{can}}^{\mathrm{Katz}} \in \omega(Y^{\mathrm{Ig}}).$$

Note that from (4.37), we have

$$\begin{pmatrix} a & b \\ 0 & d \end{pmatrix}^* \omega_{\mathrm{can}}^{\mathrm{Katz}} = d\omega_{\mathrm{can}}^{\mathrm{Katz}}.$$

We have the identification (using the notations from Definition 4.18)

$$\mathbb{Z}_p^{\times} \cong B/C \cong \mathrm{Gal}(Y^{\mathrm{Ig}}/Y^{\mathrm{ord}}),$$

where the first isomorphism is given by

$$d \mapsto \left[\begin{pmatrix} 1 & 0 \\ 0 & d \end{pmatrix}\right] \in B/C$$

where $[\cdot]$ denotes equivalence class in B/C. Hence under the above identification, we have

$$d^* \omega_{\mathrm{can}}^{\mathrm{Katz}} = d^k \omega_{\mathrm{can}}^{\mathrm{Katz}}.$$

Now we have

$$M_k^{\mathrm{Katz}} \overset{\cdot \omega_{\mathrm{can}}^{\mathrm{Katz},\otimes k}}{\underset{\sim}{\longrightarrow}} M_k^{\mathrm{Katz}} \cdot \omega_{\mathrm{can}}^{\mathrm{Katz},\otimes k} \cong \omega^{\otimes k}(Y^{\mathrm{Ig}}) \subset \omega^{\otimes k}(\mathcal{Y}^{\mathrm{Ig}}).$$

Now the proposition follows from Propositions 4.38, 5.14 and 5.38. □

Corollary 5.44. *Under the inclusions from Proposition 5.43, we have*

$$\theta_k^j|_{\mathcal{Y}^{\mathrm{Ig}}} = \delta_{k,\Delta}^j|_{M_k^{\mathrm{Katz}}} = \delta_{k,\mathrm{dR}}^j|_{M_k^{\mathrm{Katz}}} = \theta_{\mathrm{AS}}^j.$$

Proof. This follows immediately from Theorem 5.41 and Proposition 5.43. □

Hence, we can regard our theory of p-adic analysis on $\mathcal{Y}_x$ using d_k^j or $(d_k^\dagger)^j$ as an extension of Katz's theory of p-adic analysis on $\mathcal{Y}^{\mathrm{Ig}}$ using $d_{\mathrm{AS},k}^j$.

5.9   COMPARISON BETWEEN THE COMPLEX AND p-ADIC MAASS-SHIMURA OPERATORS AT CM POINTS

Let $K/\mathbb{Q}$ be an imaginary quadratic field, with ring of integers $\mathcal{O}_K$. Let $i_p : \overline{\mathbb{Q}} \hookrightarrow \mathbb{C}_p$ denote our previously fixed embedding (3.1), and now fix an embedding $i_\infty : \overline{\mathbb{Q}} \hookrightarrow \mathbb{C}$. Let $A/\overline{\mathbb{Q}}$ be an elliptic curve with CM by an order $\mathcal{O} \subset \mathcal{O}_K$. In the complex counterpart of the Hodge–de Rham filtration, we have

$$H^{1,0}(A/\mathbb{C}) = \Omega_{A/\mathbb{C}}^1, \qquad H^{0,1}(A/\mathbb{C}) = \mathrm{Lie}(A/\mathbb{C}),$$

and in fact these respectively correspond to the γ and $\overline{\gamma}$-eigenspaces for any $\gamma \in \mathcal{O} = \mathrm{End}(A/\overline{\mathbb{Q}})$, acting as scalars on these eigenspaces under the above embedding i_∞.

Let τ denote the standard coordinate on the complex upper half-plane $\mathcal{H}^+$. The real analytic Hodge splitting of the complex Hodge–de Rham filtration gives rise to the complex Maass-Shimura operator (see [33] or the introduction of [5, Section 1.2] for detailed discussions of this operator)

$$\eth_k = \frac{1}{2\pi i} \left(\frac{d}{d\tau} + \frac{k}{\tau - \bar{\tau}} \right).$$

The real analytic Hodge splitting is induced by complex conjugation acting on the complex structure of de Rham cohomology, in particular interchanging the Hodge pieces, i.e.,

$$\overline{H^{p,q}(A/\mathbb{C})} = H^{q,p}(A/\mathbb{C}),$$

and giving rise to "opposing filtrations" in Deligne's sense. Thus, as for a CM curve $A/\overline{\mathbb{Q}}$ we have

$$H^{1,0}(A/\mathbb{C}) := \Omega_{A/\mathbb{C}}^1 = \Omega_{A/\overline{\mathbb{Q}}}^1 \otimes_{\overline{\mathbb{Q}}} \mathbb{C} = H^{1,0}(A/\overline{\mathbb{Q}}) \otimes_{\overline{\mathbb{Q}}} \mathbb{C},$$

the splitting of the Hodge–de Rham filtration induced by the eigendecomposition under the $\mathcal{O}_K$-action coincides with the Hodge splitting. In fact, we have the following coincidence of the values of our p-adic and complex Maass-Shimura operators at CM points (Theorem 5.48). Let

$$\eth_k^j = \eth_{k+2j-2} \circ \eth_{k+2j-4} \circ \cdots \circ \eth_{k+2} \circ \eth_k.$$

Definition 5.45. *Let $A/\overline{\mathbb{Q}}$ be an elliptic curve with complex multiplication by an order $\mathcal{O} \subset \mathcal{O}_K$, and let t be any Γ-level structure, $\alpha : \mathbb{Z}_p^{\oplus 2} \xrightarrow{\sim} T_p A$ be any $\Gamma(p^\infty)$-level structure compatible with t. Fix any $\omega_0 \in \Omega^1_{A/\overline{\mathbb{Z}}}$, and fix an isomorphism $\mathbb{C}/\mathfrak{a}_0 \cong A$ for some ideal $\mathfrak{a}_0 \subset \mathcal{O}$ (determined up to its class in $\mathcal{C}\ell(\mathcal{O})$) which defines a differential $2\pi i dz \in \Omega^1_{A/\overline{\mathbb{Q}}}$, where z is the standard coordinate on $\mathbb{C}$. Define*

$$\Omega_\infty(A, t) \in \mathbb{C}^\times, \qquad \Omega_p(A, t, \alpha) \in \mathcal{O}_\Delta(A, t, \alpha)^\times$$

by

$$\Omega_\infty(A, t) \cdot 2\pi i dz(A, t) = \omega_0, \qquad \Omega_p(A, t, \alpha) \cdot \omega_{\mathrm{can}}(A, t, \alpha) = \omega_0. \qquad (5.60)$$

Definition 5.46. *We recall the Shimura reciprocity law on $\mathcal{Y}$. Fix an order $\mathcal{O} \subset \mathcal{O}_K$. Let $\mathbb{I}^{\mathfrak{f}}$ be the semigroup of ideals of $\mathcal{O}_K$ prime to a fixed ideal $\mathfrak{f} \subset \mathcal{O}_K$. Let $\mathfrak{a} \in \mathbb{I}^{(pN)}$ with $\mathfrak{a} \subset \mathcal{O}$. With A as in Definition 5.45, define*

$$\mathfrak{a} \star A := A/A[\mathfrak{a}]$$

where $A[\mathfrak{a}]$ denotes the $\mathfrak{a}$-torsion subgroup of A using the identification $\mathrm{End}_{\overline{\mathbb{Q}}}(A) = \mathcal{O}$. Then the natural projection

$$\pi_{\mathfrak{a}} : A \twoheadrightarrow \mathfrak{a} \star A$$

takes Γ-level structures to Γ-level structures, and $\Gamma(p^\infty)$-level structures to $\Gamma(p^\infty)$-level structures, so that we have a natural action

$$\mathfrak{a} \star (A, t) = (\mathfrak{a} \star A, \pi_{\mathfrak{a}}(t)), \qquad \mathfrak{a} \star (A, t, \alpha) = (A/A[\mathfrak{a}], \pi_{\mathfrak{a}}(t), \pi_{\mathfrak{a}}(\alpha)).$$

Proposition 5.47. *Let*
$$\omega_{0,\mathfrak{a}} \in \Omega^1_{\mathfrak{a} \star A / K_p}$$

be the unique differential such that

$$\pi_{\mathfrak{a}}^* \omega_{0,\mathfrak{a}} = \omega_0$$

(viewing $\Omega^1_{A/\overline{\mathbb{Z}}} \subset \Omega^1_{A/\mathbb{C}_p}$ using i_p), and define, for any $\mathfrak{a}$ as in Definition 5.46,

$$\Omega_\infty(\mathfrak{a} \star (A, t)) \in \mathbb{C}^\times \quad and \quad \Omega_p(\mathfrak{a} \star (A, t, \alpha)) \in \mathcal{O}_\Delta(\mathfrak{a} \star (A, t, \alpha))$$

by

$$\Omega_\infty(\mathfrak{a} \star (A, t)) \cdot 2\pi i dz(\mathfrak{a} \star (A, t)) = \omega_{0,\mathfrak{a}},$$
$$\Omega_p(\mathfrak{a} \star (A, t, \alpha)) \cdot \omega_{\mathrm{can}}(\mathfrak{a} \star (A, t, \alpha)) = \omega_{0,\mathfrak{a}}. \qquad (5.61)$$

Then

$$\Omega_\infty(A, t) = \Omega_\infty(\mathfrak{a} \star (A, t)), \qquad \Omega_p(A, t, \alpha) = \Omega_p(\mathfrak{a} \star (A, t, \alpha)).$$

Proof. This follows immediately from (5.60) and (5.61), after noting that by definitions (in particular, the functoriality of the complex uniformization

$$\mathbb{C}/(\mathbb{Z} + \mathbb{Z}\tau) \cong A,$$

which defines

$$2\pi i dz \in \omega(\mathcal{H}^+),$$

and the functoriality of the construction of ω_{can} as defined in Definition 4.35), we have

$$\pi_\mathfrak{a}^* 2\pi i dz(\mathfrak{a} \star (A, t)) = 2\pi i dz(A, t), \qquad \pi_\mathfrak{a}^* \omega_{\mathrm{can}}(\mathfrak{a} \star (A, t, \alpha)) = \omega_{\mathrm{can}}(A, t, \alpha).$$

$\square$

Theorem 5.48. *Let $A/\overline{\mathbb{Q}}$ be an elliptic curve with CM by an order $\mathcal{O} \subset \mathcal{O}_K$, let $t \in A[N]$ any $\Gamma_1(N)$-level structure, t_p any Γ-level structure, and let*

$$\alpha : \mathbb{Z}_p^{\oplus 2} \xrightarrow{\sim} T_p A$$

be any p^∞-level structure such that $(A, t, t_p, \alpha) \in \mathcal{Y}$ lies above $(A, t, t_p) \in Y$ (recall $Y = Y(\Gamma_1(N) \cap \Gamma)$ parametrizes elliptic curves with $\Gamma_1(N) \cap \Gamma$-level structure), and fix $\omega_0 \in \Omega^1_{A/\overline{\mathbb{Z}}}$ as in Definition 5.45. Let $w \in \omega(Y)$, and write

$$w|_{\mathcal{H}^+} = F \cdot (2\pi i dz)^{\otimes k}$$

and

$$\omega|_{\lambda^{-1}(Y) \cap \mathcal{Y}_x} = f \cdot \omega_{\mathrm{can}}^{\otimes k}.$$

Then for any $j \in \mathbb{Z}$, we have the following coincidence of values (in $\overline{\mathbb{Q}}$):

$$\begin{aligned}
&i_p^{-1}(\theta(\Omega_p)(A, t, t_p, \alpha))^{-(k+2j)} \cdot \theta_k^j f(A, t, t_p, \alpha) \\
&= i_p^{-1}(\Omega_p(A, t, t_p, \alpha)^{-(k+2j)} \cdot \partial_{k,\Delta}^{\diamond, j} f(A, t, t_p, \alpha)) \\
&= i_p^{-1}\left(\theta(\Omega_p)(A, t, t_p, \alpha)^{-(k+2j)} \cdot \theta \circ \partial_{k,\Delta}^{\diamond, j} f(A, t, t_p, \alpha)\right) \\
&= i_\infty^{-1}(\Omega_\infty(A, t, t_p)^{-(k+2j)} \cdot \partial_k^j F(A, t, t_p)).
\end{aligned} \tag{5.62}$$

Proof. Let

$$\overline{\mathrm{split}_\mathbb{C}} : \mathcal{H}^1_{\mathrm{dR}}(A) \otimes_{\mathcal{O}^{\mathrm{hol}}} \mathcal{O}^{\mathrm{r.an.}}$$

denote the antiholomorphic splitting of complex relative de Rham cohomology. Using $2\pi idz$, view $\mathfrak{d}_k^j$ as an operator

$$\mathfrak{d}_k^j : (\omega \otimes_{\mathcal{O}_Y} \mathcal{O}_Y^{\mathrm{r.an.}})^{\otimes k} \to (\omega \otimes_{\mathcal{O}_Y} \mathcal{O}_Y^{\mathrm{r.an.}})^{\otimes k+2j}.$$

Since the Gauss-Manin connection is holomorphically horizontal, i.e., for any holomorphic differential ω,

$$\nabla_\omega(\omega') = 0$$

for any antiholomorphic ω', then we have

$$\mathfrak{d}_k^j = \overline{\mathrm{split}}_{\mathbb{C}} \circ \nabla^j.$$

This is analogous with the horizontalness of our p-adic Maass-Shimura operator

$$\partial_k^j = \overline{\mathrm{split}} \circ \nabla^j \circ \iota_{\mathrm{dR}}$$

as proven in Proposition 5.18.

We first show

$$i_p^{-1} \left(\theta \left((\overline{\mathrm{split}} \circ \nabla^j f)(A, t, t_p, \alpha) \right) \right) = i_p^{-1} \left(\left(\mathrm{split}_{\Delta}^{\dagger} \circ \nabla^j f \right)(A, t, t_p, \alpha) \right) \tag{5.63}$$
$$= i_\infty^{-1} \left((\overline{\mathrm{split}}_{\mathbb{C}} \circ \nabla^j f)(A, t, t_p) \right)$$

for any section

$$f \in \mathrm{Sym}^k \mathcal{H}_{\mathrm{dR}}^1(\mathcal{A})(U)$$

where U is as in the statement of the theorem, where (A, t, t_p, α) and (A, t, t_p) denote specialization to those points. First note that by functoriality of θ, we have

$$\theta \left((\overline{\mathrm{split}} \circ \nabla^j f)(A, t, t_p, \alpha) \right) = (\theta \circ \overline{\mathrm{split}} \circ \nabla^j f)(A, t, t_p, \alpha)$$

and so (5.63) is equivalent to

$$i_p^{-1} \left((\theta \circ \overline{\mathrm{split}} \circ \nabla^j f)(A, t, t_p, \alpha) \right) = i_p^{-1} \left(\left(\mathrm{split}_{\Delta}^{\dagger} \circ \nabla^j f \right)(A, t, t_p, \alpha) \right) \tag{5.64}$$
$$= i_\infty^{-1} \left((\overline{\mathrm{split}}_{\mathbb{C}} \circ \nabla^j f)(A, t, t_p) \right).$$

By the functoriality of $\theta \circ \overline{\mathrm{split}}$ and $\overline{\mathrm{split}}_{\mathbb{C}}$, for any section f' in

$$\mathrm{Sym}_{\mathcal{O}_{\Delta, y_x}}^{\otimes k+2j} \left(\mathcal{H}_{\mathrm{dR}}^1(\mathcal{A}) \otimes_{\mathcal{O}_Y} \mathcal{O}_{\Delta} \right)(\lambda^{-1}(U) \cap \mathcal{Y}_x)$$
$$\cap \mathrm{Sym}_{\mathcal{O}_{\Delta, y_x}}^{\otimes k+2j} \left(T_p \mathcal{A} \otimes_{\hat{\mathbb{Z}}_{p,Y}} \mathcal{O}_{\Delta} \right)(\lambda^{-1}(U) \cap \mathcal{Y}_x),$$

where the intersection is taken in

$$\mathrm{Sym}_{\mathcal{O}_{\Delta,y_x}}^{\otimes k+2j}\left(T_p\mathcal{A}\otimes_{\hat{\mathbb{Z}}_{p,Y}}\mathcal{O}_{\Delta,y_x}\cdot t^{-1}\right)(\lambda^{-1}(U)\cap\mathcal{Y}_x)$$

using ι_{dR}, letting $f'(A,t,t_p,\alpha)$ denote its specialization to (A,t,t_p,α), we have that

$$((\theta\circ\overline{\mathrm{split}})\,(f'))\,(A,t,t_p,\alpha)=(\theta\circ\overline{\mathrm{split}})\,(f'(A,t,t_p,\alpha)).$$

Similarly, we have for any section in

$$\mathrm{Sym}_{\mathcal{O}_{\Delta^\dagger}}^{\otimes k+2j}\left(\mathcal{H}_{\mathrm{dR}}^1(\mathcal{A})\otimes_{\mathcal{O}_Y}\mathcal{O}_\Delta^\dagger\right)(\lambda^{-1}(U)\cap\mathcal{Y}_x)$$

$$\cap\,\mathrm{Sym}_{\mathcal{O}_\Delta}^{\otimes k+2j}\left(T_p\mathcal{A}\otimes_{\hat{\mathbb{Z}}_{p,Y}}\mathcal{O}_\Delta^\dagger\right)(\lambda^{-1}(U)\cap\mathcal{Y}_x),$$

where the intersection is taken in

$$\mathrm{Sym}_{\mathcal{O}_\Delta^\dagger}^{\otimes k+2j}\left(T_p\mathcal{A}\otimes_{\hat{\mathbb{Z}}_{p,Y}}\mathcal{O}_{\Delta^\dagger}\cdot t^{-1}\right)(\lambda^{-1}(U)\cap\mathcal{Y}_x)$$

using ι_{dR}, we have that

$$\left(\mathrm{split}_\Delta^\dagger(f')\right)(A,t,t_p,\alpha)=\left(\mathrm{split}_\Delta^\dagger\right)(f'(A,t,t_p,\alpha)).$$

Finally, for any section

$$f\in\left(\mathcal{H}_{\mathrm{dR}}^1(\mathcal{A})\otimes_{\mathcal{O}^{\mathrm{hol}}}\mathcal{O}^{\mathrm{r.an.}}\right)(U),$$

letting $f(A,t,t_p)$ denote its specialization to (A,t,t_p), that

$$\left(\overline{\mathrm{split}}_{\mathbb{C}}(f')\right)(A,t,t_p)=\overline{\mathrm{split}}(f'(A,t,t_p)).$$

Thus taking $f'=\nabla^j f$, can rewrite (5.64) as

$$i_p^{-1}\left((\theta\circ\overline{\mathrm{split}})_{(A,t,t_p,\alpha)}\left((\nabla^j f)\,(A,t,t_p,\alpha)\right)\right)$$

$$=i_p^{-1}\left(\left(\mathrm{split}_\Delta^\dagger\right)_{(A,t,t_p,\alpha)}\left((\nabla^j f)\,(A,t,t_p,\alpha)\right)\right)\qquad(5.65)$$

$$=i_\infty^{-1}\left(\overline{\mathrm{split}}_{\mathbb{C},(A,t,t_p)}\left((\nabla^j f)\,(A,t,t_p)\right)\right)$$

where

$$\left(\theta\circ\overline{\mathrm{split}}\right)_{(A,t,t_p,\alpha)}\quad\text{and}\quad\left(\mathrm{split}_\Delta^\dagger\right)_{(A,t,t_p,\alpha)}\quad\text{and}\quad\overline{\mathrm{split}}_{\mathbb{C},(A,t,t_p)}$$

denote the specializations of the splittings

$$\theta \circ \overline{\text{split}} \quad \text{and} \quad \text{split}_\Delta^\dagger \quad \text{and} \quad \overline{\text{split}}_\mathbb{C}$$

to the points (A, t, t_p, α) and (A, t, t_p), respectively. Now by (5.7), we have

$$\theta \circ \overline{\text{split}} = \theta \circ \text{split}_\Delta^\dagger$$

where split is defined as in Definition 5.19. Hence (5.65) is in turn equivalent to

$$
\begin{aligned}
i_p^{-1} &\left(\left(\theta \circ \text{split}_\Delta^\dagger \right)_{(A,t,t_p,\alpha)} \left((\nabla^j f)(A, t, t_p, \alpha) \right) \right) \\
&= i_p^{-1} \left(\left(\text{split}_\Delta^\dagger \right)_{(A,t,t_p,\alpha)} \left((\nabla^j f)(A, t, t_p, \alpha) \right) \right) \qquad (5.66) \\
&= i_\infty^{-1} \left(\overline{\text{split}}_{\mathbb{C},(A,t,t_p)} \left((\nabla^j f)(A, t, t_p) \right) \right)
\end{aligned}
$$

where

$$\left(\theta \circ \text{split}_\Delta^\dagger \right)_{(A,t,t_p,\alpha)}$$

denotes the specialization of the splitting $\theta \circ \text{split}_\Delta^\dagger$ to the point (A, t, t_p, α). Since $\nabla^j f$ is a section defined over $U \to Y$, then

$$(\nabla^j f)(A, t, t_p) = (\nabla^j f)(A, t, t_p, \alpha);$$

moreover, we have

$$\nabla^j f(A, t, t_p) \in H_{\mathrm{dR}}^1(A/\overline{\mathbb{Q}})$$

since $(A, t, t_p) \in U(\overline{\mathbb{Q}})$ by the theory of complex multiplication. So now it suffices to show that

$$
\begin{aligned}
i_p^{-1} \left(\left(\theta \circ \text{split}_\Delta^\dagger \right)_{(A,t,t_p,\alpha)} \right) &= i_p^{-1} \left(\text{split}_\Delta^\dagger \right)_{(A,t,t_p,\alpha)} \\
&= i_\infty^{-1} \left(\overline{\text{split}}_{\mathbb{C},(A,t,t_p)} \right). \qquad (5.67)
\end{aligned}
$$

For this, we first note that the $\mathcal{O} = \mathrm{End}(A/\overline{\mathbb{Q}})$-action induces a splitting

$$H_{\mathrm{dR}}^1(A/\overline{\mathbb{Q}}) \cong \Omega_{A/\overline{\mathbb{Q}}}^1 \oplus H^{0,1}(A/\overline{\mathbb{Q}}) \qquad (5.68)$$

functorial in $\overline{\mathbb{Q}}$-algebras, where the first factor is the subspace on which $\mathcal{O}$ acts through multiplication, and the second factor is the subspace on which $\mathcal{O}$ acts through the complex conjugation of multiplication; by the compatibility of complex structures with the CM action, (5.68) induces, upon tensoring with $\otimes_{\overline{\mathbb{Q}}}\mathbb{C}$, the complex Hodge decomposition

$$\mathcal{H}^1_{\mathrm{dR}}(\mathcal{A})(A,t,t_p)=H^1_{\mathrm{dR}}(A/\mathbb{C})\cong\left(H^{1,0}(A)\otimes_{\mathbb{Z}}\mathbb{C}\right)\oplus\left(H^{0,1}(A)\otimes_{\mathbb{Z}}\mathbb{C}\right).\quad(5.69)$$

Hence the projection onto the first factor in (5.68) is just $\overline{\mathrm{split}}_{\mathbb{C},(A,t,t_p)}$, and projection onto the first factor in (5.68) is just $\overline{\mathrm{split}}_{\mathbb{C},(A,t,t_p)}$ restricted to

$$H^1_{\mathrm{dR}}(A/\overline{\mathbb{Q}})\subset\mathcal{H}^1_{\mathrm{dR}}(\mathcal{A})(A,t,t_p).$$

The CM action also induces a splitting

$$\left(T_p\mathcal{A}\otimes_{\hat{\mathbb{Z}}_{p,Y}}\mathcal{O}_\Delta\right)(A,t,t_p,\alpha)\cong\omega_\Delta(A,t,t_p,\alpha)\oplus\omega_\Delta^{-1}(1)(A,t,t_p,\alpha)\quad(5.70)$$

where the first factor is the subspace on which $\mathcal{O}$ acts through multiplication, and the second factor is the subspace on which $\mathcal{O}$ acts through the complex conjugation of multiplication.

Consider the action of $\mathcal{O}$ on the tangent space of A. Then the Hodge-Tate filtration

$$\omega_\Delta^{-1}(1)(A,t,t_p,\alpha)\subset\left(T_p\mathcal{A}\otimes_{\hat{\mathbb{Z}}_{p,Y}}\mathcal{O}_\Delta\right)(A,t,t_p,\alpha)$$

is the subspace on which $\mathrm{End}(A/\overline{\mathbb{Q}})$ acts through the complex conjugation of multiplication, which implies that the projection onto the first factor in (5.70) is given by $\left(\mathrm{split}_\Delta^\dagger\right)_{(A,t,t_p,\alpha)}$, and the projection onto the first factor in (5.68) is just the restriction of $\left(\mathrm{split}_\Delta^\dagger\right)_{(A,t,t_p,\alpha)}$ to

$$H^1_{\mathrm{dR}}(A/\overline{\mathbb{Q}})\subset\mathcal{H}^1_{\mathrm{dR}}(\mathcal{A})(A,t,t_p).$$

Hence in all, we have

$$i_p^{-1}\left(\left(\mathrm{split}_\Delta^\dagger\right)_{(A,t,t_p,\alpha)}\right)=i_\infty^{-1}\left(\overline{\mathrm{split}}_{\mathbb{C},(A,t,t_p)}\right).$$

To finish off showing (5.67), we need the following lemma.

Lemma 5.49. *We have*

$$\mathcal{O}_\Delta(A,t,t_p,\alpha)=\mathcal{O}_\Delta^\dagger(A,t,t_p,\alpha)$$

inside the ambient ring $\hat{\mathcal{O}}_\Delta(A,t,t_p,\alpha)$.

There is a unique embedding

$$\overline{\mathbb{Q}}_p \hookrightarrow \mathcal{O}_\Delta(A, t, t_p, \alpha),$$

and moreover the composition

$$\overline{\mathbb{Q}}_p \hookrightarrow \mathcal{O}_\Delta(A, t, t_p, \alpha) = \mathcal{O}_\Delta^\dagger(A, t, t_p, \alpha) \xrightarrow{\theta} \hat{\mathcal{O}}_\mathcal{Y}(A, t, t_p, \alpha) = \mathbb{C}_p$$

is the natural inclusion.

Proof. From [29, Theorem 1.1], since (A, t, t_p, α) is a CM point we have $\hat{\mathbf{z}}(A, t, t_p, \alpha) \in \overline{\mathbb{Q}}_p$. Since $\mathbf{z}_{dR}(A, t, t_p, \alpha)$ parametrizes the complex conjugate line (the Hodge filtration), and A is defined over $\overline{\mathbb{Q}}$, then $\mathbf{z}_{dR}(A, t, t_p, \alpha) \in \overline{\mathbb{Q}}_p$. Note also that

$$\mathbf{z}_{dR}(A, t, t_p, \alpha) - \hat{\mathbf{z}}(A, t, t_p, \alpha) \neq 0$$

since the Hodge-Tate and Hodge filtrations generate the $\mathcal{O}_\Delta^\dagger(A, t, t_p, \alpha)$-module $T_p A \otimes_{\mathbb{Z}_p} \mathcal{O}_\Delta^\dagger(A, t, t_p, \alpha)$ by (5.42), and the latter has rank 2 and so the Hodge-Tate and Hodge filtrations cannot be contained in the same line. So

$$\frac{1}{\mathbf{z}_{dR}(A, t, t_p, \alpha) - \hat{\mathbf{z}}(A, t, t_p, \alpha)}, \quad \frac{1}{\hat{\mathbf{z}}(A, t, t_p, \alpha)} \in \overline{\mathbb{Q}}_p. \tag{5.71}$$

So since

$$\mathcal{O}_\Delta^\dagger(A, t, t_p, \alpha) = \mathcal{O}_\Delta(A, t, t_p, \alpha) \left[\frac{1}{\mathbf{z}_{dR}(A, t, t_p, \alpha) - \hat{\mathbf{z}}(A, t, t_p, \alpha)}, \frac{1}{\hat{\mathbf{z}}(A, t, t_p, \alpha)} \right]$$

$$\subset \hat{\mathcal{O}}_\Delta(A, t, t_p, \alpha), \tag{5.72}$$

from (5.71) and (5.72), we have

$$\mathcal{O}_\Delta(A, t, t_p, \alpha) = \mathcal{O}_\Delta^\dagger(A, t, t_p, \alpha)$$

inside the ambient ring $\hat{\mathcal{O}}_\Delta(A, t, t_p, \alpha)$, which gives the first part of the lemma.

For the second part, note that the proétale stalk $\mathcal{O}_{\Delta, y}$ is a power series ring over the complete local ring $\mathcal{O}_{\mathcal{Y}, y}$ by Proposition 4.1 (reduced modulo t), and so in particular it is Henselian. Let

$$\mathfrak{m}_y \subset \mathcal{O}_{\mathcal{Y}, y}$$

be the maximal ideal of the local ring $\mathcal{O}_{\mathcal{Y}, y}$. Then

$$\mathcal{O}_\Delta(A, t, t_p, \alpha) = \mathcal{O}_{\Delta, y} \otimes_{\mathcal{O}_{\mathcal{Y}, y}} \mathcal{O}_{\mathcal{Y}, y} / \mathfrak{m}_y$$

is also a Henselian local ring, with maximal ideal given by $\ker \theta$ (generated by an indeterminate in the above power series description of $\mathcal{O}_{\Delta,y}$) and residue field $\mathbb{C}_p$. Here, recall that the map $\mathcal{O}_{\mathcal{Y},y} \to \mathcal{O}_{\Delta,y}$ in the above tensor product is just given by the natural embedding

$$\mathcal{O}_{\mathcal{Y},y} \subset \mathcal{O}_{\Delta,y} \subset \mathcal{O}_{\Delta,y}.$$

We can thus describe the maximal ideal $\mathfrak{m}_{\Delta,y}$ of the local ring $\mathcal{O}_{\Delta,y}$ as the ideal generated by $\ker \theta$ and the image of $\mathfrak{m}_y \subset \mathcal{O}_{\mathcal{Y},y}$ under the above embedding. Now since $\overline{\mathbb{Q}}_p \subset \mathbb{C}_p = \mathcal{O}_{\Delta,y}/\mathfrak{m}_{\Delta,y}$, by Hensel's lemma the inclusion

$$\mathbb{Q}_p \subset \mathcal{O}_\Delta(A, t, t_p, \alpha)$$

lifts *uniquely* to an inclusion

$$\overline{\mathbb{Q}}_p \subset \mathcal{O}_\Delta(A, t, t_p, \alpha)$$

whose composition with θ (the reduction map to the residue field) is the natural inclusion $\overline{\mathbb{Q}}_p \subset \mathbb{C}_p$.

Alternatively, we see from the above discussion that $\mathcal{O}_\Delta(A, t, t_p, \alpha)$ is a complete local ring, and in fact complete discrete valuation ring, with maximal ideal $\ker \theta$. Such a ring satisfies Hensel's lemma, and hence the separable extension $\overline{\mathbb{Q}}_p$ of $\mathbb{Q}_p$ lifts uniquely from the residue field $\mathcal{O}_\Delta(A, t, t_p, \alpha)/\ker \theta = \mathbb{C}_p$ to $\mathcal{O}_\Delta(A, t, t_p, \alpha)$. $\qquad\square$

Since $\mathrm{split}_{(A,t,t_p,\alpha)}$ is defined over $\overline{\mathbb{Q}}_p$ and the map

$$\overline{\mathbb{Q}}_p \subset \mathcal{O}_\Delta(A, t, t_p, \alpha) \subset \mathcal{O}_\Delta^\dagger(A, t, t_p, \alpha) \xrightarrow{\theta} \mathbb{C}_p$$

is just the natural inclusion, we have

$$\theta\left(i_p^{-1}\left(\left(\mathrm{split}_\Delta^\dagger\right)_{(A,t,t_p,\alpha)}\right)\right) = i_p^{-1}\left(\left(\mathrm{split}_\Delta^\dagger\right)_{(A,t,t_p,\alpha)}\right)$$

and so

$$i_p^{-1}\left(\left(\theta \circ \mathrm{split}_\Delta^\dagger\right)_{(A,t,t_p,\alpha)}\right) = \theta\left(i_p^{-1}\left(\left(\mathrm{split}_\Delta^\dagger\right)_{(A,t,t_p,\alpha)}\right)\right)$$

$$= i_p^{-1}\left(\left(\mathrm{split}_\Delta^\dagger\right)_{(A,t,t_p,\alpha)}\right) = i_\infty^{-1}\left(\overline{\mathrm{split}}_{\mathbb{C},(A,t,t_p)}\right)$$

which gives (5.67). So we have proven (5.63). Now from the definitions of Ω_p and Ω_∞ (taken with respect to the fixed $\omega_0 \in \Omega^1_{A/\overline{\mathbb{Z}}}$), we have the following equality of elements of $\overline{\mathbb{Q}}$.

$$
\begin{aligned}
i_p^{-1}(\theta(\Omega_p)(A, t, t_p, \alpha)^{-(k+2j)} \cdot \theta_k^j f(A, t, t_p, \alpha)) \\
= i_p^{-1}(\Omega_p(A, t, t_p, \alpha)^{-(k+2j)} \cdot \partial_{k,\Delta}^{\diamond,j} f(A, t, t_p, \alpha)) \\
= i_\infty^{-1}(\Omega_\infty(A, t, t_p)^{-(k+2j)} \cdot \mathfrak{d}_k^j F(A, t, t_p)).
\end{aligned}
\tag{5.73}
$$

So now (5.62) will follow from (5.73) once we establish

$$
\begin{aligned}
\theta(\Omega_p(A, t, t_p, \alpha))^{-(k+2j)} \cdot \theta \circ \partial_{k,\Delta}^{\diamond,j} f(A, t, t_p, \alpha) \\
= \Omega_p(A, t, t_p, \alpha)^{-(k+2j)} \cdot \partial_{k,\Delta}^{\diamond,j} f(A, t, t_p, \alpha).
\end{aligned}
\tag{5.74}
$$

However, note that from the above, we have

$$
\Omega_p(A, t, t_p, \alpha)^{-(k+2j)} \cdot \partial_{k,\Delta}^{\diamond,j} f(A, t, t_p, \alpha) \in \overline{\mathbb{Q}} \overset{\iota_p}{\hookrightarrow} \overline{\mathbb{Q}}_p \subset \mathcal{O}_\Delta(A, t, t_p, \alpha).
$$

Hence from Lemma (5.49), we have

$$
\begin{aligned}
\Omega_p(A, t, t_p, \alpha)^{-(k+2j)} \cdot \partial_{k,\Delta}^{\diamond,j} f(A, t, t_p, \alpha) \\
= \theta\left(\Omega_p(A, t, t_p, \alpha)^{-(k+2j)} \cdot \partial_{k,\Delta}^{\diamond,j} f(A, t, t_p, \alpha)\right) \\
= \theta(\Omega_p(A, t, t_p, \alpha))^{-(k+2j)} \cdot \theta \circ \partial_{k,\Delta}^{\diamond,j} f(A, t, t_p, \alpha)
\end{aligned}
$$

which gives (5.74). $\qquad\square$

5.10 COMPARISON OF ALGERAIC MAASS-SHIMURA DERIVATIVES ON DIFFERENT LEVELS

We conclude this chapter with a short discussion on the relation between values of Maass-Shimura derivatives at CM points on different levels $Y(\Gamma_1(N) \cap \Gamma)$ for varying $\Gamma \subset GL_2(\mathbb{Z}_p)$. Later, we will specifically be interested in $\Gamma = \Gamma_0(p^2)$, and relating Maass-Shimura derivatives at CM points on level $\Gamma_1(N) \cap \Gamma$ to values at CM points on level $\Gamma_1(N)$, the latter being related to Rankin-Selberg central L-values. Suppose we are given an algebraic modular form

$$
w \in \omega^{\otimes k}(Y(\Gamma_1(N))),
$$

which we can then pull back to

$$
w|_{Y(\Gamma_1(N) \cap \Gamma)} \in \omega^{\otimes k}(Y(\Gamma_1(N) \cap \Gamma)).
$$

Write as before

$$w|_{\mathcal{H}^+} = F \cdot (2\pi i dz)^{\otimes k},$$

and consider the values $\eth_k^j F(A, t)$ and $\eth_k^j F(A, t, t_p)$, where $(A, t) \in Y(\Gamma_1(N))$ is a CM point as before and $(A, t, t_p) \in Y(\Gamma_1(N) \cap \Gamma)$ lies above (A, t) along the natural projection

$$Y(\Gamma_1(N) \cap \Gamma) \to Y(\Gamma_1(N)).$$

The main theorem of this section is the following.

Theorem 5.50. *Let $\Gamma \subset GL_2(\mathbb{Z}_p)$ be a congruence subgroup (i.e., of the form $\Gamma_0(p^m), \Gamma_1(p^m)$ or $\Gamma(p^m)$). Let $(A, t, t_p) \in Y(\Gamma_1(N) \cap \Gamma)$ be a CM point lying above a CM point $(A, t) \in Y(\Gamma_1(N))$ and let w be as above, and let $\mathfrak{a}$ be a fractional ideal of $\mathcal{O}_K$ prime to pN, so that we get CM points $\mathfrak{a} \star (A, t, t_p) \in Y(\Gamma_1(N) \cap \Gamma)$ and $\mathfrak{a} \star (A, t) \in Y(\Gamma_1(N))$. There is an algebraic number*

$$\Omega_\Gamma(A, t) \in \overline{\mathbb{Q}}^\times, \tag{5.75}$$

independent of w, such that for all $j \geq 0$,

$$\eth_k^j F(\mathfrak{a} \star (A, t)) = \Omega_\Gamma^j \cdot \eth_k^j F(\mathfrak{a} \star (A, t, t_p)). \tag{5.76}$$

Proof. The proof is the same for (A, t, t_p) and (A, t) as it is for $\mathfrak{a} \star (A, t, t_p)$ and $\mathfrak{a} \star (A, t)$, so we only write out the former case. Let us introduce some shorthand notation for this proof: let $Y_\Gamma = Y(\Gamma_1(N) \cap \Gamma)$, $\mathcal{A}_\Gamma \to Y_\Gamma$ and $\mathcal{A} \to Y(\Gamma_1(N))$ denote the universal objects, with associated Hodge bundles ω_Γ and ω. Let $\omega_0 \in \Omega^1_{A/\overline{\mathbb{Z}}}$ be as in Theorem 5.48. From [33, (2.3.15) and (2.4.8)], we see that $\eth_k^j F(A, t, t_p)$ is computed as

$$\eth_k^j F(A, t, t_p) = \Omega_\infty^{k+2j}(A, t) \cdot \frac{\mathrm{split}(A, t, t_p) \circ \nabla^j(w)}{\omega_0^{\otimes k+2j}} \tag{5.77}$$

where

$$\mathrm{split}(A, t, t_p) : \mathrm{Sym}^{\otimes k+2j} \mathcal{H}^1_{\mathrm{dR}}(\mathcal{A}_\Gamma)(A, t, t_p) \to \omega_{Y_\Gamma}^{\otimes k+2j}(A, t, t_p)$$

denotes the CM splitting of the Hodge filtration, and ∇^j is the j-fold composition of

$$\nabla : \mathrm{Sym}^{\otimes k} \mathcal{H}^1_{\mathrm{dR}}(\mathcal{A}_\Gamma) \to \mathrm{Sym}^{\otimes k} \mathcal{H}^1_{\mathrm{dR}}(\mathcal{A}_\Gamma) \otimes_{\mathcal{O}_{Y_\Gamma}} \Omega_{Y_\Gamma} \overset{\sigma_\Gamma}{\cong} \mathrm{Sym}^{\otimes k} \mathcal{H}^1_{\mathrm{dR}}(\mathcal{A}_\Gamma) \otimes_{\mathcal{O}_{Y_\Gamma}} \omega_{Y_\Gamma}^{\otimes 2}$$
$$\subset \mathrm{Sym}^{\otimes k+2} \mathcal{H}^1_{\mathrm{dR}}(\mathcal{A}_\Gamma),$$

where $\sigma_\Gamma : \Omega^1_{Y_\Gamma} \cong \omega_\Gamma^{\otimes 2}$ is the Kodaira-Spencer isomorphism for $\mathcal{A}_\Gamma \to Y_\Gamma$. Here, ω_0 is viewed as a generator of the fiber $\omega_\Gamma(A, t, t_p)$, and hence $\sigma_\Gamma^{-1}(\omega_0^{\otimes 2})$ is viewed as a generator of $\Omega^1_{Y_\Gamma}(A, t, t_p)$.

Similarly

$$\mathfrak{d}_k^j F(A, t) = \Omega_\infty^{k+2j}(A, t) \cdot \frac{\mathrm{split}(A, t) \circ \nabla^j(w)}{\omega_0^{\otimes k+2j}} \tag{5.78}$$

for

$$\mathrm{split}(A, t) : \mathrm{Sym}^{\otimes k+2j} \mathcal{H}^1_{\mathrm{dR}}(\mathcal{A})(A, t) \to \omega_{Y(\Gamma_1(N))}^{\otimes 2j}(A, t)$$

the CM splitting and defined using the Kodaira-Spencer isomorphism $\sigma : \Omega^1_{Y_1(N)} \cong \omega^{\otimes 2}$ is the Kodaira-Spencer isomorphism for $\mathcal{A} \to Y_1(N)$, and ω_0 is viewed as a generator of the fiber $\omega(A, t)$, and hence $\sigma^{-1}(\omega_0^{\otimes 2})$ is viewed as a generator of $\Omega^1_{Y_1(N)}(A, t)$.

The natural projection $Y_\Gamma \to Y_1(N)$ induces a map $\Omega^1_{Y_1(N)}(A, t) \to \Omega^1_{Y_\Gamma}(A, t, t_p)$, and we still denote by $e_{\Gamma_1(N)}$ the image of $e_{\Gamma_1(N)}$ under this map. Now letting

$$\Omega_\Gamma(A, t) := \sigma_\Gamma^{-1}(\omega_0^{\otimes 2})/\sigma^{-1}(\omega_0^{\otimes 2}),$$

we have $\Omega_\Gamma \in \overline{\mathbb{Q}}^\times$, since (A, t, t_p) and (A, t) are $\overline{\mathbb{Q}}$-points of their respective modular curves. Now (5.76) follows from (5.77) and (5.78). $\qquad\square$

Chapter Six

---

p-adic analysis of the p-adic Maass-Shimura operators

In this section, we analyze p-adic analytic properties of d_k^j, θ_k^j and $\delta_{k,\Delta}^{\diamond,j}$, in particular how $\theta_k^j F$ and $\delta_{k,\Delta}^{\diamond,j}$ behave as a function of j for elements $F \in \mathcal{O}_{\Delta,\mathcal{Y},y}$ in the stalk at a geometric point $y \in \mathcal{Y}_x(\mathbb{C}_p, \mathcal{O}_{\mathbb{C}_p})$ of the period ring $\mathcal{O}_\Delta$, which we view as the space of "germs at y of nearly holomorphic functions." Crucial to this study will be the "q_{dR}-expansion map"

$$\mathcal{O}_{\mathcal{Y}} \overset{q_{\mathrm{dR}}-\exp}{\hookrightarrow} \hat{\mathcal{O}}_{\mathcal{Y}} \llbracket q_{\mathrm{dR}} - 1 \rrbracket \subset \mathcal{O}_\Delta,$$

which on the supersingular locus factors the natural inclusion

$$\mathcal{O}_{\mathcal{Y}^{\mathrm{ss}}} \subset \mathcal{O}_{\Delta,\mathcal{Y}^{\mathrm{ss}}}$$

(for a precise statement, see Theorem 6.7). The q_{dR}-expansion map can be viewed naturally as analogue (or an extension) of the Serre-Tate T-expansion map, and similarly is injective and so satisfies a "q_{dR}-expansion principle." Indeed, on the natural cover $\mathcal{Y}^{\mathrm{Ig}}$ of Y^{ord}, the q_{dR}-expansion map recovers the Serre-Tate T-expansion (Theorem 6.8). The coordinate descriptions (5.25) and (5.28) allow us to compute the action of the p-adic Maass-Shimura operators d_k^j and θ_k^j on q_{dR}-expansions, and hence study their p-adic analytic properties.

Understanding the analytic properties of $\theta_k^j F$ will be important in Chapter 8 for constructing our p-adic L-function and establishing our p-adic Waldspurger formula.

6.1 q_{dR}-EXPANSIONS

We retain the notation of the previous sections and let $\mathcal{U}$ denote an affinoid subdomain of $\mathcal{Y}$. Let F be a complete nonarchimedean field with ring of integers $\mathcal{O}_F$. Recall that for $(F, \mathcal{O}_F) = (W(\overline{\mathbb{F}}_p)[1/p], W(\overline{\mathbb{F}}_p))$, the Grothendieck-Messing period map is a surjective étale map of adic spaces over $\mathrm{Spa}(F, \mathcal{O}_F)$

$$\pi_{\mathrm{GM}} : LT_\infty \to \mathbb{P}^1$$

where $LT_\infty \in Y_{\text{proét}}$ is the infinite-level Lubin-Tate tower. As Y^{ss} is covered by a finite disjoint union of LT_0's by the Rapoport-Zink uniformization theorem ([52, Theorem 6.30]), henceforth, let LT denote this finite union, so that for some i, we have surjective proétale (in fact profinite-étale) map

$$LT_i \to Y^{\text{ss}}$$

with $LT_i \in Y_{\text{proét}}$, and a surjective étale map

$$\pi_{\text{GM}} : LT_i \to \mathbb{P}^1$$

which induces a map

$$\pi_{\text{GM}} : LT_\infty \to LT_i \to \mathbb{P}^1.$$

For brevity, we denote by

$$\mathcal{LT} := LT_\infty$$

throughout this chapter.

Now let $y' \in LT(\overline{F}, \mathcal{O}_{\overline{F}})$ be any geometric point, which under the natural proétale maps

$$\lambda : LT_i \to Y^{\text{ss}} \quad \text{and} \quad \pi_{\text{GM}} : LT_i \to \mathbb{P}^1$$

induces geometric points $y := \lambda(y') \in Y(\overline{F}, \mathcal{O}_{\overline{F}})$ and $\pi_{\text{GM}}(y') \in \mathbb{P}^1(\overline{F}, \mathcal{O}_{\overline{F}})$. Since π_{GM} is proétale, given any sheaf $\mathcal{F}$ on $Y_{\text{proét}}$ and $\mathcal{G}$ on $\mathbb{P}^1_{\text{proét}}$, we have an identification of proétale stalks

$$\mathcal{F}_{y'} = \mathcal{F}_y, \qquad \mathcal{G}_{y'} = \mathcal{G}_{\pi_{\text{GM}}(y')}.$$

If further $\mathcal{F}|_{\pi_{\text{GM}}^{-1}(\mathcal{V})} = \pi_{\text{GM}}^* \mathcal{G}|_{\mathcal{V}}$ for any proétale neighborhood $\mathcal{V} \to \mathbb{P}^1$ of $\pi_{\text{GM}}(y')$, then

$$\mathcal{F}_{y'} = \mathcal{F}_y = \mathcal{G}_{\pi_{\text{GM}}(y')}. \tag{6.1}$$

Since π_{GM} is étale, by the first fundamental exact sequence for Ω^1 ([30, Proposition 1.6.3, 1.6.8]), we have

$$\Omega^1_{LT_i} = \pi_{\text{GM}}^* \Omega^1_{\mathbb{P}^1, \pi_{\text{GM}}(y')} \tag{6.2}$$

and so by (6.1), we have

$$\Omega^1_{Y,y} = \Omega^1_{LT_i, y'} = \Omega^1_{\mathbb{P}^1, \pi_{\text{GM}}(y')}. \tag{6.3}$$

Let $Y^+/\mathbb{Z}_p$ denote the adicification of the Katz-Mazur integral model for the algebraic modular curve giving rise to Y, and the exterior differential maps

$$d : \mathcal{O}_{Y^+} \to \Omega^1_{Y^+} \tag{6.4}$$

which induces a map on any stalk

$$d : \mathcal{O}_{Y^+,x} \to \Omega^1_{Y^+,x}. \tag{6.5}$$

Hence we have

$$\mathcal{O}_{Y,y} = \mathcal{O}_{LT_i,y'} = \mathcal{O}_{\mathbb{P}^1,\pi_{\mathrm{GM}}(y')}$$

and

$$\mathcal{O}^+_{Y,y} = \mathcal{O}^+_{LT_i,y'} = \mathcal{O}^+_{\mathbb{P}^1,\pi_{\mathrm{GM}}(y')}. \tag{6.6}$$

We have, from Theorem 2.53,

$$\begin{aligned}
\mathcal{O}_{Y^+,y} &= \mathcal{O}^+_{Y,y} = \mathcal{O}^+_{LT_i,y'} = \mathcal{O}^+_{\mathbb{P}^1,\pi_{\mathrm{GM}}(y')} \\
&= \{ f \in \mathcal{O}_{Y,y} = \mathcal{O}_{LT_i,y'} = \mathcal{O}_{\mathbb{P}^1,\pi_{\mathrm{GM}}(y')} : |f(y')| \le 1 \}
\end{aligned} \tag{6.7}$$

and

$$\begin{aligned}
\hat{\mathcal{O}}_{Y^+,y} &= \hat{\mathcal{O}}^+_{Y,y} = \hat{\mathcal{O}}^+_{LT_i,y'} = \hat{\mathcal{O}}^+_{\mathbb{P}^1,\pi_{\mathrm{GM}}(y')} \\
&= \{ f \in \hat{\mathcal{O}}_{Y,y} = \hat{\mathcal{O}}_{LT_i,y'} = \hat{\mathcal{O}}_{\mathbb{P}^1,\pi_{\mathrm{GM}}(y')} : |f(y')| \le 1 \}
\end{aligned} \tag{6.8}$$

where

$$f(y') := f \pmod{\mathfrak{p}_{y'}},$$

and where $\mathfrak{p}_{y'}$ is the prime ideal corresponding to the equivalence class of valuations associated with y'.

Definition 6.1. *Henceforth, let*

$$LT' = \{ y \in LT_i : \pi_{\mathrm{GM}}(y) \ne \infty \} = \{ y \in LT_i : \mathbf{z}_{\mathrm{LT}}(y) \ne \infty \}.$$

Recall that π_{GM} is not quite canonical, but depends on a choice of trivialization of the Dieudonné module $M(H)$ associated with each supersingular residue disc, where H is the p-divisible group over $\overline{\mathbb{F}}_p$ of the underlying point in $Y^+(\overline{\mathbb{F}}_p)$ (see Definition 3.12). In particular, choosing different trivialization results in changing $\mathbf{z}_{\mathrm{LT}}$ by an element of PGL_2. Hence, given any point $y \in Y^{\mathrm{ss}}(\mathbb{C}_p, \mathcal{O}_{\mathbb{C}_p})$, we can choose a trivialization $M(H) \cong \mathcal{O}_F^{\oplus 2}$ on the residue disc containing y so that $\mathbf{z}_{\mathrm{LT}}(y) \ne \infty$. Since the fibers of π_{GM} correspond exactly to isogeny classes of deformations, this also ensures that each y' in the isogeny class of y also has $\mathbf{z}_{\mathrm{LT}}(y') \ne \infty$. We further specify that the trivializations

$$i_H : M(H) \cong \mathcal{O}_F^{\oplus 2}$$

for each H (corresponding to a supersingular residue disc) be Frobenius-compatible in the following sense: letting $\varphi : \mathcal{O}_F \to \mathcal{O}_F$ denote the p-power Frobenius *automorphism, under the canonical identification*

$$M(H^{(p)}) = M(H) \otimes_{\mathcal{O}_F, \varphi} \mathcal{O}_F,$$

we have

$$i_{H^{(p)}} = i_H \otimes \varphi.$$

Moreover, let

$$\mathcal{LT}_0 = \mathcal{LT} \times_{LT_i} LT',$$

and let

$$\bar{z}_{\mathrm{LT}} = j(\mathbf{z}_{\mathrm{LT}}) \in \mathcal{O}_{\Delta, \mathcal{LT}}(\mathcal{LT})$$

where j is defined as in Definition 5.4.

Letting z denote the affine coordinate on $\mathbb{A}^1 \subset \mathbb{P}^1$ with $\mathbf{z}_{\mathrm{LT}} = \pi_{\mathrm{GM}}^* z$, we have the closed disc of radius $|z(\pi_{\mathrm{GM}}(y'))|$

$$D_{|z(\pi_{\mathrm{GM}}(y'))|} = \mathrm{Spa}(F\langle z/z(\pi_{\mathrm{GM}}(y'))\rangle, \mathcal{O}_F\langle z/z(\pi_{\mathrm{GM}}(y'))\rangle) \subset \mathbb{A}^1,$$

and a generator of its differentials

$$\frac{dz}{z(\pi_{\mathrm{GM}}(y'))} \in \Omega^1_{\mathbb{P}^1}(\mathrm{Spa}(F\langle z/z(\pi_{\mathrm{GM}}(y'))\rangle, \mathcal{O}_F\langle z/z(\pi_{\mathrm{GM}}(y'))\rangle))$$

$$= \frac{dz}{z(\pi_{\mathrm{GM}}(y'))} F\langle z/z(\pi_{\mathrm{GM}}(y'))\rangle.$$

Note that

$$\frac{1}{n!}\left(\frac{z(\pi_{\mathrm{GM}}(y'))d}{dz}\right)^n : \mathcal{O}_F\langle z/z(\pi_{\mathrm{GM}}(y'))\rangle \to \mathcal{O}_F\langle z/z(\pi_{\mathrm{GM}}(y'))\rangle,$$

i.e., the differential operator $\frac{1}{n!}\left(\frac{dz}{z(\pi_{\mathrm{GM}}(y'))}\right)^n$ preserves integrality of functions on this disc.

Recall $d\mathbf{z}_{\mathrm{LT}} = \pi_{\mathrm{GM}}^* dz$. Let

$$\mathcal{D}_{|\mathbf{z}_{\mathrm{LT}}(y')|} \subset \mathcal{LT}$$

denote the preimage of $D_{|z(\pi_{\mathrm{GM}}(y'))|} \subset \mathbb{P}^1$ under $\pi_{\mathrm{GM}} : \mathcal{LT} \to \mathbb{P}^1$. So assuming $y' \in LT'$ (i.e., $\pi_{\mathrm{GM}}(y') \neq 0$), by the étaleness of π_{GM} (Proposition 2.31 applied in the limit to the proétale presentation $\mathcal{O}_Y^+(\mathcal{LT}) = \varinjlim \mathcal{O}_Y^+(LT_i)$, noting that all the LT_i are indeed strongly noetherian (Huber) adic spaces), we have

$$\frac{1}{n!}\left(\frac{\mathbf{z}_{\mathrm{LT}}(y')d}{d\mathbf{z}_{\mathrm{LT}}}\right)^n : \mathcal{O}^+_{LT}(\mathcal{D}) \to \mathcal{O}^+_{LT}(\mathcal{D}). \tag{6.9}$$

Proposition 6.2. *We have a natural equality of sheaves on $LT_{proét}$:*

$$\mathcal{O}_{\Delta,\mathcal{L}\mathcal{T}_0} = \hat{\mathcal{O}}_{\mathcal{L}\mathcal{T}_0}[\![\mathbf{z}_{\mathrm{LT}} - \bar{z}_{\mathrm{LT}}]\!],$$

which is compatible with connections (given the natural connection on $\mathcal{O}_{\Delta,\mathcal{L}\mathcal{T}_0}$ and the connection $\frac{d}{d\mathbf{z}_{\mathrm{LT}}} \otimes d\mathbf{z}_{\mathrm{LT}}$ on the right-hand side).

Proof. We have a natural inclusion

$$\hat{\mathcal{O}}_{\mathcal{L}\mathcal{T}_0}[\![\mathbf{z}_{\mathrm{LT}} - \bar{z}_{\mathrm{LT}}]\!] \subset \mathcal{O}_{\Delta,\mathcal{L}\mathcal{T}_0} \tag{6.10}$$

where $\hat{\mathcal{O}}_{\mathcal{L}\mathcal{T}_0} \xrightarrow{j} \mathcal{O}_{\Delta,\mathcal{L}\mathcal{T}_0}$ where j is defined as in (5.4), and $\mathbf{z}_{\mathrm{LT}} - \bar{z}_{\mathrm{LT}} \mapsto \mathbf{z}_{\mathrm{LT}} - \bar{z}_{\mathrm{LT}}$.

To construct the inverse of (6.10), we again follow the argument of ([57, Proposition 6.10]). We claim it suffices to show that there is a unique map

$$\mathcal{O}_{\mathcal{L}\mathcal{T}_0} \hookrightarrow \hat{\mathcal{O}}_{\mathcal{L}\mathcal{T}_0}[\![\mathbf{z}_{\mathrm{LT}} - \bar{z}_{\mathrm{LT}}]\!] \tag{6.11}$$

sending

$$\mathbf{z}_{\mathrm{LT}} \mapsto \bar{z}_{\mathrm{LT}} + (\mathbf{z}_{\mathrm{LT}} - \bar{z}_{\mathrm{LT}})$$

which is compatible with connections, and such that the composition with the map

$$\theta : \hat{\mathcal{O}}_{\mathcal{L}\mathcal{T}_0}[\![\mathbf{z}_{\mathrm{LT}} - \bar{z}_{\mathrm{LT}}]\!] \to \hat{\mathcal{O}}_{\mathcal{L}\mathcal{T}_0}$$

is the natural inclusion. Given the existence of (6.11), we have a natural map

$$\mathcal{O}_{\mathcal{L}\mathcal{T}_0} \otimes_{W(\kappa)} \hat{\mathcal{O}}_{\mathcal{L}\mathcal{T}_0} \to \hat{\mathcal{O}}_{\mathcal{L}\mathcal{T}_0}[\![\mathbf{z}_{\mathrm{LT}} - \bar{z}_{\mathrm{LT}}]\!]$$

by extending $\otimes_{W(\kappa)}\hat{\mathcal{O}}_{\mathcal{L}\mathcal{T}_0}$-linearly, which is easily checked to induce the inverse of (6.10) on the $\ker \theta$-adic completion.

To prove the existence of (6.11), note that the Grothendieck-Messing period map restricts to an étale map

$$\pi_{\mathrm{GM}} : LT' \to \mathbb{A}^1 = \bigcup_{r>0} \mathrm{Spa}(F\langle p^{-r}z\rangle, \mathcal{O}_F\langle p^{-r}z\rangle)$$

where $\pi^*_{\mathrm{GM}}z = \mathbf{z}_{\mathrm{LT}}$. This induces a proétale map

$$\mathcal{L}\mathcal{T} \to \mathbb{A}^1.$$

Now note that we have a map

$$\bigcap_{r>0} W(\kappa)[1/p]\langle p^{-r}z\rangle \to \hat{\mathcal{O}}_{LT_x}[\![\mathbf{z}_{\mathrm{LT}} - \bar{z}_{\mathrm{LT}}]\!]$$

given by $z \mapsto \bar{z}_{\mathrm{I,T}} + (\mathbf{z}_{\mathrm{LT}} - \bar{z}_{\mathrm{LT}})$. Now by the same argument as in Claim 4.2 using Hensel's lemma (*mutatis mutandis*), this extends uniquely along the proétale cover $\mathcal{LT}_0 \to \mathbb{A}^1$ to a homomorphism

$$\mathcal{O}_{\mathcal{LT}_0} \to \hat{\mathcal{O}}_{LT'}[\![\mathbf{z}_{\mathrm{LT}} - \bar{z}_{\mathrm{LT}}]\!]$$

such that the composition

$$\mathcal{O}_{\mathcal{LT}_0} \dashrightarrow \hat{\mathcal{O}}_{\mathcal{LT}_0}[\![\mathbf{z}_{\mathrm{LT}} - \bar{z}_{\mathrm{I,T}}]\!] \xrightarrow{\theta} \hat{\mathcal{O}}_{\mathcal{LT}_0}$$

is the natural inclusion. So we have shown (0.11), and so we are done.   □

We now switch notations slightly, letting y denote a point on $\mathcal{LT} \cup Y^{\mathrm{aa}}$, rather than a point on Y^{aa}.

Proposition 6.3. *Assume* $y \in \mathcal{LT}_0$, *so that* $\bar{z}_{\mathrm{I,T}}(y) \neq 0$ *and hence* $\bar{z}_{\mathrm{LT},y} \neq 0 \in \hat{\mathcal{O}}_{LT,y}$. *The natural map induced by Proposition 6.2*

$$\mathcal{O}^+_{\mathcal{LT}}(\mathcal{LT}) \to \mathcal{O}^+_{\mathcal{LT},y} \subset \mathcal{O}_{\Delta,\mathcal{LT},y} = \hat{\mathcal{O}}_{LT,y}[\![\mathbf{z}_{\mathrm{LT}} - \bar{z}_{\mathrm{LT}}]\!] = \hat{\mathcal{O}}_{LT,y}[\![\mathbf{z}_{\mathrm{LT}}/\bar{z}_{\mathrm{LT}} - 1]\!] \tag{6.12}$$

factors through

$$\mathcal{O}^+_{\mathcal{LT}}(\mathcal{LT}) \to \hat{\mathcal{O}}^+_{LT,y}[\![\mathbf{z}_{\mathrm{LT}}/\bar{z}_{\mathrm{LT}} - 1]\!]. \tag{6.13}$$

Proof. Given $f \in \mathcal{O}^+_{\mathcal{Y},y}$, let

$$f = \sum_{n=0}^{\infty} a_n(f)(\mathbf{z}_{\mathrm{LT}}/\bar{z}_{\mathrm{LT}} - 1)^n, \quad a_n(f) \in \hat{\mathcal{O}}_{LT,y}$$

denote its image under the map (6.12). We have a commutative diagram

$$\begin{array}{ccc}
\mathcal{O}_{LT} & \xrightarrow{\nabla} & \Omega^1_{LT} \\
\downarrow & & \downarrow \\
\mathcal{O}\mathbb{B}^+_{\mathrm{dR},LT} & \xrightarrow{\nabla} & \mathcal{O}\mathbb{B}^+_{\mathrm{dR},LT} \otimes_{\mathcal{O}_{LT}} \Omega^1_{LT}
\end{array} \tag{6.14}$$

which, reducing modulo (t), trivializing $\Omega^1_{LT,y} \cong \mathcal{O}_{LT,y}$ using the section dq^{p^a}, implies that

$$\begin{array}{ccc}
\mathcal{O}_{LT,y} & \xrightarrow{\ \frac{d}{d(\mathbf{z}_{LT}/\bar{z}_{LT})}\ } & \mathcal{O}_{LT,y} \\
\downarrow & & \downarrow \\
\hat{\mathcal{O}}_{LT,y}[\![\mathbf{z}_{LT}/\bar{z}_{LT}-1]\!] & \xRightarrow{\ \frac{d}{d(\mathbf{z}_{LT}/\bar{z}_{LT})}\ } & \hat{\mathcal{O}}_{\mathcal{L}\mathcal{T},y}[\![\mathbf{z}_{LT}/\bar{z}_{LT}-1]\!]
\end{array} \tag{6.15}$$

is commutative. Note that since $\mathcal{O}_{LT}\subset\mathcal{O}\mathbb{B}^+_{\mathrm{dR},LT}\xrightarrow{\theta}\hat{\mathcal{O}}_{LT}$ is the natural inclusion, then (6.15) implies that

$$a_n(f)=\theta\left(\frac{1}{n!}\left(\frac{d}{d(\mathbf{z}_{LT}/\bar{z}_{LT})}\right)^n f\right).$$

Now using the fact that $\theta(\bar{z}_{LT})=\theta(\mathbf{z}_{LT})$ and $d\bar{z}=0$, we have

$$a_n(f)(y)=\theta\left(\frac{1}{n!}\left(\frac{d}{d(\mathbf{z}_{LT}/\bar{z}_{LT})}\right)^n f\right)(y)=\theta\left(\frac{\mathbf{z}_{LT}(y)^n}{n!}\left(\frac{d}{d\mathbf{z}_{LT}}\right)^n f\right)\overset{(6.9)}{\in}\hat{\mathcal{O}}^+_{LT,y},$$

which gives (6.13). $\hspace{1cm}\square$

Definition 6.4. *Let $y\in\mathcal{L}\mathcal{T}_0$ be as above. Now let*

$$\bar{z}_{\mathrm{dR}}=j\circ\theta(z_{\mathrm{dR}})$$

where j is defined in Definition 5.4. Let

$$q_{\mathrm{dR}}:=\exp(z_{\mathrm{dR}}-\bar{z}_{\mathrm{dR}})\in\mathcal{O}_\Delta(\mathcal{Y}_x)$$

where the fact that q_{dR} defines a section on all of $\mathcal{Y}_x$ follows from Proposition 4.34. We also denote

$$dq_{\mathrm{dR}}:=\nabla(q_{\mathrm{dR}}).$$

Note that

$$dq_{\mathrm{dR}}=q_{\mathrm{dR}}dz_{\mathrm{dR}}.$$

Furthermore, for any $b\in\mathbb{Q}$ (and a choice of an element $p^b\in\mathbb{C}_p$ with p-adic absolute value $|p|^b=(1/p)^b$), we can define

$$q_{\mathrm{dR}}^{1/p^b}:=\exp\left(\frac{z_{\mathrm{dR}}-\bar{z}_{\mathrm{dR}}}{p^b}\right).$$

Note that $\theta(z_{dR} - \bar{z}_{dR}) = 0$, and so

$$\theta(q_{dR}^{1/p^b}) = 1.$$

Lemma 6.5. *We have*

$$y_{dR}^2 \cdot \frac{dz_{dR}}{d(\mathbf{z}_{LT}/\bar{\mathbf{z}}_{LT})} \in \bar{z}_{LT}\hat{\mathcal{O}}_{\mathcal{LT}}^{\times}(\mathcal{LT}). \tag{6.16}$$

Proof. By the calculation (4.52) and the diagram (4.42), we have

$$\sigma(\mathfrak{s}^{\otimes 2}) = y_{dR}^2 \cdot \frac{dz_{dR}}{d\mathbf{z}_{LT}} \otimes d\mathbf{z}_{LT}. \tag{6.17}$$

Then since $\mathfrak{s}$ trivializes $\hat{\omega}|_{\mathcal{Y}^{ss}} \cong \hat{\mathcal{O}}_{\mathcal{Y}^{ss}}$, $\mathcal{LT} \subset \mathcal{Y}^{ss}$ and $d(\mathbf{z}_{LT})$ trivializes $\Omega_{\mathcal{Y}}^1 \cong \mathcal{O}_{\mathcal{Y}}$, we have that (6.16) holds.   □

Definition 6.6. *Let $b \in \mathbb{Q}$. We define the "q_{dR}^{1/p^b}-expansion map" as*

$$q_{dR}^{1/p^b} - \exp : \mathcal{O}_{\Delta, \mathcal{Y}_x} \to \hat{\mathcal{O}}_{\mathcal{Y}_x}[\![q_{dR}^{1/p^b} - 1]\!] \tag{6.18}$$

given by

$$f \mapsto f(q_{dR}^{1/p^b}) := \sum_{n=0}^{\infty} \theta\left(\frac{1}{n!}\left(\frac{d}{dq_{dR}^{1/p^b}}\right)^n f\right)(q_{dR}^{1/p^b} - 1)^n.$$

Note that the restriction

$$q_{dR}^{1/p^b} - \exp : \mathcal{O}_{\mathcal{Y}_x} \to \hat{\mathcal{O}}_{\mathcal{Y}_x}[\![q_{dR}^{1/p^b} - 1]\!]$$

is an injection, since $\theta \circ q_{dR}^{1/p^b} - \exp$ is the natural inclusion $\mathcal{O}_{\mathcal{Y}^{ss}} \subset \hat{\mathcal{O}}_{\mathcal{Y}^{ss}}$, and in particular injective.

Note that we have a natural inclusion

$$\hat{\mathcal{O}}_{\mathcal{Y}_x}[\![q_{dR}^{1/p^b} - 1]\!] \subset \mathcal{O}_{\Delta, \mathcal{Y}_x} \tag{6.19}$$

where

$$\hat{\mathcal{O}}_{\mathcal{Y}_x} \subset \mathcal{O}_{\Delta, \mathcal{Y}_x}$$

via (5.4), and $q_{dR}^{1/p^b} - 1$ is naturally an element of $\mathcal{O}_{\Delta, \mathcal{Y}_x}$ (and in fact is in $\ker \theta$) by (6.4), since dz_{dR} generates $\Omega_{\Delta, \mathcal{Y}_x}^1$ by Theorem 4.36.

One checks that by construction, the map (6.18) commutes with connections, where the connection on $\mathcal{O}_{\mathcal{Y}_x}$ is the natural one defined earlier, and the connection on $\hat{\mathcal{O}}_{\mathcal{Y}_x}[\![q_{\mathrm{dR}}^{1/p^b} - 1]\!]$ is induced by the connection on $\mathcal{O}_{\Delta, \mathcal{Y}_x}$ via the inclusion (6.19).

When $b = 0$, so that $q_{\mathrm{dR}}^{1/p^b} = q_{\mathrm{dR}}$, we abbreviate

$$q_{\mathrm{dR}} - \exp = q_{\mathrm{dR}}^{1/p^b} - \exp, \qquad f(q_{\mathrm{dR}}) = f(q_{\mathrm{dR}}^{1/p^b}).$$

Theorem 6.7. *On $\mathcal{Y}^{\mathrm{ss}}$, we have that the natural inclusion (6.19) is in fact an equality*

$$\mathcal{O}_{\Delta, \mathcal{Y}^{\mathrm{ss}}}[\![q_{\mathrm{dR}}^{1/p^b} - 1]\!] = \mathcal{O}_{\Delta, \mathcal{Y}^{\mathrm{ss}}} \tag{6.20}$$

and moreover that

$$q_{\mathrm{dR}}^{1/p^b} - \exp|_{\mathcal{Y}^{\mathrm{ss}}} = \mathrm{id} \tag{6.21}$$

is the identity under the identification (6.20).

   The natural inclusion

$$\mathcal{O}_{\mathcal{Y}^{\mathrm{ss}}} \subset \mathcal{O}_{\Delta, \mathcal{Y}^{\mathrm{ss}}} \tag{6.22}$$

factors as

$$\mathcal{O}_{\mathcal{Y}^{\mathrm{ss}}} \overset{q_{\mathrm{dR}}^{1/p^b} - \exp}{\hookrightarrow} \hat{\mathcal{O}}_{\mathcal{Y}^{\mathrm{ss}}}[\![q_{\mathrm{dR}}^{1/p^b} - 1]\!] \overset{(6.20)}{=} \mathcal{O}_{\Delta, \mathcal{Y}^{\mathrm{ss}}}.$$

Proof. We first show (6.20); from this, (6.21) immediately follows since the composition

$$\hat{\mathcal{O}}_{\mathcal{Y}^{\mathrm{ss}}}[\![q_{\mathrm{dR}}^{1/p^b} - 1]\!] \subset \mathcal{O}_{\Delta, \mathcal{Y}^{\mathrm{ss}}} \overset{q_{\mathrm{dR}}^{1/p^b} - \exp}{\to} \hat{\mathcal{O}}_{\mathcal{Y}^{\mathrm{ss}}}[\![q_{\mathrm{dR}}^{1/p^b} - 1]\!]$$

is compatible with natural connections, and so we can check coefficient-wise that

$$\sum_{n=0}^{\infty} a_n (q_{\mathrm{dR}}^{1/p^b} - 1)^n \mapsto \sum_{n=0}^{\infty} a_n (q_{\mathrm{dR}}^{1/p^b} - 1)^n,$$

i.e., the composition is the identity.

   First, we show that, as subsheaves of $\mathcal{O}_{\Delta, \mathcal{LT}}$, we have

$$\hat{\mathcal{O}}_{\mathcal{LT}}[\![\mathbf{z}_{\mathrm{LT}} - \bar{z}_{\mathrm{LT}}]\!] = \hat{\mathcal{O}}_{\mathcal{LT}}[\![z_{\mathrm{dR}} - \bar{z}_{\mathrm{dR}}]\!] = \hat{\mathcal{O}}_{\mathcal{LT}}[\![q_{\mathrm{dR}}^{1/p^b} - 1]\!]. \tag{6.23}$$

The first equality follows from performing a change of variables $\mathbf{z}_{\mathrm{LT}} - \bar{z}_{\mathrm{LT}}$ which uses the fact that $\frac{dz_{\mathrm{dR}}}{dz_{\mathrm{LT}}}$ is invertible on $\mathcal{LT}$ by Proposition 4.39. The second equality is completely formal, given by a "change of variables"

$$z_{\mathrm{dR}} - \bar{z}_{\mathrm{dR}} \mapsto \exp\left(\frac{z_{\mathrm{dR}} - \bar{z}_{\mathrm{dR}}}{p^b}\right).$$

Now, from Proposition 6.2, we have

$$\mathcal{O}_{\Delta,\mathcal{LT}} = \hat{\mathcal{O}}_{\mathcal{LT}}[\![\mathbf{z}_{\mathrm{LT}} - \bar{z}_{\mathrm{LT}}]\!]$$

compatible with connections, and so from (6.23) we have

$$\mathcal{O}_{\Delta,\mathcal{LT}} = \hat{\mathcal{O}}_{\mathcal{LT}}[\![q_{\mathrm{dR}}^{1/p^b} - 1]\!] \tag{6.24}$$

compatible with connections. From the compatibility with connections, we see that the natural map

$$\mathcal{O}_{\Delta,\mathcal{Y}^{\mathrm{ss}}} \to \mathcal{O}_{\Delta,\mathcal{LT}} \overset{(6.24)}{=} \hat{\mathcal{O}}_{\mathcal{LT}}[\![q_{\mathrm{dR}}^{1/p^b} - 1]\!] \tag{6.25}$$

factors through an equality

$$\mathcal{O}_{\Delta,\mathcal{Y}^{\mathrm{ss}}} = \hat{\mathcal{O}}_{\mathcal{Y}^{\mathrm{ss}}}[\![q_{\mathrm{dR}}^{1/p^b} - 1]\!]$$

as follows. Given a section f of $\mathcal{O}_{\Delta,\mathcal{Y}^{\mathrm{ss}}}$, its image under (6.25) is given by

$$\sum_{n=0}^{\infty} \theta\left(\frac{1}{n!}\left(\frac{d}{dq_{\mathrm{dR}}^{1/p^b}}\right)^n f\right)(q_{\mathrm{dR}}^{1/p^b} - 1)^n. \tag{6.26}$$

However, we see that $\theta\left(\frac{1}{n!}\left(\frac{d}{dq_{\mathrm{dR}}^{1/p^b}}\right)^n\right)$ is a section of $\hat{\mathcal{O}}_{\mathcal{Y}^{\mathrm{ss}}}$, and so we have that (6.26) is a section of $\hat{\mathcal{O}}_{\mathcal{Y}^{\mathrm{ss}}}[\![q_{\mathrm{dR}}^{1/p^b} - 1]\!]$, as desired.

Now to prove the final statement of the theorem, we recall that

$$\mathcal{O}_{\mathcal{Y}^{\mathrm{ss}}} \subset \mathcal{O}_{\Delta,\mathcal{Y}^{\mathrm{ss}}} = \hat{\mathcal{O}}_{\mathcal{Y}^{\mathrm{ss}}}[\![q_{\mathrm{dR}}^{1/p^b} - 1]\!] \tag{6.27}$$

(6.22) commutes with connections, and that

$$\mathcal{O}_{\mathcal{Y}^{\mathrm{ss}}} \subset \mathcal{O}_{\Delta,\mathcal{Y}^{\mathrm{ss}}} \overset{\theta}{\twoheadrightarrow} \hat{\mathcal{O}}_{\mathcal{Y}^{\mathrm{ss}}}$$

is the natural inclusion, hence given a section f of $\mathcal{O}_{\mathcal{Y}^{\mathrm{ss}}}$, each coefficient of its image under (6.27) is given by Taylor's formula, and so its image is just $f(q_{\mathrm{dR}}^{1/p^b})$. $\qquad\square$

6.2   RELATION BETWEEN q_{dR}-EXPANSIONS AND SERRE-TATE EXPANSIONS

We have the following theorem relating $q_{dR} - \exp$ to the Serre-Tate expansion. Given a q_{dR}-expansion

$$f(q_{dR}) = (q_{dR} - \exp)(f)$$

or T-expansion

$$f(T) = (T - \exp)(f),$$

let

$$f(y)(q_{dR}) = (q_{dR} - \exp)(f)(y)$$

and

$$f(T)(y) = (T - \exp)(f)(y)$$

denote the corresponding power series obtained by evaluating each coefficient in the fiber at y (i.e., reducing modulo $\mathfrak{p}_y$, the prime ideal corresponding to y).

Theorem 6.8. *For a point* $y = (A, \alpha) \in \mathcal{Y}^{\mathrm{Ig}}$ *where A is an ordinary elliptic curve with complex multiplication, we have*

$$q_{dR} - \exp(\cdot)(y) = T - \exp|_{D(y)}(\cdot)(y)$$

where $T - \exp$ *is the Serre-Tate expansion map (see Definition 4.24), and $D(y)$ is the preimage under the projection* $\lambda : \mathcal{Y}^{\mathrm{Ig}} \to Y^{\mathrm{ord}}$ *of the ordinary residue disc in Y^{ord} centered around the geometric point of Y corresponding to $\lambda(y)$.*

Proof. First, note that by a strictly formal computation, for a formal variable Y, we have that

$$\theta\left(\frac{d}{d\exp(Y)}\right)^n \Big|_{\exp Y = 1} = P_n\left(\left(\frac{d}{dY}\right)^n \Big|_{Y=0}\right)$$

for some easily computable polynomial $P_n(X) \in \mathbb{Q}[X]$. On $\mathcal{Y}^{\mathrm{Ig}}$, by Theorem 5.9, we then have

$$\theta\left(\frac{1}{n!}\left(\frac{d}{dq_{dR}}\right)^n f\right) = \frac{1}{n!}P_n\left(\theta\left(\left(\frac{d}{dz_{dR}}\right)^n f\right)\right) = \frac{1}{n!}P_n\left(\frac{d}{d\log T}\Big|_{\log T = 0}f(y)\right)$$

$$= \frac{1}{n!}\left(\frac{d}{dT}\right)^n \Big|_{T=1}f(y),$$

where for a section s of $\mathcal{O}|_{\mathcal{Y}^{\mathrm{Ig}}}$, $s(y)$ denotes evaluation at y (i.e., the image of s in the residue field y). Now we are done by (4.35). $\square$

Let

$$Y_{0r} = Y(\Gamma_1(N) \cap \Gamma_0(p^r))$$

and recall that $Y = Y_1(N) = Y_{00}$. Recall the U_p operator which can be viewed as a correspondence

$$Y_{0r} \leftarrow Y_{0r} \rightarrow Y$$

and V_p viewed as a correspondence

$$Y_{0r} \leftarrow Y_{0r} \rightarrow Y.$$

Hence they can be viewed as Hecke operators which act on modular forms $w \in \omega^{\otimes k}(Y_{0r})$ by considering the pullback $w|_{Y_{0r+1}}$, so that

$$U_p^*(w|_{Y_{0,r+1}}), V_p^*(w|_{Y_{0r+1}}) \in \omega^{\otimes k}(Y_{0r+1}).$$

Given a modular form $w \in \omega^{\otimes k}(Y)$, we can define the *p-stabilization*

$$w|_{U_p V_p - V_p U_p} = (U_p V_p - V_p U_p)^* w \in \omega^{\otimes k}(Y_{02}). \tag{6.28}$$

We can find lifts of them in terms of the $GL_2(\mathbb{Q}_p)$ acting on the infinite-level modular curve $\mathcal{Y}$; in other words, we can find correspondences $\tilde{U}_p, \tilde{V}_p$ on $\mathcal{Y}$ defined in terms of the $GL_2(\mathbb{Q}_p)$-action such that

$$\tilde{U}_p^* w|_{\mathcal{Y}_x} = (U_p^* w)|_{\mathcal{Y}_x}, \qquad \tilde{V}_p^* w|_{\mathcal{Y}_x} = (V_p^* w)|_{\mathcal{Y}_x}.$$

In fact, one such choice of $\tilde{U}_p, \tilde{V}_p$ is given by

$$\tilde{V}_p = \begin{pmatrix} 1 & 0 \\ 0 & p \end{pmatrix}, \qquad \tilde{U}_p = \frac{1}{p} \sum_{j=0}^{p-1} \begin{pmatrix} p & j \\ 0 & 1 \end{pmatrix}.$$

We recall here that $GL_2(\mathbb{Q}_p)$ acts on functions f by $\gamma^* f(x) = f(x\gamma)$.

Again, given a T-expansion $f(T) \in W[\![T]\!]$ and a geometric point $y \in \mathcal{Y}$ $(\mathbb{C}_p, \mathcal{O}_{\mathbb{C}_p})$, recall we denote

$$f(T)(y) \in \mathbb{C}_p$$

the specialization of $f(T)$ at (i.e., image in the fiber at) y.

Theorem 6.9 ([8] Lemma 8.2, [46] Proposition A.0.1). *Suppose that $y \in \mathcal{Y}^{\mathrm{Ig}}$ is an ordinary CM point. Let $w \in \omega^{\otimes k}(Y)$, so that*

$$w^{(p)} := (U_p V_p - V_p U_p)^* w \in \omega^{\otimes k}(Y_{02}). \tag{6.29}$$

Write

$$w|_{\mathcal{Y}^{\mathrm{Ig}}} = f \cdot \left(\omega_{\mathrm{can}}^{\mathrm{Katz}}\right)^{\otimes k}, \qquad w^{(p)}|_{\mathcal{Y}^{\mathrm{Ig}}} = f^{(p)} \cdot \left(\omega_{\mathrm{can}}^{\mathrm{Katz}}\right)^{\otimes k}$$

in $\omega^{\otimes k}(\mathcal{Y}^{\mathrm{Ig}})$. Then

$$f^{(p)}|_{\mathcal{Y}^{\mathrm{Ig}}} := ((U_p V_p - V_p U_p)^* f)|_{\mathcal{Y}^{\mathrm{Ig}}}(T)(y) = (((\tilde{U}_p \tilde{V}_p)^* - (\tilde{V}_p \tilde{U}_p)^*) f|_{\mathcal{Y}^{\mathrm{Ig}}})(T)(y)$$

$$= f|_{\mathcal{Y}^{\mathrm{Ig}}}(T)(y) - \frac{1}{p} \sum_{j=0}^{p-1} f|_{\mathcal{Y}^{\mathrm{Ig}}}(\zeta_p^j T)(y).$$

Furthermore, letting F_w denote the Coleman primitive associated with w which takes the value 0 at the cusp (∞) (see Section 9.1 and Theorem 9.9), and $F_{w^{(p)}}$ denote the Coleman primitive associated with w, we have for any $-\frac{k}{2} \le r \le -1$ the following identity:

$$\lim_{m\to\infty} \left(\frac{Td}{dT}\right)^{r+p^m(p-1)} \left(f^{(p)}|_{\mathcal{Y}^{\mathrm{Ig}}}(T)\right)$$

$$= \lim_{m\to\infty} \left(\frac{Td}{dT}\right)^{r+p^m(p-1)} (((U_p V_p - V_p U_p)^* f)|_{\mathcal{Y}^{\mathrm{Ig}}}(T))$$

$$= \left(\lim_{m\to\infty} \theta_{\mathrm{AS}}^{r+p^m(p-1)} (U_p V_p - V_p U_p)^* f\right)|_{\mathcal{Y}^{\mathrm{Ig}}}(T) \qquad (6.30)$$

$$= \theta_{\mathrm{AS}}^{r+1} ((U_p V_p - V_p U_p)^* F_w)|_{\mathcal{Y}^{\mathrm{Ig}}}(T)$$

$$= (\theta_{\mathrm{AS}}^{r+1} F_w)|_{\mathcal{Y}^{\mathrm{Ig}}}(T) - \frac{1}{p} \sum_{j=0}^{p-1} \theta_{\mathrm{AS}}^{r+1} F_w|_{\mathcal{Y}^{\mathrm{Ig}}}(\zeta_p^j T) = \theta_{\mathrm{AS}}^{r+1} F_{w^{(p)}}|_{\mathcal{Y}^{\mathrm{Ig}}}(T).$$

6.3   INTEGRALITY PROPERTIES OF q_{dR}-EXPANSIONS: THE DIEUDONNÉ-DWORK LEMMA

The goal of this section is to prove a "Key Lemma," which will be essential for our p-adic analytic computations, in particular for showing the convergence of the sequence of p-adic Maass-Shimura derivatives of a rigid function in stalks at certain supersingular CM points.

Lemma 6.10 (Key Lemma). *View $\hat{z} \in \hat{\mathcal{O}}_{\mathcal{Y}}(\mathcal{LT})$. Suppose that*

$$\left| \theta \left(\frac{d(\mathbf{z}_{\mathrm{LT}}/\bar{z}_{\mathrm{LT}})}{dq_{\mathrm{dR}}^{1/p^b}} \right)(y) \right| \le 1. \qquad (6.31)$$

Then we have that the q_{dR}^{1/p^b}-expansion of $(\mathbf{z}_{\mathrm{LT}}/\bar{z}_{\mathrm{LT}}) \in \mathcal{O}_{\Delta,y}$ satisfies

$$(\mathbf{z}_{\mathrm{LT}}/\bar{z}_{\mathrm{LT}})_y(q_{\mathrm{dR}}^{1/p^b}) \in \hat{\mathcal{O}}_{\mathcal{LT},y}^+[\![q_{\mathrm{dR}}^{1/p^b} - 1]\!]. \qquad (6.32)$$

As a consequence of (6.32) and (6.13), we have

$$\mathcal{O}^+_{\mathcal{LT}}(\mathcal{LT}) \overset{q_{\mathrm{dR}}^{1/p^b}-\exp}{\longrightarrow} \hat{\mathcal{O}}^+_{\mathcal{LT},y}\llbracket q_{\mathrm{dR}}^{1/p^b} - 1 \rrbracket. \qquad (6.33)$$

The argument essentially comes down to an application of the Dieudonné-Dwork lemma. We recall a result of Chojecki-Hansen-Johansson ([13]) that in particular implies that the subgroup

$$\langle \alpha_1 \pmod{p} \rangle \subset A[p],$$

called the *pseudocanonical subgroup* in loc. cit., coincides with the canonical subgroup when $|\mathbf{z}| > p^{p/(p^2-1)}$.

Lemma 6.11 ([13], Lemma 2.14). *Assume that*

$$|\hat{\mathbf{z}}(A,\alpha)| > p^{p/(p^2-1)}.$$

Then the pseudocanonical subgroup

$$\langle \alpha_1 \pmod{p} \rangle \subset A[p]$$

is in fact the canonical subgroup of A.

Let $\mathcal{Y}_n$ denote the adic space over $\mathrm{Spa}(\mathbb{C}_p, \mathcal{O}_{\mathbb{C}_p})$ associated with the modular curve parametrizing elliptic curves with $\Gamma_1(p^n N)$-level structure, that is, the modular curve represents the moduli space of isomorphism classes of quadruples (E, t, β) with embeddings of group schemes $t : \mathbb{Z}/N \hookrightarrow E[N]$ and $\beta : \mathbb{Z}/p^n \hookrightarrow E[p^n]$. Let $\mathcal{Y}_n^{\mathrm{ss}} = \{|\mathrm{Ha}| < 1\} \subset \mathcal{Y}_n$ denote the associated supersingular locus, cut out by (locally defined lifts of) the Hasse invariant Ha.

Recall that by the theory of canonical subgroup ([36, Chapter 3]), for every $n \in \mathbb{Z}_{\geq 0}$, there exists $1/p < r < 1$ such that the rigid analytic subdomain $\mathcal{Y}_{0,r}^{\mathrm{ss}} = \{r < |\mathrm{Ha}| < 1\} \subset \mathcal{Y}_0^{\mathrm{ss}}$ has a rigid analytic section to the projection $\mathcal{Y}_n^{\mathrm{ss}} \to \mathcal{Y}_0^{\mathrm{ss}}$, given by choosing a generator of the level n canonical subgroup (i.e., the unique, when it exists, order p^n cyclic subgroup of $E[p^n]$ which lifts p^n-power Frobenius). Therefore, on the locus of $\mathcal{Y}_n^{\mathrm{ss},r}$ parametrizing curves with level $r > n$ canonical subgroup, the classifying map corresponding to division of the universal object by the universal (level 1) canonical subgroup corresponds to a Frobenius morphism $\mathcal{Y}_n^{\mathrm{ss},r} \to \mathcal{Y}_n^{\mathrm{ss},r-1}$.

Proof of Lemma 6.10. Let

$$\gamma_r = \begin{pmatrix} 1 & 0 \\ 0 & p^r \end{pmatrix}$$

for the rest of the lemma. Note that by (4.44) and (3.15), and by the definition $z_{\mathrm{dR}} = \mathbf{z}_{\mathrm{dR}} \pmod{t}$, for any integer $r \geq 0$ we have

$$\gamma_r^* z_{\mathrm{dR}} = p^r z_{\mathrm{dR}}, \tag{6.34}$$

and from Definition 6.4, we see that

$$\gamma_r^* q_{\mathrm{dR}} = q_{\mathrm{dR}}^{p^r}. \tag{6.35}$$

Write

$$\mathbf{z}_{\mathrm{LT}} / \bar{z}_{\mathrm{LT}} = 1 + \sum_{n=1}^{\infty} a_n (q_{\mathrm{dR}}^{1/p^b} - 1)^n. \tag{6.36}$$

To prove the lemma, we will apply a Dieudonné-Dwork-style inductive argument. The rough idea of the argument is this. We will use γ_r^* to "transport structure" to $y_r := y \cdot \gamma_r^{-1}$, which has $|\hat{\mathbf{z}}(y_r)| = p^r |\hat{\mathbf{z}}(y)|$. As $r \to \infty$, then $|\hat{\mathbf{z}}(y_r)| \to \infty$, and we move the situation to smaller and smaller neighborhoods of $\mathcal{Y}^{\mathrm{Ig}}$. By [56, Lemma III.3.8], as $|\hat{\mathbf{z}}| \to \infty$, the Hodge height $\epsilon = \mathrm{Hdg}$ (defined as the valuation of the Hasse invariant, i.e., $|\mathrm{Ha}| = p^{\mathrm{Hdg}}$) tends to 0, and so the canonical subgroup lifts the kernel of p-power Frobenius modulo $p^{1-\epsilon}$ for smaller and smaller ϵ. Using the canonical subgroup as $r \to \infty$ (and hence $\epsilon \to 0$), as well as an intermediate lemma to descend to finite level modulo p, we show that each $a_{n,y} \in p^{-\epsilon} \hat{\mathcal{O}}_{\mathcal{LT},y}^+$ for all $\epsilon > 0$, and so $a_{n,y} \in \hat{\mathcal{O}}_{\mathcal{LT},y}^+$.

First note that, tautologically, on stalks and fibers,

$$(\gamma_r^* a_n)_{y_r} = a_{n,y}, \qquad (\gamma_r^* a_n)(y_r) = a_n(y). \tag{6.37}$$

Now consider the image of $\gamma_r^*(\mathbf{z}_{\mathrm{LT}} / \bar{z}_{\mathrm{LT}}) \in \mathcal{O}_\Delta$ in the stalk $\mathcal{O}_{\Delta, y_r}$ at y_r

$$
\begin{aligned}
(\gamma_r^*(\mathbf{z}_{\mathrm{LT}} / \bar{z}_{\mathrm{LT}}))_{y_r} &= 1 + \sum_{n=1}^{\infty} (\gamma_r^* a_n)_{y_r} ((\gamma_r^* q_{\mathrm{dR}}^{1/p^b}) - 1)^n \\
&\overset{(6.35)}{=} 1 + \sum_{n=1}^{\infty} (\gamma_r^* a_n)_{y_r} (q_{\mathrm{dR}}^{1/p^{b-r}} - 1)^n \in \mathcal{O}_{\Delta, y_r}.
\end{aligned}
\tag{6.38}
$$

We need the following intermediate lemma, which shows that for every r and every n, there exists an $m \in \mathbb{Z}_{\geq 0}$ such that we can descend each $(\gamma_r^* a_n)_{y_r}$ $\pmod{p}$ from (6.36) to the finite level curves $\mathcal{Y}_m$ "modulo p." This will be used in order to show that γ_1^* acts on $(\gamma_r^* a_n)_{y_r}$ $\pmod{p}$ as p-power Frobenius, using the fact that γ_1 acts on the universal object by division by the canonical subgroup at y_r.

Lemma 6.12. *For every r and every n, for all sufficiently large $m \in \mathbb{Z}_{\geq 0}$ we have that $(\gamma_r^* a_n)_{y_r} \in \hat{\mathcal{O}}_{\mathcal{LT}, y_r} = \hat{\mathcal{O}}_{\mathcal{Y}, y_r}$ satisfies*

$$\delta^*(\gamma_r^* a_n)_{y_r} \equiv (\gamma_r^* a_n)_{y_r} \pmod{p \hat{\mathcal{O}}_{\mathcal{Y}, y_r}^+} \tag{6.39}$$

for every $\delta = \begin{pmatrix} a & b \\ c & d \end{pmatrix} \in SL_2(\mathbb{Z}_p)$ with $a \equiv 1 \pmod{p^m}$, $c \equiv 0 \pmod{p^m}$. In particular,

$$(\gamma_r^* a_n)_{y_r} \equiv b_{n,m,r} \pmod{p \hat{\mathcal{O}}_{\mathcal{L}T,y_r}^+} \tag{6.40}$$

for some $b_{n,m,r} \in \hat{\mathcal{O}}_{\mathcal{Y}_m^{ss}, \bar{y}_r}$, where $\bar{y}_r$ is the image of y_r under the projection $\mathcal{Y}^{ss} \to \mathcal{Y}_m^{ss}$. The $b_{n,m,r}$ are all compatible for all $m \gg 0$ with respect to the projections $\mathcal{Y}_{m+1} \to \mathcal{Y}_m$.

Proof of Lemma 6.12. Using the elementary identity

$$\frac{1}{n!} q^n \left(\frac{d}{dq} \right)^n = \binom{\frac{qd}{dq}}{n},$$

we have the formula

$$\gamma_r^* a_n \overset{(6.38)}{=} \frac{1}{n!} \theta \left(\frac{d}{dq_{dR}^{1/p^{b-r}}} \right)^n (\gamma_r^*(\mathbf{z}_{LT}/\bar{z}_{LT})) = \theta \circ \binom{p^{b-r} \frac{d}{dz_{dR}}}{n} (\gamma_r^*(\mathbf{z}_{LT}/\bar{z}_{LT})). \tag{6.41}$$

We need to check (6.39), for which we note that, using (4.44), we have

$$\delta^*(dz_{dR}) = (cz_{dR} + a)^{-2} dz_{dR}$$

(recalling $(ad - bc) = 1$ for δ). Since $\delta \in SL_2(\mathbb{Z}_p) \subset GL_2(\mathbb{Z}_p)$, we have $\delta^* \mathbf{z}_{LT} = \mathbf{z}_{LT}$ and $\delta^* \bar{z}_{LT} = \bar{z}_{LT}$, since π_{GM} is invariant under $GL_2(\mathbb{Z}_p)$ acting on $\mathbb{Q}_p^{\oplus 2}$-level structure (note that the $GL_2(\mathbb{Z}_p)$-action on $\mathcal{Y}$ does not change the elliptic curve in the test triple corresponding to any geometric point). These identities, combined with (6.41), imply

$$\delta^*(\gamma_r^* a_n) = \theta \circ \binom{(cz_{dR} + a)^2 p^{b-r} \frac{d}{dz_{dR}}}{n} (\gamma_r^*(\mathbf{z}_{LT}/\bar{z}_{LT})). \tag{6.42}$$

Now choose any $m \in \mathbb{Z}_{\geq 0}$ such that

$$p^m \theta(z_{dR})_{y_r} \frac{1}{n!} \left(p^{b-r} \frac{d}{dz_{dR}} \right)^i (\gamma_r^*(\mathbf{z}_{LT}/\bar{z}_{LT})) \in p \hat{\mathcal{O}}_{\mathcal{Y}, y_r}^+$$

for all $0 \leq i \leq n$ (one can check this inclusion by evaluating at the fiber at y_r, using (6.8)). Then for $c \equiv 0 \pmod{p^m}$, $a \equiv 1 \pmod{p^m}$, one checks by computation, using $\frac{d}{dz_{dR}}(cz_{dR} + a)^2 = 2c(cz_{dR} + a)$, that

$$\theta \circ \binom{(cz_{dR} + a)^2 p^{b-r} \frac{d}{dz_{dR}}}{n} (\gamma_r^*(\mathbf{z}_{LT}/\bar{z}_{LT}))_{y_r} \equiv \binom{p^{b-r} \frac{d}{dz_{dR}}}{n} (\gamma_r^*(\mathbf{z}_{LT}/\bar{z}_{LT}))_{y_r}$$

$$= (\gamma_r^* a_n)_{y_r} \pmod{p \hat{\mathcal{O}}_{\mathcal{Y}, y_r}^+},$$

which with (6.42), implies (6.39). Now using profinite descent (see Lemma 4.27) on the proétale sheaf $\hat{\mathcal{O}}^+/p$ along the geometric connected component of $\mathcal{Y} \to \mathcal{Y}_r$ containing y_r, we get (6.40). The compatibility of the $b_{n,m,r}$ follows immediately from the descent. $\qquad\square$

We now begin the Dieudonné-Dwork-style induction, showing that $a_{n,y} \in \hat{\mathcal{O}}^+_{\mathcal{LT},y}$ for all $n \in \mathbb{Z}_{\geq 1}$. By hypothesis (6.31) and (6.8), this is true for $n = 1$. Now assume $n > 1$ and $a_{m,y} \in \hat{\mathcal{O}}^+_{\mathcal{LT},y}$ for all $1 \leq m < n$. Then by our above observation, for any $r \in \mathbb{Z}_{\geq 0}$, $(\gamma_r^* a_m)_{y_r} \in \hat{\mathcal{O}}^+_{\mathcal{LT},y_r}$ for all $1 \leq m < n$. Applying γ_1^* to (6.38) gives

$$\gamma_1^*(\gamma_r^*(\mathbf{z}_{\mathrm{LT}}/\bar{z}_{\mathrm{LT}}))_{y_r} \overset{(6.35)}{=} 1 + \sum_{n=1}^{\infty} \gamma_1^*(\gamma_r^* a_n)_{y_r} (q_{\mathrm{dR}}^{1/p^{b-r-1}} - 1)^n \in \mathcal{O}_{\Delta,y_r}. \quad (6.43)$$

Let $\epsilon_r = \mathrm{Hdg}(y_r)$, so that as mentioned above, $\epsilon_r \to 0$ as $r \to \infty$ by Lemma III.3.8 of loc. cit. Moreover, by Lemma 6.11, we have that for all $r \gg 0$, the action by γ_1^* corresponds to division by the canonical subgroup, and so is a lifting of p-power Frobenius modulo $p^{1-\epsilon_r}$ on $\hat{\mathcal{O}}^+_{\mathcal{Y}_m,\tilde{y}_r}$, in the notation of Lemma 6.12. In particular, by (6.40), for all large enough $m \gg 0$ we have

$$\gamma_1^*(\gamma_r^* a_j)_{y_r} \equiv \gamma_1^*(b_{j,m,r}) \equiv b_{j,m,r}^p \quad (\mathrm{mod}\ p^{1-\epsilon_r})\hat{\mathcal{O}}^+_{\mathcal{LT},y_r}$$

for $0 < j < n-1$. Moreover, since $(\gamma_r^* \mathbf{z}_{\mathrm{LT}}) \in \mathcal{O}_{\mathcal{Y}_m,y_r}$, i.e., $\mathbf{z}_{\mathrm{LT}}$ is in the proétale stalk on finite level, and γ_1^* is a lifting of p-power Frobenius modulo $p^{1-\epsilon_r}$, we have the identity in $\mathcal{O}^+_{\mathcal{Y}_m,\tilde{y}_y}$

$$\gamma_1^*\left((\gamma_r^*(\mathbf{z}_{\mathrm{LT}}))_{y_r}/(\gamma_r^*(\mathbf{z}_{\mathrm{LT}}))(y_r)\right) \equiv ((\gamma_r^*(\mathbf{z}_{\mathrm{LT}}))_{y_r}/(\gamma_r^*(\mathbf{z}_{\mathrm{LT}}))(y_r))^p$$
$$(\mathrm{mod}\ p^{1-\epsilon_r}\mathcal{O}^+_{\mathcal{Y}_m,\tilde{y}_y}),$$

and analogously in $\mathcal{O}^+_{\mathcal{Y}_m,\tilde{y}_y}$,

$$\gamma_1^*\left((\gamma_r^*(\bar{z}_{\mathrm{LT}}))_{y_r}/(\gamma_r^*(\bar{z}_{\mathrm{LT}}))(y_r)\right) \equiv ((\gamma_r^*(\bar{z}_{\mathrm{LT}}))_{y_r}/(\gamma_r^*(\bar{z}_{\mathrm{LT}}))(y_r))^p$$
$$(\mathrm{mod}\ p^{1-\epsilon_r}\mathcal{O}^+_{\mathcal{Y}_m,\tilde{y}_y}),$$

and so dividing these two identities and using that $\mathbf{z}_{\mathrm{LT}}(y_r) = \bar{z}_{\mathrm{LT}}(y_r)$, we get

$$\gamma_1^*(\gamma_r^*(\mathbf{z}_{\mathrm{LT}}/\bar{z}_{\mathrm{LT}}))_{y_r} \equiv (\gamma_r^*(\mathbf{z}_{\mathrm{LT}}/\bar{z}_{\mathrm{LT}}))_{y_r}^p \quad (\mathrm{mod}\ p^{1-\epsilon_r})\hat{\mathcal{O}}^+_{\mathcal{LT},y_r}.$$

For the rest of the induction, let

$$X = q_{\mathrm{dR}}^{1/p^{b-r}} - 1, \qquad b_j = (\gamma_r^* a_j)_{y_r}$$

for brevity. Note that

$$\gamma_1^* X = (1+X)^p - 1 = X(X^{p-1} + pg(X))$$

where $g(X) = \sum_{i=0}^{p-2} \binom{p}{i+1} X^i \in p\mathbb{Z}[X]$. Hence comparing terms of degree at most n in (6.43) and reducing modulo $p^{1-\epsilon_r}$, we have

$$p(\gamma_r^* a_n)_{y_r} X^n + \sum_{j=1}^{n-1} b_j^p X^{jp} \equiv \sum_{j=1}^{n} \gamma_1^*(b_j) X^j (X^{p-1} + g(X))^j \tag{6.44}$$

$$(\text{mod } (p^{1-\epsilon_r}, X^{n+1}) \mathcal{O}_{\mathcal{LT},y_r}^+[\![X]\!]).$$

Since $b_j \in \hat{\mathcal{O}}_{\mathcal{LT},y_r}^+$ for $1 \leq j \leq n-1$ by induction hypothesis, we have

$$\gamma_1^*(b_j) X^j (X^{p-1} + g(X))^j \equiv b_j^p X^{jp} \quad (\text{mod } p^{1-\epsilon_r} \hat{\mathcal{O}}_{\mathcal{LT},y_r}^+[\![X]\!])$$

for $1 \leq j \leq n-1$. Thus we can clear all terms indexed by $1 \leq j \leq n-1$ from both sides of (6.44), and we get

$$pa_n X^n \equiv \gamma_1^*(b_n) X^n (X^{p-1} + g(X))^n$$
$$\equiv \gamma_1^*(b_n) g(0)^n X^n$$
$$= \gamma_1^*(b_n) p^n X^n \quad (\text{mod } (p^{1-\epsilon_r}, X^{n+1}) \hat{\mathcal{O}}_{\mathcal{LT},y_r}^+[\![X]\!]).$$

Hence we have

$$b_n \equiv \gamma_1^*(b_n) p^{n-1} \quad (\text{mod } (p^{-\epsilon_r}) \hat{\mathcal{O}}_{\mathcal{LT},y_r}^+).$$

By (6.8), we have

$$|b_n(y_r) - \gamma_1^*(b_n)(y_r) p^{n-1}| \leq p^{\epsilon_r}.$$

Since as before γ_1^* is a Frobenius lift modulo $p^{1-\epsilon_r}$ on $\hat{\mathcal{O}}_{y_m,\tilde{y}_y}$ by (6.40), we have $|\gamma_1^*(b_n)(y_r)| \leq |b_n(y_r)|$, and so since $n > 1$, we have

$$|b_n(y_r)| > |\gamma_1^*(b_n) p^{n-1}(y_r)|$$

and so

$$|a_n(y)| \stackrel{(6.37)}{=} |b_n(y_r)| \leq p^{\epsilon_r}.$$

Letting $r \to \infty$ and hence $\epsilon_r \to 0$, we get $|a_n(y)| \leq 1$, and so (6.8) implies $a_{n,y} \in \hat{\mathcal{O}}_{\mathcal{LT},y}^+$. This completes the induction. $\qquad \square$

Remark 6.13. We note that there is no analogue of the assumption (6.31) on a_1 made in the original Dieudonné-Dwork lemma [22, Lemma 1]. It is made here because we use a different Frobenius lifting considered in loc. cit. Namely, instead of the lifting

$$\phi(X) = X^p,$$

we use

$$\phi(X) = (1 + X)^p - 1.$$

6.4   INTEGRAL STRUCTURES ON STALKS OF INTERMEDIATE PERIOD SHEAVES BETWEEN $\mathcal{O}_\Delta$ AND $\hat{\mathcal{O}}_\Delta$

In this section we consider integral subrings of geometric stalks of the various period sheaves

$$\mathcal{O}_\Delta \subset \mathcal{O}_\Delta^\diamond \subset \mathcal{O}_\Delta^\dagger \subset \hat{\mathcal{O}}_\Delta$$

defined in Definitions 5.22 and 5.23. For modular forms on Y, which arise as sections of the generic fiber of powers of $\pi_* \Omega^1_{\mathcal{A}/Y^+}$ on the Katz-Mazur integral model Y^+ of Y, the stalks of their q_{dR}-expansions at certain geometric points in $\mathcal{Y}^{\mathrm{ss}}$ (including CM points) will belong to the generic fibers of the integral stalks of these intermediate period rings. Hence the study of these integral structures will be crucial to establishing p-adic analytic properties of stalks of p-adic Maass-Shimura derivatives of modular forms on Y, which when specialized to stalks at CM points, will be crucial for the construction of p-adic L-functions.

Definition 6.14. *As before, make the identification*

$$\mathcal{O}_{\Delta,\mathcal{Y}^{\mathrm{ss}}} = \hat{\mathcal{O}}_{\mathcal{Y}^{\mathrm{ss}}}[\![q_{\mathrm{dR}}^{1/p^b} - 1]\!]$$

using (6.20); by the definition of $\mathcal{O}_\Delta^\diamond$ given in Definition 5.23, this induces an identification

$$\mathcal{O}_{\Delta,\mathcal{Y}^{\mathrm{ss}}}^\diamond = \hat{\mathcal{O}}_{\mathcal{Y}^{\mathrm{ss}}}[\![q_{\mathrm{dR}}^{1/p^b} - 1]\!]\left[\frac{1}{z_{\mathrm{dR}} - \hat{\mathbf{z}}}\right] = \hat{\mathcal{O}}_{\mathcal{Y}^{\mathrm{ss}}}[\![q_{\mathrm{dR}}^{1/p^b} - 1]\!]\left[\frac{y_{\mathrm{dR}}}{\hat{\mathbf{z}}}\right].$$

We define

$$\mathcal{O}_\Delta^+ := \hat{\mathcal{O}}_{\mathcal{Y}^{\mathrm{ss}}}^+[\![q_{\mathrm{dR}}^{1/p^b} - 1]\!]$$

and

$$\mathcal{O}_\Delta^{\dagger,+} = \mathcal{O}_\Delta^+\left[\frac{p^b}{z_{\mathrm{dR}} - \hat{\mathbf{z}}}, \hat{\mathbf{z}}, \frac{1}{\hat{\mathbf{z}}}\right] = \mathcal{O}_\Delta^+\left[p^b\frac{y_{\mathrm{dR}}}{\hat{\mathbf{z}}}, \hat{\mathbf{z}}, \frac{1}{\hat{\mathbf{z}}}\right],$$

$$\mathcal{O}_\Delta^{\diamond,+} := \mathcal{O}_\Delta^+\left[\frac{p^b}{z_{\mathrm{dR}} - \hat{\mathbf{z}}}\right] = \mathcal{O}_\Delta^+\left[p^b\frac{y_{\mathrm{dR}}}{\hat{\mathbf{z}}}\right].$$

When $y \in \mathcal{Y}^{ss}(\mathbb{C}_p, \mathcal{O}_{\mathbb{C}_p})$ is a point satisfying the conditions of Lemma 6.10, we recall that

$$\mathcal{O}_{\mathcal{Y},y}^+ \overset{q_{dR}^{1/p^b}-\exp}{\subset} \mathcal{O}_{\Delta,y}^+$$

and in this situation, we define

$$\hat{\mathcal{O}}_{\Delta,y}^+ := \mathcal{O}_{\Delta,y}^+ \otimes_{\mathcal{O}_{\mathcal{Y},y}^+} \hat{\mathcal{O}}_{\mathcal{Y},y}^+ \tag{6.45}$$

so that

$$\hat{\mathcal{O}}_{\Delta,y}^+[1/p] = \mathcal{O}_{\Delta,y}^+ \otimes_{\mathcal{O}_{\mathcal{Y},y}^+} \hat{\mathcal{O}}_{\mathcal{Y},y}.$$

Note that we have the inclusions

$$\mathcal{O}_{\Delta,y}^{\diamond,+} \subset \hat{\mathcal{O}}_{\Delta,y}^+ \subset \hat{\mathcal{O}}_{\Delta,y}. \tag{6.46}$$

Note that $\hat{\mathcal{O}}_{\mathcal{Y},y}^+$ has a natural topology of a pseudometric space, which is given by the residue seminorm

$$|f| = |f(y)|.$$

We give $\mathcal{O}_\Delta^+$ the topology of uniform convergence with respect to the natural topology on $\hat{\mathcal{O}}_{\mathcal{Y}}^+$, i.e., the topology induced by the Gauss norm. We give $\mathcal{O}_\Delta^{\diamond,+}$ the topology of uniform convergence with respect to the Gauss norm topology on $\mathcal{O}_\Delta^+$. Note that both

$$\theta : \mathcal{O}_\Delta^+ \to \hat{\mathcal{O}}_{\mathcal{Y}^{ss}}^+$$

and

$$\theta : \mathcal{O}_\Delta^{\diamond,+} \to \hat{\mathcal{O}}_{\mathcal{Y}^{ss}}^+$$

are continuous with respect to these topologies.

Proposition 6.15. We have

$$\frac{d}{dz_{dR}}\left(\frac{y_{dR}}{\hat{\mathbf{z}}}\right)^k = k\left(\frac{y_{dR}}{\hat{\mathbf{z}}}\right)^{k+1}, \quad \frac{q_{dR}^{1/p^b} d}{dq_{dR}^{1/p^b}}\left(p^b\frac{y_{dR}}{\hat{\mathbf{z}}}\right)^k = k\left(p^b\frac{y_{dR}}{\hat{\mathbf{z}}}\right)^{k+1}. \tag{6.47}$$

We also have that

$$\mathcal{O}_{\Delta,y}^{\diamond,\nabla=0} = \hat{\mathcal{O}}_{\mathcal{Y}_x,y}, \quad \mathcal{O}_{\Delta,y}^{\diamond,+}[1/p]^{\nabla=0} = \hat{\mathcal{O}}_{\mathcal{Y}^{ss},y}, \quad \mathcal{O}_{\Delta,y}^{\diamond,+,\nabla=0} = \hat{\mathcal{O}}_{\mathcal{Y}^{ss},y}^+. \tag{6.48}$$

Proof. From (4.17) we have

$$\frac{y_{dR}}{\hat{\mathbf{z}}} = -\frac{1}{z_{dR} - \hat{\mathbf{z}}},$$

from which one easily computes (6.47) (recalling that $\frac{d}{dz_{\mathrm{dR}}}\hat{\mathbf{z}}=0$). The identities (6.48) immediately follow from the fact that

$$O_{\Delta,y}^{\nabla=0} = \hat{O}_{\mathcal{Y}_x,y}, \quad O_{\Delta,y}^{+,\nabla=0} = \hat{O}_{\mathcal{Y}^{\mathrm{ss}},y}^{+} \tag{6.49}$$

and from (6.47). $\square$

6.5   THE p-ADIC MAASS-SHIMURA OPERATOR θ_k^j IN q_{dR}-COORDINATES

Now we can rewrite (5.25) as

$$\delta_k^j = \sum_{i=0}^{j} \binom{j+k-1}{i}\binom{j}{i}i!\frac{1}{(z_{\mathrm{dR}}-\bar{z})^i}\left(\frac{1}{p^b}\frac{q_{\mathrm{dR}}^{1/p^b}d}{dq_{\mathrm{dR}}^{1/p^b}}\right)^{j-i}, \tag{6.50}$$

and (5.28) as

$$
\begin{aligned}
\theta_k^j &= \sum_{i=0}^{j} \binom{j+k-1}{i}\binom{j}{i}i!\frac{1}{(\bar{z}_{\mathrm{dR}}-\theta(\bar{z}))^i}\,\theta\circ\left(\frac{1}{p^b}\frac{q_{\mathrm{dR}}^{1/p^b}d}{dq_{\mathrm{dR}}^{1/p^b}}\right)^{j-i}\\
&= \sum_{i=0}^{j} \binom{j+k-1}{i}\binom{j}{i}i!\left(-\frac{\theta(y_{\mathrm{dR}})}{\mathbf{z}}\right)^i\left(\frac{1}{p^b}\right)^{j-i}\theta\circ\left(\frac{q_{\mathrm{dR}}^{1/p^b}d}{dq_{\mathrm{dR}}^{1/p^b}}\right)^{j-i} \\
&= \sum_{i=0}^{\infty} c_i(j)\left(-\frac{\theta(y_{\mathrm{dR}})}{\mathbf{z}}\right)^i\left(\frac{1}{p^b}\right)^{j-i}\theta\circ\left(\frac{q_{\mathrm{dR}}^{1/p^b}d}{dq_{\mathrm{dR}}^{1/p^b}}\right)^{j-i}
\end{aligned}
\tag{6.51}
$$

where the second equality follows from the p-adic Legendre relation (4.17), and

$$c_i(j) := \begin{cases} \binom{j+k-1}{i}\binom{j}{i}i! & i \le j \\ 0 & i > j. \end{cases} \tag{6.52}$$

Note that $c_i(j)$ extends to a p-adic continuous function in $j\in\mathbb{Z}/(p-1)\times\mathbb{Z}_p\to\mathbb{Z}_p$. Here

$$\mathbb{Z}\subset\mathbb{Z}/(p-1)\times\mathbb{Z}_p$$

is embedded diagonally and is dense with respect to the p-adic topology (by the Chinese remainder theorem). Similarly, we have from Theorem 5.34

$$(\delta^\diamond_{\Delta,k})^j = \delta^{\diamond,j}_{\Delta,k} = \sum_{i=0}^\infty \frac{c_i(j)}{(z_{\mathrm{dR}} - \hat{\mathbf{z}})^i} \left(\frac{1}{p^b} \frac{q_{\mathrm{dR}}^{1/p^b} d}{dq_{\mathrm{dR}}^{1/p^b}} \right)^{j-i}$$

$$= \sum_{i=0}^\infty c_i(j) \left(-\frac{y_{\mathrm{dR}}}{\hat{\mathbf{z}}} \right)^i \left(\frac{1}{p^b} \frac{q_{\mathrm{dR}}^{1/p^b} d}{dq_{\mathrm{dR}}^{1/p^b}} \right)^{j-i} \qquad (6.53)$$

where the last equality again follows from the *p*-adic Legendre relation (4.17).

Note that $\frac{q_{\mathrm{dR}}^{1/p^b} d}{dq_{\mathrm{dR}}^{1/p^b}}$ has the following simple action on power series

$$\frac{q_{\mathrm{dR}}^{1/p^b} d}{dq_{\mathrm{dR}}^{1/p^b}} \left(\sum_{n=0}^\infty a_n (q_{\mathrm{dR}}^{1/p^b} - 1)^n \right) = \sum_{n=1}^\infty n a_n q_{\mathrm{dR}}^{1/p^b} (q_{\mathrm{dR}}^{1/p^b} - 1)^{n-1}$$

where $a_n \in \hat{\mathcal{O}}^+_{\mathcal{Y},y}$. In particular, on polynomials, we can write

$$\sum_{n=0}^d a_n (q_{\mathrm{dR}}^{1/p^b} - 1)^n = \sum_{n-0}^d b_n \left(q_{\mathrm{dR}}^{1/p^b} \right)^n$$

for appropriate b_n, and hence

$$\left(\frac{q_{\mathrm{dR}}^{1/p^b} d}{dq_{\mathrm{dR}}^{1/p^b}} \right)^j \left(\sum_{n=0}^d a_n (q_{\mathrm{dR}}^{1/p^b} - 1)^n \right) = \left(\frac{q_{\mathrm{dR}}^{1/p^b} d}{dq_{\mathrm{dR}}^{1/p^b}} \right)^j \left(\sum_{n=0}^d b_n \left(q_{\mathrm{dR}}^{1/p^b} \right)^n \right)$$

$$= \sum_{n=0}^d n^j b_n \left(q_{\mathrm{dR}}^{1/p^b} \right)^n .$$

Now endow $\hat{\mathcal{O}}_{\mathcal{Y},y}[\![q_{\mathrm{dR}}^{1/p^b} - 1]\!]$ with the *p*-adic uniform convergence topology, i.e., the metric topology induced by the "*p*-adic Gauss norm"

$$\left| \sum_{n=0}^\infty a_n (q_{\mathrm{dR}}^{1/p^b} - 1)^n \right| := \sup_n |a_n|.$$

Note that θ takes a power series $F(q_{\mathrm{dR}}^{1/p^b})$ to its constant term, and gives a continuous map $\theta : \mathcal{O}_{\Delta,\mathcal{Y}^{ss}} \to \hat{\mathcal{O}}_{\mathcal{Y}^{ss}}$. Hence in order to show *p*-adic analytic properties of an element of

$$F \in \mathcal{O}_{\Delta,\mathcal{Y},y}$$

where $y \in \mathcal{Y}$, it suffices to show such properties for $F(q_{\mathrm{dR}}^{1/p^b})$.

6.6   INTEGRALITY OF q_{dR}-EXPANSIONS AND THE $\flat$-OPERATOR

Lemma 6.16. *Suppose $y \in \mathcal{Y}^{\mathrm{ss}}(\mathbb{C}_p, \mathcal{O}_{\mathbb{C}_p})$ satisfies the assumptions of Lemma 6.10, and that $f \in \hat{\mathcal{O}}_{\mathcal{Y}^{\mathrm{ss}}}(\mathcal{Y}^{\mathrm{ss}})$, so that $f_y \in \hat{\mathcal{O}}_{\mathcal{Y},y} = \hat{\mathcal{O}}^+_{\mathcal{Y},y}[1/p]$, $y^k_{\mathrm{dR}}f \in \mathcal{O}_\Delta(\mathcal{Y}^{\mathrm{ss}})$, and $(y^k_{\mathrm{dR}}f)_y \in \mathcal{O}_{\Delta,y}$. Then under the inclusion (6.33), we have*

$$(y^k_{\mathrm{dR}}f)_y(q_{\mathrm{dR}}^{1/p^b}) \in \mathcal{O}^+_{\Delta,y}[1/p] = \hat{\mathcal{O}}^+_{\mathcal{Y},y}[\![q_{\mathrm{dR}}^{1/p^b} - 1]\!][1/p] \tag{6.54}$$

for any even integer $k \geq 0$. Moreover, if $p > 2$, then (6.54) holds for any integer $k \geq 0$.

Proof. By (6.16), we have

$$y^2_{\mathrm{dR}} \in \bar{z}_{\mathrm{LT}}\hat{\mathcal{O}}^\times_{\mathcal{Y}}(\mathcal{LT}) \cdot \left(\frac{d(\mathbf{z}_{\mathrm{LT}}/\bar{z}_{\mathrm{LT}})}{d\mathbf{z}_{\mathrm{dR}}/p^b} \right) = \bar{z}_{\mathrm{LT}}\hat{\mathcal{O}}^\times_{\mathcal{Y}}(\mathcal{LT}) \cdot q_{\mathrm{dR}}^{1/p^b} \left(\frac{d(\mathbf{z}_{\mathrm{LT}}/\bar{z}_{\mathrm{LT}})}{dq_{\mathrm{dR}}^{1/p^b}} \right)$$

$$\overset{(6.32)}{\mapsto} \bar{z}_{\mathrm{LT}}\hat{\mathcal{O}}^\times_{\mathcal{Y}}(\mathcal{LT}) \cdot \mathcal{O}^+_{\Delta,y} \subset \hat{\mathcal{O}}_{\mathcal{Y}}(\mathcal{LT}) \cdot \mathcal{O}^+_{\Delta,y}$$

$$\tag{6.55}$$

where the map "$\mapsto$" denotes mapping to the stalk at y, and the entire equality in (6.55) takes place in the stalk $\hat{\mathcal{O}}_{\Delta,y}$. When k is even, raising to the $k/2^{\mathrm{th}}$ power implies

$$y_{\mathrm{dR},y} \in \hat{\mathcal{O}}^\times_{\mathcal{Y}}(\mathcal{LT}) \cdot \mathcal{O}^+_{\Delta,y},$$

and when $p > 2$ and k is odd, we can first take the square root of (6.55) and then raise to the kth power to get the same conclusion. Hence for $f \in \hat{\mathcal{O}}_{\mathcal{Y}}(\mathcal{Y}^{\mathrm{ss}})$, we have the equality in $\hat{\mathcal{O}}_{\Delta,y}$

$$(y^k_{\mathrm{dR}}f)_y \in \hat{\mathcal{O}}^\times_{\mathcal{Y},y} \cdot \hat{\mathcal{O}}_{\mathcal{Y}}(\mathcal{LT}) \cdot \mathcal{O}^+_{\Delta,y}.$$

By assumption, $(y^k_{\mathrm{dR}}f) \in \mathcal{O}_\Delta(\mathcal{Y}^{\mathrm{ss}})$, and hence

$$(y^k_{\mathrm{dR}}f)_y \in (\hat{\mathcal{O}}_{\mathcal{Y}}(\mathcal{LT}) \cdot \mathcal{O}^+_{\Delta,y}[1/p]) \cap \mathcal{O}_\Delta(\mathcal{Y}^{\mathrm{ss}}) = \mathcal{O}^+_{\Delta,y}[1/p],$$

where the intersection takes place in $\hat{\mathcal{O}}_{\Delta,y} = \mathcal{O}_{\Delta,y} \otimes_{\mathcal{O}_y} \hat{\mathcal{O}}_y$. $\square$

From Lemma 6.16, we see that integral properties of modular forms from finite level are somewhat preserved upon passing to infinite level (i.e., they are preserved at *stalks* at geometric points y satisfying the conditions of Lemma 6.10). Suppose now that $w \in \omega^{\otimes k}(Y^+)$ (where we recall Y^+ is the Katz-Mazur model of Y) is a normalized eigenform. Then since

$$\mathcal{Y}_x \twoheadrightarrow Y$$

and

$$\omega_{\mathrm{can}} = \mathfrak{s}/y_{\mathrm{dR}}$$

and $\mathfrak{s}$ trivializes $\omega_\Delta(\mathcal{Y}_x)$, we can write

$$w|_{\mathcal{Y}_x} = y_{\mathrm{dR}}^k f \cdot \omega_{\mathrm{can}}^{\otimes k}$$

where $y_{\mathrm{dR}}^k f \in \mathcal{O}_\Delta(\mathcal{Y}_x)$ and $(y_{\mathrm{dR}}^k f)_y \in \mathcal{O}_{\Delta,y}^+(\mathcal{Y}_x)[1/p]$ by Lemma 6.16.

Definition 6.17. *Let*

$$(y_{\mathrm{dR}}^k f)_y^\flat(q_{\mathrm{dR}}^{1/p^b}) \subset \hat{\mathcal{O}}_{\mathcal{Y}_x}^+[\![q_{\mathrm{dR}}^{1/p^b} - 1]\!][1/p]$$

denote the $\flat$-*stabilization of* $(y_{\mathrm{dR}}^k f)_y$, *which is*

$$(y_{\mathrm{dR}}^k f)_y^\flat(q_{\mathrm{dR}}^{1/p^b}) := (y_{\mathrm{dR}}^k f)_y(q_{\mathrm{dR}}^{1/p^b}) - \frac{1}{p}\sum_{j=0}^{p-1}(y_{\mathrm{dR}}^k f)_y(\zeta_p^j q_{\mathrm{dR}}^{1/p^b}). \tag{6.56}$$

We now define a global version of $\flat$-*stabilization, which we will use in Chapter 8, Section 8.3. Given* $F \in \mathcal{O}_{\Delta,\mathcal{Y}_x}(U)$ *with* q_{dR}^{1/p^b}-*expansion (see (6.6)) for any* $b \in \mathbb{Q}$ *given by*

$$F(q_{\mathrm{dR}}^{1/p^b}) = \sum_{n=0}^\infty a_n(F)(q_{\mathrm{dR}}^{1/p^b} - 1)^n \in \hat{\mathcal{O}}_{\mathcal{Y}_x}(U)[\![q_{\mathrm{dR}}^{1/p^b} - 1]\!],$$

we define

$$U_N(F)^+ := \{y \in U : |\theta(a_n(F))(y)| \le 1, \quad \forall n \le N\}$$

which defines a rational subset $U_N(F)^+ \subset U$, *we have*

$$U(F)^+ := \varprojlim_N U_N(F)^+ \in \mathcal{Y}_{x,pro\acute{e}t} = Y_{pro\acute{e}t}/\mathcal{Y}_x.$$

Definition 6.18. *We then define the* $\flat$-*stabilization of* F *to be*

$$F^\flat(q_{\mathrm{dR}}^{1/p^b}) := F(q_{\mathrm{dR}}^{1/p^b}) - \frac{1}{p}\sum_{j=0}^{p-1}\sum_{n=0}^\infty a_n(F)|_{U(F)^+}(\zeta_p^j q_{\mathrm{dR}}^{1/p^b} - 1)^n$$

$$\in \hat{\mathcal{O}}_{\mathcal{Y}_x}^+(U(F)^+)[\![q_{\mathrm{dR}}^{1/p^b} - 1]\!]$$

where one easily checks the inclusion

$$F^\flat(q_{\mathrm{dR}}^{1/p^b}) \in \hat{\mathcal{O}}_{\mathcal{Y}_x}^+(U(F)^+)[\![q_{\mathrm{dR}}^{1/p^b} - 1]\!]$$

using the definition of $F^\flat(q_{\mathrm{dR}}^{1/p^b})$ and the fact that

$$a_n|_{U(F)^+} \in \hat{\mathcal{O}}_{\mathcal{Y}_x}^+(U(F)^+)$$

for all $n \geq 0$.

It is clear that the stalk of $(y_{\mathrm{dR}}^k f)^\flat$ at y is just $(y_{\mathrm{dR}}^k f)_y(q_{\mathrm{dR}}^{1/p^b})$, since $y \in U(y_{\mathrm{dR}}^k f)^+(\mathbb{C}_p, \mathcal{O}_{\mathbb{C}_p})$ by Lemma 6.16.

Definition 6.19. *For any $j \in \mathbb{Z}/(p-1) \times \mathbb{Z}_p$, let $\{j_m\} \subset \mathbb{Z}_{\geq 0}$ be a sequence with $j_m \to j$. We define*

$$\delta_k^j F^\flat(q_{\mathrm{dR}}^{1/p^b}) := \lim_{m \to \infty} \delta_k^{j_m} F^\flat(q_{\mathrm{dR}}^{1/p^b}), \quad \delta_k^{\diamond, j} F^\flat(q_{\mathrm{dR}}^{1/p^b}) := \lim_{m \to \infty} \delta_k^{\diamond, j_m} F^\flat(q_{\mathrm{dR}}^{1/p^b}).$$

By the definition of $U(F)^+$ given in Definition 6.18, and the formulas in (5.25) and Theorem 5.34, it is clear that the above limits exist.

6.7 p-ADIC ANALYTIC PROPERTIES OF p-ADIC MAASS-SHIMURA OPERATORS

Definition 6.20. *Let $r \in \mathbb{Q}$ be minimal such that*

$$p^r(y_{\mathrm{dR}}^k f)_y^\flat(q_{\mathrm{dR}}^{1/p^b}) \in \hat{\mathcal{O}}_{\mathcal{Y}, y}^+[\![q_{\mathrm{dR}}^{1/p^b} - 1]\!], \tag{6.57}$$

where the subscript y, as usual, denotes the image in the stalk at y. The fact that such $r \in \mathbb{Q}$ exists follows from Lemma 6.16 and the definition of the stabilization (6.56), and r can be seen from the proof of Lemma 6.16 and (6.8) to only depend on y and $|f(y)|$.

Now let

$$p^r(y_{\mathrm{dR}}^k f)_y(q_{\mathrm{dR}}^{1/p^b}) = \sum_{n=0}^{\infty} a_n(q_{\mathrm{dR}}^{1/p^b} - 1)^n$$

where $a_n \in \hat{\mathcal{O}}_{\mathcal{Y}, y}^+$.

Theorem 6.21. *Let f, y, y_{dR} be as above satisfying the assumptions of Lemma 6.16, and*

$$|\theta(y_{\mathrm{dR}})(y) p^b / \hat{\mathbf{z}}(y)| < p^{1/(p-1)} \tag{6.58}$$

for y.

Then we have:

1. *The function* $\mathbb{Z}_{\geq 0} \to \hat{\mathcal{O}}_{\mathcal{Y},y}^+$ *defined by*

$$j \mapsto (p^b \theta_k)^j ((y_{\mathrm{dR}}^k f)_y^\flat (q_{\mathrm{dR}}^{1/p^b})),$$

where the subscript y denotes the image in the stalk at y, extends to a continuous function in $j \in \mathbb{Z}/(p-1) \times \mathbb{Z}_p \to \hat{\mathcal{O}}_{\mathcal{Y},y}$ by defining

$$(p^b \theta_k)^j ((y_{\mathrm{dR}}^k f)_y^\flat (q_{\mathrm{dR}}^{1/p^b})) := \lim_{m \to \infty} (p^b \theta_k)^{j_m} ((y_{\mathrm{dR}}^k f)_y^\flat (q_{\mathrm{dR}}^{1/p^b})) \qquad (6.59)$$

where writing

$$j - \sum_{n=0}^{\infty} \alpha_n p^n$$

uniquely with $0 \leq \alpha_n \leq p-1$, we let

$$j_m = \sum_{n=0}^{m} \alpha_n p^n.$$

(In particular, the above limit exists.)
For any $j_0 \in \mathbb{Z}_{\geq 0}$, we have

$$\lim_{m \to \infty} (p^b \theta_k)^{j_0 + (p-1)p^m} ((y_{\mathrm{dR}}^k f)_y (q_{\mathrm{dR}}^{1/p^b})) = (p^b \theta_k)^{j_0} ((y_{\mathrm{dR}}^k f)^\flat (q_{\mathrm{dR}}^{1/p^b})). \qquad (6.60)$$

2. *The function* $\mathbb{Z}_{\geq 0} \to \mathcal{O}_{\Delta,y}^{\diamond,+} = \mathcal{O}_{\Delta,y}^+ \left[\left(\frac{y_{\mathrm{dR}} p^b}{\hat{\mathbf{z}}} \right)_y \right]$ *defined by*

$$j \mapsto (p^b \delta_{\Delta,k}^\diamond)^j ((y_{\mathrm{dR}}^k f)_y^\flat (q_{\mathrm{dR}}^{1/p^b})),$$

where the subscript y denotes the image in the stalk at y, extends to a continuous function in $j \in \mathbb{Z}/(p-1) \times \mathbb{Z}_p \to \mathcal{O}_{\Delta,y}^{\diamond,+} = \mathcal{O}_{\Delta,y}^+ \left[\left(\frac{y_{\mathrm{dR}} p^b}{\hat{\mathbf{z}}} \right)_y \right]$ (where the target has the uniform convergence topology with respect to the topology on $\mathcal{O}_{\Delta,y}^+$) by defining

$$((p^b \delta_{\Delta,k})^\diamond)^j ((y_{\mathrm{dR}}^k f)_y^\flat (q_{\mathrm{dR}}^{1/p^b})) := \lim_{m \to \infty} (p^b \delta_{\Delta,k}^\diamond)^{j_m} ((y_{\mathrm{dR}}^k f)_y^\flat (q_{\mathrm{dR}}^{1/p^b})) \qquad (6.61)$$

where writing

$$j = \sum_{n=0}^{\infty} \alpha_n p^n$$

uniquely with $0 \leq \alpha_n \leq p-1$, we let

$$j_m = \sum_{n=0}^{m} \alpha_n p^n.$$

(In particular, the above limit exists.)
For any $j_0 \in \mathbb{Z}_{\geq 0}$, we have

$$\lim_{m \to \infty} (p^b \delta_{\Delta,k}^{\diamond})^{j_0+(p-1)p^m}((y_{\mathrm{dR}}^k f)_y(q_{\mathrm{dR}}^{1/p^b})) = (p^b \delta_{\Delta,k}^{\diamond})^{j_0}((y_{\mathrm{dR}}^k f)^{\flat}(q_{\mathrm{dR}}^{1/p^b})). \quad (6.62)$$

3. *The function $\mathbb{Z}_{\geq 0} \to \hat{\mathcal{O}}_{\mathcal{Y},y}^{+}$ defined by*

$$j \mapsto (p^b \delta_{\Delta,k}^{\diamond})^j((y_{\mathrm{dR}}^k f)_y^{\flat}(q_{\mathrm{dR}}^{1/p^b})),$$

also extends to a continuous function in $j \in \mathbb{Z}/(p-1) \times \mathbb{Z}_p \to \hat{\mathcal{O}}_{\mathcal{Y},y}$ by defining

$$\theta \circ ((p^b \delta_{\Delta,k}^{\diamond}))^j((y_{\mathrm{dR}}^k f)_y^{\flat}(q_{\mathrm{dR}}^{1/p^b})) := \lim_{m \to \infty} \theta \circ (p^b \delta_{\Delta,k}^{\diamond})^{j_m}((y_{\mathrm{dR}}^k f)_y^{\flat}(q_{\mathrm{dR}}^{1/p^b})). \quad (6.63)$$

Proof. (1): By (6.50), we have

$$(p^b \delta_k)^j((y_{\mathrm{dR}}^k f)_y^{\flat}(q_{\mathrm{dR}}^{1/p^b})) = \sum_{i=0}^{\infty} c_i(j) \frac{(p^b)^i}{(z_{\mathrm{dR}} - \bar{z})_y^i} \left(\frac{q_{\mathrm{dR}}^{1/p^b} d}{dq_{\mathrm{dR}}^{1/p^b}} \right)^{j-i} ((y_{\mathrm{dR}}^k f)_y^{\flat}(q_{\mathrm{dR}}^{1/p^b}))$$

$$(6.64)$$

for $j \in \mathbb{Z}_{\geq 0}$, where again the subscript y denotes the image in the stalk at y.
By (6.51), we have

$$(p^b \theta_k)^j((y_{\mathrm{dR}}^k f)_y^{\flat}(q_{\mathrm{dR}}^{1/p^b}))$$

$$= \sum_{i=0}^{\infty} c_i(j) \left(-\frac{\theta(y_{\mathrm{dR}})p^b}{\hat{\mathbf{z}}} \right)_y^i \theta \circ \left(\frac{q_{\mathrm{dR}}^{1/p^b} d}{dq_{\mathrm{dR}}^{1/p^b}} \right)^{j-i} ((y_{\mathrm{dR}}^k f)_y^{\flat}(q_{\mathrm{dR}}^{1/p^b})) \quad (6.65)$$

for $j \in \mathbb{Z}_{\geq 0}$. Note that since the q_{dR}^{1/p^b}-expansion map (6.18) is injective and commutes with derivations, it suffices to prove all statements on q_{dR}^{1/p^b}-expansions.

First, note that for any polynomial

$$g(q_{\mathrm{dR}}^{1/p^b}) = \sum_{n=0,\,p\nmid n}^{d} b_n \left(q_{\mathrm{dR}}^{1/p^b} \right)^n \in \hat{\mathcal{O}}_{\mathcal{Y},y}^+ [q_{\mathrm{dR}}^{1/p^b}] = \hat{\mathcal{O}}_{\mathcal{Y},y}^+ [q_{\mathrm{dR}}^{1/p^b} - 1],$$

we have

$$\left(\frac{q_{\mathrm{dR}}^{1/p^b} d}{dq_{\mathrm{dR}}^{1/p^b}} \right)^j g(q_{\mathrm{dR}}^{1/p^b}) = \sum_{n=0,\,p\nmid n}^{d} n^j b_n \left(q_{\mathrm{dR}}^{1/p^b} \right)^n$$

which, by Fermat's little theorem, is a continuous function in

$$j \in \mathbb{Z}/(p-1) \times \mathbb{Z}_p \to \hat{\mathcal{O}}_{\mathcal{Y},y}^+ [q_{\mathrm{dR}}^{1/p^b}] - \hat{\mathcal{O}}_{\mathcal{Y},y}^+ [q_{\mathrm{dR}}^{1/p^b} - 1],$$

where $\hat{\mathcal{O}}_{\mathcal{Y},y}^+ [q_{\mathrm{dR}}^{1/p^b} - 1]$ has the p-adic uniform convergence topology. (Note that it is important that $b_n \in \hat{\mathcal{O}}_{\mathcal{Y},y}^+$ for all $0 \le n \le d$ for this last assertion.)

In particular, writing

$$\left(\frac{q_{\mathrm{dR}}^{1/p^b} d}{dq_{\mathrm{dR}}^{1/p^b}} \right)^j g(q_{\mathrm{dR}}^{1/p^b}) = \sum_{n=0,\,p\nmid n}^{d} c_{n,j} (q_{\mathrm{dR}}^{1/p^b} - 1)^n,$$

by the above discussion we see that each coefficient $c_{n,j}$ is a continuous function in

$$j \in \mathbb{Z}/(p-1) \times \mathbb{Z}_p \to \hat{\mathcal{O}}_{\mathcal{Y},y}^+.$$

Moreover, note that for a more general polynomial

$$h(q_{\mathrm{dR}}^{1/p^b}) = \sum_{n=0}^{d} b_n \left(q_{\mathrm{dR}}^{1/p^b} \right)^n \in \hat{\mathcal{O}}_{\mathcal{Y},y}^+ [q_{\mathrm{dR}}^{1/p^b}] = \hat{\mathcal{O}}_{\mathcal{Y},y}^+ [q_{\mathrm{dR}}^{1/p^b} - 1],$$

we have

$$\left(\frac{q_{\mathrm{dR}}^{1/p^b} d}{dq_{\mathrm{dR}}^{1/p^b}} \right)^j h(q_{\mathrm{dR}}^{1/p^b}) = \sum_{n=0}^{d} n^j b_n \left(q_{\mathrm{dR}}^{1/p^b} \right)^n.$$

Letting $j = j_0 + (p-1)p^m$ for any $j_0 \in \mathbb{Z}$, we thus clearly see that

$$\lim_{m \to \infty} \left(\frac{q_{\mathrm{dR}}^{1/p^b} d}{dq_{\mathrm{dR}}^{1/p^b}} \right)^{j_0 + p^m(p-1)} = \sum_{n=0,\,p\nmid n}^{d} n^{j_0} b_n \left(q_{\mathrm{dR}}^{1/p^b} \right)^n. \tag{6.66}$$

We now need the following lemma.

Lemma 6.22. *Let p^r be as in (6.57) for $(y_{dR}^k f)_y^\flat (q_{dR}^{1/p^b})$. Each coefficient of*

$$\left(\frac{q_{dR}^{1/p^b} d}{dq_{dR}^{1/p^b}}\right)^j (p^r (y_{dR}^k f)_y^\flat (q_{dR}^{1/p^b}))(q_{dR}^{1/p^b}) =: \sum_{n=0}^{\infty} a_{0,n,j}(q_{dR}^{1/p^b} - 1)^n, \qquad (6.67)$$

where

$$a_{0,n,j} \in \hat{\mathcal{O}}_{\mathcal{Y},y}^+,$$

is a continuous function in $j \in \mathbb{Z}/(p-1) \times \mathbb{Z}_p \to \hat{\mathcal{O}}_{\mathcal{Y},y}^+$.
  Moreover, for any $j_0 \in \mathbb{Z}$, we have

$$\lim_{m \to \infty} \left(\frac{q_{dR}^{1/p^b} d}{dq_{dR}^{1/p^b}}\right)^{j_0 + p^m(p-1)} ((y_{dR}^k f)_y (q_{dR}^{1/p^b})) = \left(\frac{q_{dR}^{1/p^b} d}{dq_{dR}^{1/p^b}}\right)^{j_0} ((y_{dR}^k f)_y^\flat (q_{dR}^{1/p^b})).$$

$$(6.68)$$

Proof of Lemma 6.22. Note that since $a_{0,n,j}$ can be expressed recursively in terms of $a_{0,n,j-1}$ and $a_{0,n+1,j-1}$, then $a_{0,n,j}$ can be expressed entirely in terms of the coefficients

$$a_{0,n,0}, a_{0,n+1,0}, \ldots a_{0,n+j,0}.$$

Hence for any $N \geq 0$, there exists a polynomial truncation

$$(p^r (y_{dR}^k f)^\flat)_y^{(N)} (q_{dR}^{1/p^b}) = \sum_{n=0}^{N} a_n (q_{dR}^{1/p^b} - 1)^n \in \hat{\mathcal{O}}_{\mathcal{Y},y}^+ [q_{dR}^{1/p^b} - 1]$$

of $(p^r (y_{dR}^k f)_y^\flat)(q_{dR}^{1/p^b})$ such that letting

$$\left(\frac{q_{dR}^{1/p^b} d}{dq_{dR}^{1/p^b}}\right)^j (p^r (y_{dR}^k f)^\flat)_y^{(N)} (q_{dR}^{1/p^b}) = \sum_{n=0}^{N} a_{0,n,j}^{(N)} (q_{dR}^{1/p^b} - 1)^n, \qquad (6.69)$$

we have

$$a_{0,n,j} = a_{0,n,j}^{(N)}$$

for all $n + j \leq N$. By the previous paragraph, we know that each $a_{0,n,j}^{(N)}$ is a continuous function in

$$j \in \mathbb{Z}/(p-1) \times \mathbb{Z}_p \to \hat{\mathcal{O}}_{\mathcal{Y},y}^+.$$

In particular, all the $a_{0,n,j}^{(N)}$'s patch together and show that $a_{0,n,j}$ is a continuous function in

$$j \in \mathbb{Z}/(p-1) \times \mathbb{Z}_p \to \hat{\mathcal{O}}_{\mathcal{Y},y}^+.$$

Moreover, by the uniform convergence of the coefficients from the previous paragraph, we have that (6.69) is a continuous function in j, giving $\hat{\mathcal{O}}_{\mathcal{Y},y}[\![q_{\mathrm{dR}}^{1/p^b} - 1]\!]$ the uniform convergence topology.

Now write for any $j \in \mathbb{Z}_{\geq 0}$

$$\left(\frac{q_{\mathrm{dR}}^{1/p^b} d}{dq_{\mathrm{dR}}^{1/p^b}} \right)^j (y_{\mathrm{dR}}^k f)_y (q_{\mathrm{dR}}^{1/p^b}) = \sum_{n=0}^{\infty} b_{0,n,j} (q_{\mathrm{dR}}^{1/p^b} - 1)^n.$$

Then (6.68) follows by the same argument as above, expressing $b_{0,n,j}$ in terms of the coefficients $b_{0,n,0}, b_{0,n+1,0}, \ldots b_{0,n+j,0}$, then considering truncations $b_{0,n,j}^{(N)}$ and using (6.66). □

Fix any $j_0 \in \mathbb{Z}_{\geq 0}$. By (6.64), we have

$$\left(\frac{q_{\mathrm{dR}}^{1/p^b} d}{dq_{\mathrm{dR}}^{1/p^b}} \right)^j \left((p^b \delta_k)^{j_0} \left(p^r (y_{\mathrm{dR}}^k f)_y^b \right) \left(q_{\mathrm{dR}}^{1/p^b} \right) \right)$$

$$= \left(\frac{q_{\mathrm{dR}}^{1/p^b} d}{dq_{\mathrm{dR}}^{1/p^b}} \right)^j \left(\sum_{i=0}^{\infty} c_i(j_0) \frac{(p^b)^i}{(z_{\mathrm{dR}} - \bar{z})_y^i} \left(\frac{q_{\mathrm{dR}}^{1/p^b} d}{dq_{\mathrm{dR}}^{1/p^b}} \right)^{j_0 - i} \left(p^r (y_{\mathrm{dR}}^k f)_y^b \left(q_{\mathrm{dR}}^{1/p^b} \right) \right) \right).$$

(6.70)

Note that for all large $i \gg 0$, we have

$$c_i(j) \frac{q_{\mathrm{dR}}^{1/p^b} d}{dq_{\mathrm{dR}}^{1/p^b}} \frac{(p^b)^i}{(z_{\mathrm{dR}} - \bar{z})_y^i} = c_i(j) \frac{d}{dz_{\mathrm{dR}}} \frac{(p^b)^{i+1}}{(z_{\mathrm{dR}} - \bar{z})_y^i} = -c_i(j) \frac{i(p^b)^{i+1}}{(z_{\mathrm{dR}} - \bar{z})_y^{i+1}}$$

$$\overset{\theta}{\mapsto} -c_i(j) \cdot i \cdot \left(\frac{\theta(y_{\mathrm{dR}})p^b}{\hat{\mathbf{z}}} \right)_y^{i+1} \in \hat{\mathcal{O}}_{\mathcal{Y},y}^+$$

(6.71)

where the last inclusion follows since

$$c_i(j) \left(-\frac{\theta(y_{\mathrm{dR}})p^b}{\hat{\mathbf{z}}} \right)_y \in \hat{\mathcal{O}}_{\mathcal{Y},y}^+$$

(6.72)

for all large $i \gg 0$; this in turn follows from our assumption (6.58), the fact that

$$i! | c_i(j) \quad \forall j \in \mathbb{Z}_{\geq 0},$$

that $|i!|$ becomes arbitrarily close to $p^{1/p-1}$ as $i \to \infty$, and (6.7).

In particular, one can see from Lemma 6.22, (6.70) and (6.71) that

$$\theta \circ \left(\frac{q_{\mathrm{dR}}^{1/p^b} d}{d q_{\mathrm{dR}}^{1/p^b}} \right)^j (((p^b \delta_k)^{j_0} p^r (y_{\mathrm{dR}}^k f)_y^{\flat})(q_{\mathrm{dR}}^{1/p^b})) = a_{j_0,0,j} \tag{6.73}$$

is a continuous function in $j \in \mathbb{Z}/(p-1) \times \mathbb{Z}_p \to \hat{\mathcal{O}}_{\mathcal{Y},y}^+$. Now we can rewrite (6.65) as

$$\theta_{k+2j_0}^j \left(\left(p^b \delta_k \right)^{j_0} \left(p^r \left(y_{\mathrm{dR}}^k f \right)_y^{\flat} \right) \left(q_{\mathrm{dR}}^{1/p^b} \right) \right) = \sum_{i=0}^{\infty} c_i(j) \left(-\frac{\theta(y_{\mathrm{dR}}) p^b}{\hat{\mathbf{z}}} \right)_y^i a_{j_0,0,j-i} \tag{6.74}$$

where we see by the previous paragraph that each $a_{j_0,0,j-i}$ is a continuous function in

$$j \in \mathbb{Z}/(p-1) \times \mathbb{Z}_p \to \hat{\mathcal{O}}_{\mathcal{Y},y}.$$

Now we show that the extension of

$$j \mapsto (p^b \theta_k)^j \left(p^r (y_{\mathrm{dR}}^k f)_y^{\flat} \left(q_{\mathrm{dR}}^{1/p^b} \right) \right)$$

to $j \in \mathbb{Z}/(p-1) \times \mathbb{Z}_p$ is well-defined (i.e., the limit in (6.59) exists).

For this, given

$$j = \sum_{n=0}^{\infty} \alpha_n p^n \in \mathbb{Z}/(p-1) \times \mathbb{Z}_p$$

where $0 \leq \alpha_n \leq p-1$, let

$$j_m = \sum_{n=0}^{m} \alpha_n p^n.$$

Given any $\epsilon > 0$, choose $N > 0$ such that $|p^N| < \epsilon$ and such that

$$|a_{j_N,n,x} - a_{j_N,n,y}| < \epsilon \tag{6.75}$$

for any $x, y \in \mathbb{Z}/(p-1) \times \mathbb{Z}_p$ with

$$|x - y| < p^N$$

and for all $n \in \mathbb{Z}_{\geq 0}$. For each individual n, we can do this since $\mathbb{Z}/(p-1) \times \mathbb{Z}_p$ is compact and the functions

$$\mathbb{Z}/(p-1) \times \mathbb{Z}_p \ni x \mapsto a_{j_N,0,x} \in \hat{\mathcal{O}}_{\mathcal{Y},y}^{+}$$

as in (6.73) are analytic and hence continuous, and hence uniformly continuous. The fact that such N exists uniformly for all $n \in \mathbb{Z}_{\geq 0}$ follows from the statement that (6.69) is a continuous function in j when $\hat{\mathcal{O}}_{\mathcal{Y},y}^{+}[\![q_{\mathrm{dR}}^{1/p^b} - 1]\!]$ is endowed with the uniform convergence topology, which we proved above.

Note that for all $m > N$, we have for any $|x| \leq |p^N|$,

$$|c_i(x)| \leq |p^N|, \qquad |c_i(1-k+x)| \leq |p^N| \tag{6.76}$$

if $i > 0$, by (6.52). Then we have for all $m, m' > N$, since

$$|(p^b\theta_k)^{j_m}(p^r(y_{\mathrm{dR}}^k f)_y^\flat(q_{\mathrm{dR}}^{1/p^b})) - (p^b\theta_k)^{j_{m'}}(p^r(y_{\mathrm{dR}}^k f)_y^\flat(q_{\mathrm{dR}}^{1/p^b}))|$$

$$= |\theta_{k+2j_N}^{j_m-j_N} \circ (p^b\delta_k)^{j_N}(p^r(y_{\mathrm{dR}}^k f)_y^\flat(q_{\mathrm{dR}}^{1/p^b})) - \theta_{k+2j_N}^{j_{m'}-j_N} \circ (p^b\delta_k)^{j_N}(p^r(y_{\mathrm{dR}}^k f)_y^\flat(q_{\mathrm{dR}}^{1/p^b}))|$$

$$= \left| \sum_{i=0}^{\infty} c_i(j_{m'} - j_N)\left(-\frac{\theta(y_{\mathrm{dR}})}{\hat{\mathbf{z}}}\right)_y^i \theta \circ \left(\frac{q_{\mathrm{dR}}^{1/p^b} d}{dq_{\mathrm{dR}}^{1/p^b}}\right)^{j_{m'}-j_N-i}\left((p^b\delta_k)^{j_N}(p^r(y_{\mathrm{dR}}^k f)_y^\flat)(q_{\mathrm{dR}}^{1/p^b})\right)\right.$$

$$\left. - \sum_{i=0}^{\infty} c_i(j_m - j_N)\left(-\frac{\theta(y_{\mathrm{dR}})}{\hat{\mathbf{z}}}\right)_y^i \theta \circ \left(\frac{q_{\mathrm{dR}}^{1/p^b} d}{dq_{\mathrm{dR}}^{1/p^b}}\right)^{j_m-j_N-i}\left((p^b\delta_k)^{j_N}(p^r(y_{\mathrm{dR}}^k f)_y^\flat)(q_{\mathrm{dR}}^{1/p^b})\right)\right|$$

$$\overset{|j_m-j_N|\leq|p|^N,(6.76)}{\leq} \max\left(|p^N|, \left|\theta \circ \left(\frac{q_{\mathrm{dR}}^{1/p^b} d}{dq_{\mathrm{dR}}^{1/p^b}}\right)^{j_{m'}-j_N}\left((p^b\delta_k)^{j_N}(p^r(y_{\mathrm{dR}}^k f)_y^\flat)(q_{\mathrm{dR}}^{1/p^b})\right)\right.\right.$$

$$\left.\left. - \theta \circ \left(\frac{q_{\mathrm{dR}}^{1/p^b} d}{dq_{\mathrm{dR}}^{1/p^b}}\right)^{j_m-j_N}\left((p^b\delta_k)^{j_N}(p^r(y_{\mathrm{dR}}^k f)_y^\flat)(q_{\mathrm{dR}}^{1/p^b})\right)\right|\right)$$

$$\overset{(6.75)}{=} \max\left(|p^N|, |a_{j_N,0,j_{m'}} - a_{j_N,0,j_m}|\right) < \epsilon. \tag{6.77}$$

So we have shown that

$$\left\{(p^b\theta_k)^{j_m}(p^r(y_{\mathrm{dR}}^k f)_y^\flat(q_{\mathrm{dR}}^{1/p^b}))\right\}_{m\in\mathbb{Z}_{\geq 0}}$$

is a Cauchy sequence in the p-adic topology. Now the statement that (6.59) exists follows from the fact that $\hat{\mathcal{O}}_{\mathcal{Y},y}$ is p-adically complete.

We now show that the function defined by (6.59) is continuous as a function in

$$j \in \mathbb{Z}/(p-1) \times \mathbb{Z}_p \to \hat{\mathcal{O}}_{\mathcal{Y},y}.$$

Consider the sequence of truncations

$$(p^b \theta_k)^j (p^r (y_{\mathrm{dR}}^k f)_y^\flat (q_{\mathrm{dR}}^{1/p^b}))^{(N)} = \sum_{i=0}^N c_i(j) \left(-\frac{\theta(y_{\mathrm{dR}})p^b}{\hat{\mathbf{z}}} \right)_y^i a_{0,0,j-i} \qquad (6.78)$$

each of which is a continuous function in $j \in \mathbb{Z}/(p-1) \times \mathbb{Z}_p \to \hat{\mathcal{O}}_{\mathcal{Y},y}^+$, since the $a_{0,0,j-i}$ as in (6.73) are continuous.

We claim that the sequence

$$\{(p^b \theta_k)^j (p^r (y_{\mathrm{dR}}^k f)_y^\flat (q_{\mathrm{dR}}^{1/p^b}))^{(N)}\}_{N \in \mathbb{Z}_{\geq 0}}$$

converges uniformly to

$$(p^b \delta_k)^j (p^r (y_{\mathrm{dR}}^k f)_y^\flat (q_{\mathrm{dR}}^{1/p^b}))$$

as a function in

$$j \in \mathbb{Z}/(p-1) \times \mathbb{Z}_p \to \hat{\mathcal{O}}_{\mathcal{Y},y},$$

and hence

$$(p^b \delta_k)^j (p^r (y_{\mathrm{dR}}^k f)_y^\flat (q_{\mathrm{dR}}^{1/p^b}))$$

is a continuous function in

$$j \in \mathbb{Z}/(p-1) \times \mathbb{Z}_p \to \hat{\mathcal{O}}_{\mathcal{Y},y}$$

by the standard argument from analysis.

Let us now show the uniform convergence. Given $j \in \mathbb{Z}/(p-1) \times \mathbb{Z}_p$, suppose we are given any $\epsilon > 0$. Choose $t \in \mathbb{Z}_{\geq 0}$ so that $|p^t| < \epsilon$. Note that for all $i \geq p^t$

$$c_i(j) \in p^t \mathbb{Z}_p \qquad (6.79)$$

for all $j \in \mathbb{Z}/(p-1) \times \mathbb{Z}_p$ (since one sees from the definition of $c_i(j)$ that for every residue class $\mathbb{Z}_p/(p^t)$, $c_i(j)$ is divisible by some representative of that residue class). Then for all $N \geq p^t$, we have

$$|(p^b \theta_k)^j (p^r (y_{\mathrm{dR}}^k f)_y^\flat (q_{\mathrm{dR}}^{1/p^b})) - (p^b \theta_k)^j (p^r (y_{\mathrm{dR}}^k f)_y^\flat (q_{\mathrm{dR}}^{1/p^b}))^{(N)}|$$

$$= \left| \sum_{n=N}^{\infty} c_i(j) \left(-\frac{\theta(y_{\mathrm{dR}})p^b}{\hat{\mathbf{z}}} \right)_y^i a_{0,0,j-i} \right|$$

$$\leq \max_{N \leq i < \infty} \left(c_i(j) \left(-\frac{\theta(y_{\mathrm{dR}})p^b}{\hat{\mathbf{z}}} \right)_y^i a_{0,0,j-i} \right) \overset{(6.72),(6.73),(6.79),}{\leq} |p^t| < \epsilon.$$

This gives the desired uniform convergence.

Now (6.60) follows from (6.68) in Lemma 6.22 and the calculation (6.71).

(2): The proof of (2) is almost entirely the same as that of (1). By (6.53), we have

$$(p^b \delta_{\Delta,k}^{\diamond})^j (p^r f)_y^{\flat} (q_{\mathrm{dR}}^{1/p^b}) = \sum_{i=0}^{\infty} c_i(j) \left(-\frac{y_{\mathrm{dR}}p^b}{\hat{\mathbf{z}}} \right)^i \left(\frac{q_{\mathrm{dR}}^{1/p^b} d}{dq_{\mathrm{dR}}^{1/p^b}} \right)^{j-i} (p^r f)_y^{\flat} (q_{\mathrm{dR}}^{1/p^b}).$$

By Lemma (6.22) and the continuity of $c_i(j)$, we see that each term

$$j \mapsto c_i(j) \left(-\frac{y_{\mathrm{dR}}p^b}{\hat{\mathbf{z}}} \right)_y^i \left(\frac{q_{\mathrm{dR}}^{1/p^b} d}{dq_{\mathrm{dR}}^{1/p^b}} \right)^{j-i} (p^r f)_y^{\flat} (q_{\mathrm{dR}}^{1/p^b}) \qquad (6.80)$$

extends to a continuous function $\mathbb{Z}/(p-1) \times \mathbb{Z}_p \to \mathcal{O}_{\Delta,y}^+$. Hence, for each $N \in \mathbb{Z}_{\geq 0}$, the truncation

$$j \mapsto (p^b \delta_{\Delta,k}^{\diamond})^j (p^r f)_y^{\flat} (q_{\mathrm{dR}}^{1/p^b})^N = \sum_{i=0}^{N} c_i(j) \left(-\frac{y_{\mathrm{dR}}p^b}{\hat{\mathbf{z}}} \right)_y^i \left(\frac{q_{\mathrm{dR}}^{1/p^b} d}{dq_{\mathrm{dR}}^{1/p^b}} \right)^{j-i} (p^r f)_y^{\flat} (q_{\mathrm{dR}}^{1/p^b})$$

$$(6.81)$$

extends to a continuous function $\mathbb{Z}/(p-1) \times \mathbb{Z}_p \to \mathcal{O}_{\Delta,y}^{\diamond,+} = \mathcal{O}_{\Delta,y}^+ \left[\left(\frac{y_{\mathrm{dR}}p^b}{\hat{\mathbf{z}}} \right)_y \right]$. Let $\|\cdot\|$ denote the Gauss norm on $\mathcal{O}_{\Delta,y}^+$. We have by Lemma (6.22) that

$$\left\| \left(\frac{q_{\mathrm{dR}}^{1/p^b} d}{dq_{\mathrm{dR}}^{1/p^b}} \right)^{j-i} (p^r f)_y^{\flat} (q_{\mathrm{dR}}^{1/p^b}) \right\| \leq 1$$

and so

$$\left\| c_i(j) \left(\frac{q_{\mathrm{dR}}^{1/p^b} d}{dq_{\mathrm{dR}}^{1/p^b}} \right)^{j-i} (p^r f)_y^{\flat} (q_{\mathrm{dR}}^{1/p^b}) \right\| \leq |c_i(j)| \to 0 \quad \text{as} \quad i \to \infty.$$

Hence the sequence of truncations (6.81) *uniformly* converges in the topology of uniform convergence on $\mathcal{O}_{\Delta,y}^{\diamond,+} = \mathcal{O}_{\Delta,y}^{+}\left[\left(\frac{y_{\mathrm{dR}}p^{b}}{\hat{z}}\right)_{y}\right]$ to a continuous function in $\mathbb{Z}/(p-1) \times \mathbb{Z}_{p} \to \mathcal{O}_{\Delta,y}^{\diamond,+}$. This is the desired continuous function (6.61).

Now (6.62) follows from (6.68) in Lemma 6.22 and the calculation (6.71).

(3): This follows by applying θ to (2), and noting that $\theta:\mathcal{O}_{\Delta,y}^{\diamond,+} \to \hat{\mathcal{O}}_{\mathcal{Y},y}^{+}$ is continuous. $\qquad\square$

Chapter Seven

Bounding periods at supersingular CM points

In this chapter, let K be an imaginary quadratic field. We construct a one-variable p-adic L-function for Rankin-Selberg families (w, χ^{-1}), where f is an eigenform of weight k for $\Gamma_1(N)$ and χ runs through central anticyclotomic characters of K. This will be done by applying Theorem 6.21 in the case where y is a supersingular CM point. We hence construct a p-adic L-function, which is a continuous function on a space of p-adic Hecke characters (the closure of p-adic central critical characters for f inside the space of functions

$$\mathbb{A}_K^{\times,(p\infty)} \to \mathcal{O}_{\mathbb{C}_p},$$

where $\mathbb{A}_K^{\times,(p\infty)}$ is the group of idèles prime to $p\infty$, with the uniform convergence topology). We use Theorem 5.48 to show that our p-adic L-function satisfies an interpolation property, namely that special values of our p-adic L-function are, up to normalizations by certain periods, equal to central L-values

$$L(w, \chi^{-1}, 0) := L((\pi_w)_K \times \chi^{-1}, 1/2),$$

as χ varies through a sequence of central critical characters in an "interpolation range" inside a space of characters.

We conclude by showing that in the case $k = 2$, a special value of our p-adic L-function is equal to the evaluation at a certain Heegner point of the formal logarithm of a certain differential in the same Hecke isotypic component away from p as that of f. We show that if this special value is nonzero, then the Heegner point descends to a point of infinite order in $A_f(K)$ for a GL_2-type abelian variety A_f attached to f. In particular, combined with the interpolation property described in the previous paragraph, this gives a new criterion to produce a point of infinite order in $A_f(K)$, namely showing that a sequence of central critical L-values has a nonzero limit.

7.1 PERIODS OF SUPERSINGULAR CM POINTS

Now fix an imaginary quadratic field $K = \mathbb{Q}(\sqrt{-d})$, where d is squarefree, and with fundamental discriminant d_K. Assume p is inert or ramified in $K/\mathbb{Q}$, and let $\mathfrak{p}$ denote the prime of K above p, and let K_p denote the completion of K at $\mathfrak{p}$. Let $\mathcal{O} \subset \mathcal{O}_K$ be a fixed suborder. Let H denote the ring class field over K associated with $\mathcal{O}$.

Suppose that A/H is an elliptic curve with CM by $\mathcal{O}$. For any field extension F/H, there is a canonical algebraic splitting

$$H^1_{\mathrm{dR}}(A/F) \cong H^{1,0}(A/F) \oplus H^{0,1}(A/F) = \Omega^1_{A/F} \oplus (\Omega^1_{A/F})^* \qquad (7.1)$$

of the Hodge–de Rham sequence

$$0 \to H^{1,0}(A/F) = \Omega^1_{A/F} \to H^1_{\mathrm{dR}}(A/F) \to H^{0,1}(A/F) = (\Omega^1_{A/F})^* \to 0$$

given by the action of $\mathcal{O}$: for

$$\gamma \in \mathcal{O} = \mathrm{End}(A/F),$$

γ^* acts on $H^1_{\mathrm{dR}}(A/F)$ as multiplication by γ on $H^{1,0}(A/F)$ and as multiplication by $\overline{\gamma}$ on $H^{0,1}(A/F)$.

By virtue of it splitting the Hodge filtration, this splitting is also compatible with Poincaré duality, that is, under the specialization of the Poincaré pairing (i.e., the de Rham cup product)

$$H^1_{\mathrm{dR}}(A/F) \times H^1_{\mathrm{dR}}(A/F) \to F$$

we have that

$$H^{1,0}(A/F) = \Omega^1_{A/F} \quad \text{and} \quad H^{0,1}(A/F) \quad \text{are dual.}$$

Moreover, recall the integral Hodge-Tate complex as stated in Theorem 3.13

$$0 \to \mathrm{Lie}(A/\mathcal{O}_{\mathbb{C}_p})(1) \xrightarrow{(HT_A)^\vee (1)} T_p A \otimes_{\mathbb{Z}_p} \mathcal{O}_{\mathbb{C}_p} \xrightarrow{HT_A} = \Omega^1_{A/\mathcal{O}_{\mathbb{C}_p}} \to 0. \qquad (7.2)$$

Consider $(A, \alpha) \in \mathcal{Y}(\mathbb{C}_p, \mathcal{O}_{\mathbb{C}_p})$ for any p^∞-level structure $\alpha : \mathbb{Z}_p^{\oplus 2} \xrightarrow{\sim} T_p A$. Specializing Proposition 5.3 to (A, α), we have an injective map

$$\Omega^1_{A/F} \subset \Omega^1_{A/F} \otimes_F \mathcal{O}_\Delta(A, \alpha) \xrightarrow{\iota_{\mathrm{dR}}(A, \alpha)} T_p A \otimes_{\mathbb{Z}_p} \mathcal{O}_\Delta(A, \alpha) \xrightarrow{\theta} T_p A \otimes_{\mathbb{Z}_p} \mathbb{C}_p \qquad (7.3)$$

whose composition with the natural map $T_p A \otimes_{\mathbb{Z}_p} \mathbb{C}_p \xrightarrow{HT_A} \Omega^1_{A/\mathbb{C}_p}$ is the natural inclusion $\Omega^1_{A/F} \subset \Omega^1_{A/\mathbb{C}_p}$. The next proposition gives more information on (7.3) in the case where A has complex multiplication.

Proposition 7.1. *Suppose A is an elliptic curve with CM by some order $\mathcal{O} \subset \mathcal{O}_K$ of conductor prime to p, and with p^∞-level structure α such that $|\hat{\mathbf{z}}(A, \alpha)| \geq$*

1. *Then for any generator*

$$\omega_0 \in \Omega^1_{A/\mathcal{O}_{\mathbb{C}_p}}$$

we have

$$|\mathfrak{s}(A, \alpha)/\omega_0| = p^{-\left(\frac{1}{p-1} - \frac{1}{e(\#\kappa - 1)}\right)}, \tag{7.4}$$

where $\#\kappa$ is the order of the residue field of $\mathcal{O}_{K_\mathfrak{p}}$, i.e.,

$$\#\kappa = \begin{cases} p & p \text{ ramified in } K/\mathbb{Q} \\ p^2 & p \text{ inert in } K/\mathbb{Q}, \end{cases}$$

and e is the ramification index of $K_\mathfrak{p}/\mathbb{Q}_p$.

Proof. By (7.2), we have $\mathfrak{s}(A, \alpha) \subset \Omega^1_{A/\mathcal{O}_{\mathbb{C}_p}}$. If $|\hat{\mathfrak{z}}(A, \alpha)| \geq 1$, then $\mathfrak{s}(A, \alpha)$ generates the image of HT_A in (7.2), and consequently, $\Omega = \mathfrak{s}(A, \alpha)/\omega_0$ is a Lubin-Tate period associated with the formal group of A, in the terminology of [63, Appendix]. By Theorem c of the appendix of loc. cit., we thus get (7.4). $\square$

Corollary 7.2. *Let $y = (A, \alpha) \in \mathcal{Y}(\mathbb{C}_p, \mathcal{O}_{\mathbb{C}_p})$ be a CM point (i.e., A is an elliptic curve with CM by some order $\mathcal{O} \subset \mathcal{O}_K$). Then in the notation of Definition 5.45 for y, we have*

$$|\theta(\Omega_p(A, \alpha))| = p^{\left(\frac{1}{p-1} - \frac{1}{e(\#\kappa - 1)}\right)} |\theta(y_{\mathrm{dR}}(A, \alpha))|. \tag{7.5}$$

Proof. Let $\omega_0 \in \Omega^1_{A/\mathcal{O}_{\mathbb{Q}}}$ be an integral generator as in Definition 5.45. We have

$$\frac{\omega_0}{\theta(\Omega_p(A, \alpha))} = \omega_{\mathrm{can}}(A, \alpha) = \frac{\mathfrak{s}(A, \alpha)}{y_{\mathrm{dR}}(A, \alpha)}.$$

Now the assertion follows from (7.4). $\square$

We now analyze the Hodge-Tate, de Rham and Ω_p periods of CM points. We first deal with the case where p is ramified in K.

Proposition 7.3. *Suppose p is ramified in K and A is an elliptic curve with CM by $\mathcal{O} \subset \mathcal{O}_K$, say $\mathcal{O} = \mathbb{Z} + c\mathcal{O}_K$ with $c \in \mathbb{Z}_{>0}$ prime to p, so that $\mathcal{O} \otimes_{\mathbb{Z}} \mathbb{Z}_p = \mathcal{O}_{K_p}$. Let π denote a uniformizer of $\mathcal{O}_{K_p}$.*

Let t_p denote any $\Gamma_0(p^n)$-level structure on A, i.e., the image $\langle t_p \rangle \subset A[p^n]$ is an order p^n cyclic subgroup, such that $\langle \alpha_1 \pmod{p} \rangle = A[\pi]$. Then there exists a p^∞-level structure $\alpha : \mathbb{Z}_p^{\oplus 2} \xrightarrow{\sim} T_p A$ with

$$\langle \alpha_1 \pmod{p^n} \rangle = \langle t_p \rangle, \qquad \langle \alpha_1 \pmod{p} \rangle = A[\pi]$$

and

$$\hat{\mathbf{z}}(A,\alpha), z_{\mathrm{dR}}(A,\alpha), y_{\mathrm{dR}}(A,\alpha) \in K_p,$$

and

$$|\hat{\mathbf{z}}(A,\alpha)| = p^{1/2}, \quad |z_{\mathrm{dR}}(A,\alpha)| = p^{1/2}, \quad |y_{\mathrm{dR}}(A,\alpha)| = |\theta(y_{\mathrm{dR}}(A,\alpha))| = |2|^{-1}.$$

Furthermore

$$|\theta(\Omega_p(A,\alpha))| = p^{\frac{1}{2(p-1)}}|2| \tag{7.6}$$

for $\Omega_p(A,\alpha)$ as defined in Definition 5.45.
   Moreover, letting

$$q := \begin{cases} p & p>2 \\ 8 & p=2 \end{cases}$$

and defining (A,α') by

$$(A',\alpha') = (A,\alpha) \cdot \begin{pmatrix} q & 0 \\ 0 & 1 \end{pmatrix},$$

then $\alpha' : \mathbb{Z}_p^2 \xrightarrow{\sim} T_pA$ satisfies $\langle \alpha_1' \pmod{p} \rangle$ is the canonical subgroup of A', and

$$\hat{\mathbf{z}}(A,\alpha'), z_{\mathrm{dR}}(A,\alpha'), y_{\mathrm{dR}}(A,\alpha') \in K_p,$$

and

$$|\hat{\mathbf{z}}(A,\alpha')| = qp^{1/2}, \quad |z_{\mathrm{dR}}(A,\alpha')| = qp^{1/2}, \quad |y_{\mathrm{dR}}(A,\alpha')| = |\theta(y_{\mathrm{dR}}(A,\alpha'))| = |2|^{-1}.$$

Proof. Since $\mathcal{O} \otimes_{\mathbb{Z}} \mathbb{Z}_p = \mathcal{O}_{K_p}$, then T_pA has CM by $\mathcal{O}_{K_p}$. We have by Shimura's reciprocity law that $A \to A[\pi]$ lifts p-power Frobenius, and so $A[\pi]$ is the canonical subgroup of A.
   Pick any $\alpha_2 : \mathbb{Z}_p \hookrightarrow T_pA$, and pick any $\alpha_1 : \mathbb{Z}_p \hookrightarrow T_pA$ extending the given $\Gamma_0(p^n)$-level structure t_p (i.e., with $\langle \alpha_1 \pmod{p^n} \rangle = \langle t_p \rangle$ with $\langle \alpha_1 \pmod{p} \rangle = A[\pi]$). By the main result of [29], or by directly observing, from (7.1), that $\alpha_1 - (1/z_{\mathrm{dR}}(A,\alpha))\alpha_2$ and $\alpha_1 - (1/\hat{\mathbf{z}}(A,\alpha))\alpha_2$ parametrize the eigenspaces of $T_pA \otimes_{\mathbb{Z}_p} \mathbb{C}_p$ on which the CM action $\mathcal{O}_{K_p}$ acts by multiplication on the right and conjugate multiplication on the right, respectively, we deduce

$$\hat{\mathbf{z}}(A,\alpha'), z_{\mathrm{dR}}(A,\alpha') \in K_p,$$

and hence

$$y_{\mathrm{dR}}(A,\alpha) \in K_p$$

by (4.17). In particular, $z_{\mathrm{dR}}(A,\alpha)$ and $\hat{\mathbf{z}}(A,\alpha)$ are conjugate in $\mathcal{O}_{K_p}$.

Since $\langle \alpha_1 \ (\mathrm{mod} \ p) \rangle = A[\pi]$, and the sequence of matrix actions

$$(A, \alpha) \xrightarrow{\begin{pmatrix} 0 & 1 \\ p & 0 \end{pmatrix}} (A, \alpha) \cdot \begin{pmatrix} 0 & 1 \\ p & 0 \end{pmatrix} \xrightarrow{\begin{pmatrix} 0 & 1 \\ p & 0 \end{pmatrix}} (A, \alpha) \cdot \begin{pmatrix} p & 0 \\ 0 & p \end{pmatrix} = (A, \alpha)$$

corresponds to the sequence of isogenies

$$A \to A/A[\pi] \to (A/A[\pi])/(A/A[\pi])[\pi] = A,$$

we see that

$$|p\hat{\mathbf{z}}(A, \alpha)| = |1/\hat{\mathbf{z}}(A, \alpha)|$$

and so

$$|\hat{\mathbf{z}}(A, \alpha)| = p^{1/2},$$

which implies

$$|z_{\mathrm{dR}}(A, \alpha)| = p^{1/2}.$$

Finally, denoting the CM action on A by $[\cdot]$, we see that $[1/z_{\mathrm{dR}}(A, \alpha)](\alpha_2) = \alpha_1$, and consequently letting $u \in \mathcal{O}_{K_p}^\times$ such that $u/z_{\mathrm{dR}}(A, \alpha) = \sqrt{d}$ (recall $|\sqrt{d}| = p^{1/2}$) and replacing α_1 by $[u](\alpha_1)$, we have $z_{\mathrm{dR}}(A, \alpha) = \frac{1}{\sqrt{d}}$, and consequently $\hat{\mathbf{z}}(A, \alpha) = -\frac{1}{\sqrt{d}}$. Now from (4.17), we get

$$y_{\mathrm{dR}}(A, \alpha) = \theta(y_{\mathrm{dR}}(A, \alpha)) = 1 - z_{\mathrm{dR}}(A, \alpha)/\hat{\mathbf{z}}(A, \alpha) = 2.$$

Now (7.6) follows from (7.5).

The corresponding claims for (A', α') now follow easily from (3.15), (4.10) and (4.44). The assertion that $\langle \alpha_1' \ (\mathrm{mod} \ p) \rangle$ is the canonical subgroup of A' follows from $|\hat{\mathbf{z}}(A', \alpha')| = qp^{1/2} > p^{p/(p^2-1)}$ and Lemma 6.11. $\square$

Corollary 7.4. *For (A, α) and associated (A', α') as in Proposition 7.3, we have*

$$\left| \theta \left(q \frac{d(\mathbf{z}_{\mathrm{LT}}/\bar{\mathbf{z}}_{\mathrm{LT}})}{dz_{\mathrm{dR}}} \right)(A', \alpha') \right| \leq 1, \qquad \left| \theta \left(\frac{y_{\mathrm{dR}}}{\hat{\mathbf{z}}} \right)(A', \alpha') \right| < 1. \qquad (7.7)$$

In particular, (6.31) and the assumptions of Corollary 6.21 are satisfied for the stalk at (A', α') with $b = 1$ if $p > 2$ and with $b = 3$ if $p = 2$.

Proof. The second inequality of (7.7) follows immediately from Proposition 7.3. Let $y' = (A', \alpha')$. Recall the canonical derivation $d : \mathcal{O}_Y \to \Omega_Y^1$. We want to compute the valuation

$$\left| \theta(\frac{d\hat{\mathbf{z}}}{dz_{\mathrm{dR}}})(A', \alpha') \right|$$

for any generator $d\mathfrak{z}$ of the integral differentials $d(\mathcal{O}_{Y,y'})$. By (4.52), we have

$$\sigma^{-1}(dz_{\mathrm{dR}}(A',\alpha')) = \mathfrak{s}(A',\alpha')^{\otimes 2}/y_{\mathrm{dR}}(A',\alpha')^2.$$

So to understand the p-adic valuation, we need to understand the p-adic valuation of $\mathfrak{s}(A',\alpha')$ and $y_{\mathrm{dR}}(A',\alpha')$. For the former, we will calculate this from our knowledge of $\mathfrak{s}(A,\alpha)$ (which we saw above had p-adic valuation equal to that of a Lubin-Tate period).

By Proposition 7.3 and Lemma 6.11, $\langle \alpha_1' \pmod{p} \rangle \subset A'[p]$ is the canonical subgroup of A', and the kernel of the isogeny $A' \to A$ given by the action of $\begin{pmatrix} 1 & 0 \\ 0 & q \end{pmatrix}$ sending the stalk at y' to $y = (A,\alpha)$. In particular, the dual isogeny $\phi:$ $A \to A'$ is Verschiebung modulo $p^{1-\mathrm{Hdg}(A')}$ (recall that $\mathrm{Hdg}(A) = \mathrm{ord}(\mathrm{Ha}(A'))$). In particular, the image of

$$\phi: \Omega^1_{A'/\mathcal{O}_{\mathbb{C}_p}} \to \Omega^1_{A/\mathcal{O}_{\mathbb{C}_p}}$$

is $\mathrm{Ha}(A')\Omega^1_{A/\mathcal{O}_{\mathbb{C}_p}}$. Since $\mathfrak{s}(A,\alpha)$ generates $p^{-p/(p^2-1)}\Omega^1_{A/\mathcal{O}_{\mathbb{C}_p}}$, and $\phi^*\mathfrak{s}(A',\alpha') = \mathfrak{s}(A,\alpha)$ (which follows from $\begin{pmatrix} q & 0 \\ 0 & 1 \end{pmatrix}^* \mathfrak{s} = \mathfrak{s}$, by (4.7)), $\phi^*\mathfrak{s}(A',\alpha')$ generates

$$(p^{-p/(p^2-1)}/\mathrm{Ha}(A'))\Omega^1_{A'/\mathcal{O}_{\mathbb{C}_p}}.$$

Since $\langle \alpha_1' \pmod{p} \rangle \subset A'[p]$ is the canonical subgroup, by [36, Theorem 3.10.7], $|\mathrm{Ha}(A')| = p^{-1/(p+1)}$. Hence $\mathfrak{s}(A',\alpha')$ generates

$$p^{-p/(p^2-1)+1/(p+1)}\Omega^1_{A'/\mathcal{O}_{\mathbb{C}_p}} = p^{-1/(p^2-1)}\Omega^1_{A'/\mathcal{O}_{\mathbb{C}_p}}.$$

Recall that by Proposition 7.3 we have $|\theta(y_{\mathrm{dR}})(A',\alpha')| = |2|^{-1}$. So in all, we have

$$\left|\theta(\frac{d\mathfrak{z}}{dz_{\mathrm{dR}}})(A',\alpha')\right| = p^{-1/(p^2-1)}|2|^{-2}.$$

Recalling the canonical derivation $d: \mathcal{O}_Y \to \Omega^1_Y$, note that because $\mathbf{z}_{\mathrm{LT}}/|\mathbf{z}_{\mathrm{LT}}(y)| \in \mathcal{O}^+_{Y(\Gamma),y'}$ for all y (by (6.7)),

$$\theta(d(\mathbf{z}_{\mathrm{LT}}/\bar{z}_{\mathrm{LT}}))_{y'} \in d(\mathcal{O}^+_{Y,y})$$

for all y, and so

$$\left|\theta(\frac{d((\mathbf{z}_{\mathrm{LT}}/\bar{z}_{\mathrm{LT}}))}{d\mathfrak{z}})(A',\alpha')\right| \leq 1.$$

Hence, in all

$$\left| \theta \left(q \frac{d(\mathbf{z}_{\mathrm{LT}}/\bar{z}_{\mathrm{LT}})}{dz_{\mathrm{dR}}} \right) (A', \alpha') \right| \leq |q| p^{-1/(p^2-1)} |2|^{-2} \leq 1.$$

$\square$

To deal with the p inert in K case, we need to recall another result on canonical subgroups.

Definition 7.5. *For an abelian variety A over $\mathcal{O}_{\mathbb{C}_p}$, let $\mathrm{Ha}(A)$ denote the (truncated) Hasse invariant, so that*

$$|\mathrm{Ha}(A)| \in [1/p, 1].$$

For more details, see [18, Section 1].

Proposition 7.6 (Theorem 4.2.5 of [18], see also p. 2 of [17]). *Suppose A is an elliptic curve. Then A admits a canonical subgroup (of order p^n) if and only if $|\mathrm{Ha}(A)| > p^{-1/p^{n-2}(p+1)}$.*

Proposition 7.7. *Suppose p is inert in K. Then suppose A is an elliptic curve with CM by $\mathcal{O}_K$. Then for any p^∞-level structure $\alpha : \mathbb{Z}_p^{\oplus 2} \xrightarrow{\sim} T_p A$, we have*

$$\hat{\mathbf{z}}(A, \alpha), z_{\mathrm{dR}}(A, \alpha), y_{\mathrm{dR}}(A, \alpha) \in K_p,$$

and

$$|\hat{\mathbf{z}}(A, \alpha)| = 1, \qquad |z_{\mathrm{dR}}(A, \alpha)| = 1,$$

and moreover there exists such α with

$$|y_{\mathrm{dR}}(A, \alpha)| = |\theta(y_{\mathrm{dR}}(A, \alpha))| = |2|^{-1}.$$

Furthermore

$$|\theta(\Omega_p(A, \alpha))| = p^{\frac{p}{p^2-1}} |2| \tag{7.8}$$

for $\Omega_p(A, \alpha)$ as defined in Definition 5.45.
    Moreover, letting

$$q := \begin{cases} p & p > 2 \\ 8 & p = 2 \end{cases}$$

and defining (A, α') by

$$(A', \alpha') = (A, \alpha) \cdot \begin{pmatrix} q & 0 \\ 0 & 1 \end{pmatrix},$$

then $\alpha' : \mathbb{Z}_p^2 \xrightarrow{\sim} T_p A$ *satisfies* $\langle \alpha_1' \pmod{p} \rangle$ *is the canonical subgroup of* A', *and*

$$\hat{\mathbf{z}}(A, \alpha'), z_{\mathrm{dR}}(A, \alpha'), y_{\mathrm{dR}}(A, \alpha') \in K_p,$$

and

$$|\hat{\mathbf{z}}(A, \alpha')| = q, \qquad |z_{\mathrm{dR}}(A, \alpha')| = q, \qquad |y_{\mathrm{dR}}(A, \alpha')| = |\theta(y_{\mathrm{dR}}(A, \alpha'))| = |2|^{-1}.$$

Proof. Since p is inert in K, by Deuring's theorem A is supersingular and so

$$|\mathrm{Ha}(A)| < 1.$$

Since $\mathrm{Ha}(A) \in K_p$, its p-adic valuation is an integral power of p, and so

$$|\mathrm{Ha}(A)| = 1/p.$$

Now by Proposition 7.6, we have that A does *not* have a canonical subgroup. Hence, by Lemma 6.11,

$$|z(A, \alpha)| \leq p^{p/(p^2 - 1)}$$

for any p^∞-level structure α on A.

By the discussion at the beginning of this section, the Hodge-Tate sequence splits over K_p. By [29, Theorem 1.2 (2)], we have

$$K_p = \mathbb{Q}_p(\hat{\mathbf{z}}(A, \alpha)). \tag{7.9}$$

In particular $|\hat{\mathbf{z}}(A, \alpha)|$ is an integral power of p, and so the previous paragraph implies that

$$|\hat{\mathbf{z}}(A, \alpha)| \leq 1.$$

Since this holds for all (A, α), repeating the argument after replacing α with

$$(A, \alpha) \cdot \begin{pmatrix} 0 & 1 \\ 1 & 0 \end{pmatrix}$$

and using from (3.15)

$$\mathbf{z}\left((A, \alpha) \cdot \begin{pmatrix} 0 & 1 \\ 1 & 0 \end{pmatrix} \right) = 1/\hat{\mathbf{z}}(A, \alpha), \tag{7.10}$$

we also obtain

$$|\hat{\mathbf{z}}(A, \alpha)| \geq 1,$$

so in all we have

$$|\hat{\mathbf{z}}(A,\alpha)| = 1.$$

As in the proof of Proposition 7.3, we have that $\hat{\mathbf{z}}(A,\alpha)$ and $z_{\mathrm{dR}}(A,\alpha)$ are conjugate in $\mathcal{O}_{K_p}$, and so $z_{\mathrm{dR}}(A,\alpha) \in K_p$ with

$$|z_{\mathrm{dR}}(A,\alpha)| = 1.$$

By (4.17), we have $1/y_{\mathrm{dR}}(A,\alpha) = 1 - z_{\mathrm{dR}}(A,\alpha)/\hat{\mathbf{z}}(A,\alpha)$. As in the proof of Proposition 7.7, making a change $\alpha_1 \mapsto [u](\alpha_1)$ if necessary, we may assume that $z_{\mathrm{dR}}(A,\alpha) = \frac{2}{1+\sqrt{-d}}$ and hence $\hat{\mathbf{z}}_{\mathrm{dR}}(A,\alpha) = \frac{2}{1-\sqrt{-d}}$, which then implies $y_{\mathrm{dR}} = \frac{1+\sqrt{-d}}{2\sqrt{-d}}$, which gives $|y_{\mathrm{dR}}(A,\alpha)| = |\theta(y_{\mathrm{dR}}(A,\alpha))| = |2|^{-1}$. Now (7.8) follows from (7.5).

The corresponding claims for (A',α') now follow easily from (3.15), (4.10) and (4.44). The assertion that $\langle \alpha_1' \pmod{p} \rangle$ is the canonical subgroup of A' follows from $|\hat{\mathbf{z}}(A',\alpha')| = p > p^{p/(p^2-1)}$ and Lemma 6.11. $\qquad\square$

Corollary 7.8. *For (A,α) and associated (A',α') as in Proposition 7.7, we have*

$$\left| \theta\left(\frac{d(\mathbf{z}_{\mathrm{LT}}/\bar{z}_{\mathrm{LT}})}{dz_{\mathrm{dR}}} \right) (A',\alpha') \right| < 1, \qquad \left| \theta\left(\frac{y_{\mathrm{dR}}}{\hat{\mathbf{z}}} \right) (A',\alpha) \right| < 1. \qquad (7.11)$$

In particular, (6.31) is satisfied for $y = (A',\alpha')$ with $b = 1$ if $p > 2$ and $b = 3$ if $p = 2$.

Proof. First, observe that since $A[\mathfrak{p}] \subset A[p]$ is the canonical subgroup of A and $(A/A[\mathfrak{p}])[\mathfrak{p}]$ is the canonical subgroup of $A/A[\mathfrak{p}]$. Hence by [36, Theorem 3.10.7], considering the division by the canonical subgroup isogeny $A \to A/A[\mathfrak{p}]$ we have

$$p/|\mathrm{Ha}(A)| = |\mathrm{Ha}(A)|,$$

and so $|\mathrm{Ha}(A)| = p^{-1/2}$. Now since the kernel of $A' \to A$ is $\langle \alpha_1' \pmod{p} \rangle \subset A'[p]$, which is the canonical subgroup, we have by loc. cit. that $|\mathrm{Ha}(A')| = p^{-1/(2p)}$. Using this, the assertion follows from the same argument as in the proof of Corollary 7.4. $\qquad\square$

7.2 WEIGHTS

In this section, we slightly refine our study of the notion of weights for the p-adic Maass-Shimura derivatives

$$(p^b \theta_k)^j f$$

and apply its restrictions of sections to the closed adic subspaces $\mathcal{V} \subset \mathcal{Y}_x$ of the open affinoid subdomain $\mathcal{Y}_x$ of $\mathcal{Y}$. This notion of weights will be crucial to do descent arguments along $\mathcal{Y} \to Y$ and go from infinite to finite level, on which we can work with rigid analytic sections to prove certain identities in the next section.

Given a closed immersion of adic spaces

$$i : \mathcal{V} \hookrightarrow \mathcal{Y}_x,$$

let $\mathcal{O}_\mathcal{V}$ denote the proétale structure sheaf on the adic space $\mathcal{V}$, with p-adic completion $\hat{\mathcal{O}}_\mathcal{V}$. Henceforth, denote

$$\hat{\mathcal{O}}_{\mathcal{Y}_x}|_\mathcal{V} := i^{-1}\hat{\mathcal{O}}_\mathcal{V}$$

the restriction of $\hat{\mathcal{O}}_{\mathcal{Y}_x}$ to $\mathcal{V}$.

Note that since

$$\mathcal{Y}^{ss} \subset \mathcal{Y}$$

is an open affinoid subdomain, then

$$\mathcal{Y}^{ord} = \mathcal{Y} \setminus \mathcal{Y}^{ss} \subset \mathcal{Y}$$

is a closed adic subspace, and so

$$\mathcal{Y}_x^{ord} := \mathcal{Y}_x \cap \mathcal{Y}^{ord} \subset \mathcal{Y}_x$$

is a closed adic subspace.

Definition 7.9. *Let*
$$\Gamma \subset GL_2(\mathbb{Z}_p),$$

where $\mathcal{V} \subset \mathcal{Y}_x$ is an affinoid open or closed adic subspace, be a subgroup such that $\mathcal{V} \cdot \Gamma = \Gamma$. A section

$$F \in \hat{\mathcal{O}}_{\mathcal{Y}_x}|_\mathcal{V} \ \ if \ \ \mathcal{V} \subset \mathcal{Y}_x \ \ is \ closed \ and \ \ F \in \hat{\mathcal{O}}_{\mathcal{Y}_x}(\mathcal{V}) \ \ if \ \ \mathcal{V} \ \ is \ open$$

is said to have weight k *for Γ on $\mathcal{V}$ if*

$$\begin{pmatrix} a & b \\ c & d \end{pmatrix}^* F = (ad - bc)^{-k}(c\theta(z_{dR})|_\mathcal{V} + a)^k F$$

for all

$$\begin{pmatrix} a & b \\ c & d \end{pmatrix} \in \Gamma.$$

Remark 7.10. Note that if $F \in \mathcal{O}_\Delta(\mathcal{Y}_x)$ is of weight k in the sense of Definition 5.13 for any subgroup $\Gamma \subset GL_2(\mathbb{Z}_p)$, so that $\mathcal{Y}^{ss} \cdot \Gamma = \mathcal{Y}^{ss}$, then $\theta(F) \in \hat{\mathcal{O}}(\mathcal{Y}^{ss})$ is of weight k on $\mathcal{Y}^{ss}$ for Γ in the sense of Definition 7.9. Moreover, if $\mathcal{V} \cdot \Gamma = \mathcal{V}$ then $\theta(F)|_{\mathcal{V}}$ has weight k for Γ on $\mathcal{V}$.

Proposition 7.11. *Suppose $F \in \mathcal{O}_\Delta(\mathcal{Y}_x)$ has weight k for $\Gamma \subset GL_2(\mathbb{Z}_p)$ in the sense of Definition 5.13, and suppose $\mathcal{V} \cdot \Gamma = \mathcal{V}$. Then for any $j \in \mathbb{Z}_{\geq 0}$, $\theta_k^j F|_{\mathcal{V}}$ has weight $k + 2j$ for Γ on $\mathcal{V}$.*

Proof. This follows immediately from Remark 7.10 and by Proposition 5.21. □

Proposition 7.12. *Suppose that F has weight k on $\mathcal{Y}_x$ for a subgroup $\Gamma \subset GL_2(\mathbb{Z}_p)$ as defined in Definition 5.16, and let $\mathcal{V} \subset \mathcal{Y}$ be an open affinoid subdomain or closed adic subspace $\mathcal{V} \cdot \Gamma = \mathcal{V}$, such that*

$$|c\theta(z_{dR})|_{\mathcal{V}} + a| = 1.$$

Suppose that (6.31) is satisfied for every rank 1 point in $\mathcal{V}$ for a fixed $b \in \mathbb{Q}$, as well as the assumptions of Theorem 6.21. Then the limit

$$\lim_{m \to \infty} (p^b \theta_k)^{j_0 + (p-1)p^m} F|_{\mathcal{V}}$$

exists for $j_0 \in \mathbb{Z}/(p-1) \times \mathbb{Z}_p$, then it has weight $k + 2j_0$ for Γ on $\mathcal{V}$.

Proof. We see by Theorem 6.21 that the limit exists. Now we have

$$\lim_{m \to \infty} (c\theta(z_{dR})|_{\mathcal{V}} + a)^{j_0 + 2(j_0 + (p-1)p^m)} = (c\theta(z_{dR})|_{\mathcal{V}} + a)^{k + 2j_0}.$$

For any $\begin{pmatrix} a & b \\ c & d \end{pmatrix} \in \Gamma$, we compute, using Proposition 7.11, that

$$\begin{pmatrix} a & b \\ c & d \end{pmatrix}^* \lim_{m \to \infty} (p^b \theta_k)^{j_0 + (p-1)p^m} F|_{\mathcal{V}}$$

$$= \lim_{m \to \infty} \begin{pmatrix} a & b \\ c & d \end{pmatrix}^* (p^b \theta_k)^{j_0 + (p-1)p^m} F|_{\mathcal{V}}$$

$$= \lim_{m \to \infty} (ad - bc)^{k + 2(j_0 + (p-1)p^m)} (c\theta(z_{dR})|_{\mathcal{V}} + a)^{j_0 + 2(j_0 + (p-1)p^m)}$$

$$(p^b \theta_k)^{j_0 + (p-1)p^m} F|_{\mathcal{V}}$$

$$= \lim_{m \to \infty} (ad - bc)^{k + 2(j_0 + (p-1)p^m)} \cdot \lim_{m \to \infty} (c\theta(z_{dR})|_{\mathcal{V}} + a)^{j_0 + 2(j_0 + (p-1)p^m)}$$

$$\cdot \lim_{m \to \infty} (p^b \theta_k)^{j_0 + (p-1)p^m} F|_{\mathcal{V}}$$

$$= (ad - bc)^{k + 2j_0} (c\theta(z_{dR})|_{\mathcal{V}} + a)^{k + 2j_0} \lim_{m \to \infty} (p^b \theta_k)^{j_0 + (p-1)p^m} F|_{\mathcal{V}}. \qquad \square$$

7.3   GOOD CM POINTS

In this section, we relate Maass-Shimura derivatives of $\flat$-stabilizations (1.19) and p-stabilizations (6.28), as well as their values on CM points $y = (A, t, \alpha)$ and

$$y' = y \cdot \begin{pmatrix} q & 0 \\ 0 & 1 \end{pmatrix}.$$

The upshot is that the Maass-Shimura derivatives have good continuity properties at y' (though not a priori at y), because the "good" CM points y' satisfy the assumptions of Theorem 6.21. We will relate the derivatives at y' to the derivatives at the CM points y in the final corollary of this section (Corollary 7.21). For brevity, we work with θ_k^j-operators throughout this section, though analogous results for $\theta \circ \delta_{k,\Delta}^\diamond$ also hold with essentially the same arguments.

Definition 7.13. *For any* $m, n \in \mathbb{Z}_{\geq 0}$, *let*

$$\Gamma(m, n) := \Gamma_1(m) \cap \Gamma_0(n)$$
$$= \left\{ \begin{pmatrix} a & b \\ c & d \end{pmatrix} \in GL_2(\mathbb{Z}_p) : a - 1, c \equiv 0 \pmod{p^m}, c \equiv 0 \pmod{p^n} \right\}.$$

Suppose our base modular curve $Y = Y(\Gamma_1(N) \cap \Gamma)$ *has* $\Gamma \subset \Gamma(m, n)$. *Let* $Y(m, n)$ *denote the (adic space associated with the) modular curve determined by* $\Gamma(m, n)$, *i.e., the subcover of* $\mathcal{Y} \to Y$ *with* $\mathrm{Gal}(\mathcal{Y}/Y(m, n)) = \Gamma(m, n)$. *Hence*

$$Y(m, n) = Y(\Gamma_1(p^m N) \cap \Gamma_0(p^n)),$$

and parametrize quadruples (A, t, P, C) *where* (A, t) *is an elliptic curve* A, *with* $\Gamma_1(N)$-*level structure*, $P \in A[p^m]$ *is a point of exact order* m, *and* $C \subset A[p^n]$ *is a cyclic subgroup of order* p^n *such that* $C \pmod{p^{\min(m,n)}} = \langle P \pmod{p^{\min(m,n)}} \rangle$ *(the image of* P *in* $A[p^{\min(m,n)}]$). *We sometimes write more compactly (in keeping with previous notation)* $t_p = (P, C)$ *to denote a* $\Gamma(m, n) = \Gamma_1(p^m) \cap \Gamma_0(p^n)$-*level structure. Note that the p-stabilization (see (6.28)) of a modular form on* $Y_1(N)$ *can thus naturally be viewed on* $Y(0, n)$ *for any* $n \geq 2$, *and hence on* $Y(m, n)$ *for any* $m \geq 0$, $n \geq 2$ *via pullback through* $Y(m, n) \to Y(0, n)$.

Definition 7.14. *Suppose we are given a CM point* (A, t, t_p) *as in Propositions 7.3 and 7.7 where* t *is a* $\Gamma_1(N)$-*level structure* $(p \nmid N)$ *and* t_p *is a* $\Gamma_0(p^n)$-*level structure, and fix* $y = (A, \alpha) \in \mathcal{Y}$ *any lift of* (A, t, t_p) *to* $\mathcal{Y}$ *as in those propositions (suppressing t from the notation), and let* $y' = (A', \alpha')$, *and define the set*

$$C_{m,n}^{\mathrm{good}}(y) := y' \cdot \Gamma(m, n) \subset \mathcal{Y}.$$

Proposition 7.15. *Suppose* $m \geq 2$ *if* $p > 2$ *and* $m \geq 3$ *if* $p = 2$. *Then any point* $s = (A', \alpha')$ *of* $C_{m,n}^{\mathrm{good}}(y)$ *satisfies the conclusions for* (A', α') *in Propositions 7.3* *and 7.7, and Corollaries 7.4 and 7.8. In particular, $C^{\mathrm{good}}_{m,n}(y)$ is a $\Gamma(m,n)$-invariant set of points with these properties. We also have*

$$|\theta(cz_{\mathrm{dR}}(s)+a)(ad-bc)^{-1}|=1$$

for any $\begin{pmatrix} a & b \\ c & d \end{pmatrix} \in \Gamma(m,n)$ and any $s \in C^{\mathrm{good}}_{m,n}(y)$.

In fact, there exists an affinoid neighborhood $\mathcal{U} \subset \hat{\mathcal{Y}}$ with $\mathcal{U} \cap C^{\mathrm{good}}_{m,n}(y)$ and $\mathcal{U} \cdot \Gamma(m,n) = \mathcal{U}$ such that

$$|\theta(cz_{\mathrm{dR}}+a)(ad-bc)^{-1}|=1$$

on $\mathcal{U}$, for any $\begin{pmatrix} a & b \\ c & d \end{pmatrix} \in \Gamma(m,n)$.

Proof. The first part follows immediately from the conclusions of the aforementioned propositions and corollaries, together with (4.44), (3.15) and (4.10), using that $|c\theta(z_{\mathrm{dR}})(y')| < 1$.

For the second part, by continuity of $y \mapsto |\theta(z_{\mathrm{dR}})(y)|$ for $\theta(z_{\mathrm{dR}}) \in \hat{\mathcal{O}}_{\mathcal{Y}}(\mathcal{Y}_x) = \mathcal{O}^{\mathrm{an}}_{\hat{\mathcal{Y}}}(\hat{\mathcal{Y}}_x)$ (recalling that $\mathcal{Y} \sim \hat{\mathcal{Y}}$ and $\mathcal{Y}_x$ are affinoid perfectoid), we see that an analytic neighborhood $U \subset \hat{\mathcal{Y}}$ containing y exists with the desired property, and now we use the $\Gamma(m,n)$-action along with (4.44) to translate it to $\mathcal{U} := U \cdot \Gamma(m,n)$. By continuity of the $\Gamma(m,n)$-action, we can shrink U if necessary to make it affinoid (cf. [13, Theorem 2.17]). $\qquad\square$

Given $w \in \omega^{\otimes k}(Y_1(N))$, we now define some related generalized p-adic modular forms, coming from the p-stabilization $w^{(p)}$ of w. These will be used later in our construction of p-adic L-functions.

Definition 7.16. *Recall*

$$q := \begin{cases} p & p > 2 \\ 8 & p = 2. \end{cases}$$

Given a modular form $w \in \omega^{\otimes k}(Y(\Gamma_1(N)))$, we can consider the pullback

$$w|_{\mathcal{Y}_x} = F \cdot \omega^{\otimes k}_{\mathrm{can}}$$

for some $F \in \mathcal{O}_{\Delta}(\mathcal{Y}_x)$. Let

$$w^{(p)} \in \omega^{\otimes k}(Y(0,2)) = \omega^{\otimes k}(Y(\Gamma_1(N) \cap \Gamma_0(p^2)))$$

denote the p-stabilization, as in (6.28), and write

$$w^{(p)}|_{\mathcal{Y}_x} = F^{(p)} \cdot \omega^{\otimes k}_{\mathrm{can}}.$$

Applying our p-adic Maass-Shimura derivatives ∂_k^j (Definition 5.15), we have weight $k + 2j$ modular forms

$$\partial_k^j w^{(p)} = \partial_k^j(w^{(p)}|_{\mathcal{Y}_x}) = \delta_k^j F^{(p)} \cdot \omega_{\mathrm{can}}^{\otimes k + 2j}.$$

Let

$$w_j' := \begin{pmatrix} 1 & 0 \\ 0 & q \end{pmatrix}^* \left(\partial_k^j w^{(p)}\right) = \begin{pmatrix} 1 & 0 \\ 0 & q \end{pmatrix}^* \left(\delta_k^j F^{(p)} \cdot \omega_{\mathrm{can}}^{\otimes k + 2j}\right)$$

$$\in \omega^{\otimes k}(\mathcal{Y}_x) \subset \omega^{\otimes k}(\mathcal{Y}^{\mathrm{ss}}). \tag{7.12}$$

Definition 7.17. *Henceforth, for brevity, let*

$$\Gamma = \begin{cases} \Gamma(1,2) & p > 2 \\ \Gamma(3,2) & p = 2. \end{cases}$$

Proposition 7.18. *Suppose for some affinoid $\mathcal{U} \subset \hat{\mathcal{Y}}$ (which we can then view with $\mathcal{U} \in Y(\Gamma)_{\mathrm{pro\acute{e}t}}$ using $|\mathcal{Y}| \cong |\hat{\mathcal{Y}}|$) satisfying $\mathcal{U} \cdot \Gamma = \mathcal{U}$, we are given an arbitrary section*

$$\mathfrak{w} \in \omega_\Delta^{\otimes k + 2j}(\mathcal{U}) = (\omega \otimes_{\mathcal{O}_{Y(\Gamma)}} \mathcal{O}_\Delta)^{\otimes k + 2j}(\mathcal{U})$$

such that

$$\theta(\mathfrak{w}) \in \theta(\omega_\Delta)^{\otimes k + 2j}(\mathcal{U}) = (\omega \otimes_{\mathcal{O}_{Y(\Gamma)}} \hat{\mathcal{O}}_{\mathcal{Y}}^{\otimes k + 2j})(\mathcal{U})$$

is Γ-invariant. Then

$$\mathfrak{w}' := \theta\left(\begin{pmatrix} 1 & 0 \\ 0 & q \end{pmatrix}^* \mathfrak{w}\right) \in (\omega \otimes_{\mathcal{O}_{Y(\Gamma)}} \hat{\mathcal{O}}_{\mathcal{Y}}^{\otimes k + 2j})(\mathcal{U})$$

naturally descends to a section

$$\mathfrak{w}' \in (\omega \otimes_{\mathcal{O}_{Y(\Gamma)}} \hat{\mathcal{O}}_{Y(\Gamma)})^{\otimes k + 2j}(Y(\Gamma)^{\mathrm{ss}}).$$

Proof. We want to apply proétale descent along $\mathcal{U} \to Y(\Gamma)$ to the sheaf $\omega \otimes_{\mathcal{O}_{Y(\Gamma)}} \hat{\mathcal{O}}_{\mathcal{Y}^{\mathrm{ss}}}$ along the proétale Γ-cover $\mathcal{Y}^{\mathrm{ss}} \to Y(\Gamma)^{\mathrm{ss}}$ (see Lemma 4.27), and so need to check that for any $\gamma \in \Gamma$,
$$\gamma^* \mathfrak{w}' = \mathfrak{w}'.$$

By continuity arguments, it suffices to check this on rank 1 geometric points. Write
$$\gamma = \begin{pmatrix} a & b \\ c & d \end{pmatrix} \in \Gamma,$$

so that $a-1, c \equiv 0 \pmod{q}$, and $c \equiv 0 \pmod{p^2}$ (recall our notation: $q = p$ if $p > 2$ and $q = 8$ if $p = 2$). We have, by definition of the $GL_2(\mathbb{Q}_p)$-action (see (3.5))

$$(\gamma^* \mathfrak{w}')(A, \alpha) = \mathfrak{w}'((A, \alpha) \cdot \gamma) = \theta(\mathfrak{w})((A, \alpha) \cdot \gamma \begin{pmatrix} 1 & 0 \\ 0 & q \end{pmatrix})$$

$$= \theta(\mathfrak{w})((A, \alpha) \cdot \begin{pmatrix} a & bq \\ c & dq \end{pmatrix}) = \theta(\mathfrak{w})(A/H, \alpha'),$$

where $H = \langle \alpha_1 \pmod{p} \rangle$ and α' is given by (3.6). Note that α_1' is entirely determined by a, c and H. In fact, we have

$$(A, \alpha) \cdot \begin{pmatrix} 1 & 0 \\ 0 & q \end{pmatrix} = (A/H, \alpha'')$$

where

$$\alpha_1' = a\alpha_1'' + c\alpha_2'' \equiv \alpha_1'' \pmod{q},$$

and

$$\langle \alpha_1 \pmod{p^2} \rangle = \langle a\alpha_1'' \pmod{p^2} \rangle = \langle \alpha_1'' \pmod{p^2} \rangle.$$

Hence, since $\theta(\mathfrak{w}) \in (\omega \otimes_{\mathcal{O}_{Y(\Gamma)}} \hat{\mathcal{O}}_{\mathcal{Y}})^{\otimes k + 2j}(Y(\Gamma)^{ss})$, we have that any value $\theta(\mathfrak{w})$ (B, β') only depends on $(B, \beta_1 \pmod{q}, \langle \beta_1 \pmod{p^2} \rangle)$, and so

$$\theta(\mathfrak{w})(A/H, \alpha') = \theta(\mathfrak{w})(A/H, \alpha'')$$

and hence

$$(\gamma^* \mathfrak{w}')(A, \alpha) = \theta(\mathfrak{w})(A/H, \alpha') = \theta(\mathfrak{w})(A/H, \alpha'') = \theta(\mathfrak{w}) \left((A, \alpha) \cdot \begin{pmatrix} 1 & 0 \\ 0 & q \end{pmatrix} \right)$$

$$= \mathfrak{w}'(A, \alpha). \qquad \square$$

Note that the calculation of (4.48) can be extended to show

$$\begin{pmatrix} 1 & 0 \\ 0 & q \end{pmatrix}^* \omega_{\mathrm{can}} = q\omega_{\mathrm{can}}. \tag{7.13}$$

Moreover, we have by (4.44)

$$\begin{pmatrix} 1 & 0 \\ 0 & q \end{pmatrix}^* (dz_{\mathrm{dR}}) = q dz_{\mathrm{dR}}.$$

In particular,

$$\begin{pmatrix} 1 & 0 \\ 0 & q \end{pmatrix}^* (\omega_{\text{can}}^{\otimes k} \otimes (dz_{\text{dR}})^j) = q^{k+j} \omega_{\text{can}}^{\otimes k} \otimes (dz_{\text{dR}})^j.$$

Definition 7.19. *Suppose* $w'_j \in \omega^{\otimes k}(Y(\Gamma))$ *as above and write*

$$w'_j|_{\mathcal{Y}_x} = \begin{pmatrix} 1 & 0 \\ 0 & q \end{pmatrix}^* (\delta_k^j F^{(p)} \cdot \omega_{\text{can}}^{\otimes k} \otimes (dz_{\text{dR}})^j) = F'_j \cdot \omega_{\text{can}}^{\otimes k} \otimes (dz_{\text{dR}})^j$$

where

$$F'_j = q^k \begin{pmatrix} 1 & 0 \\ 0 & q \end{pmatrix}^* \left(q^j \delta_k^j F^{(p)} \right) \in \mathcal{O}_\Delta(\mathcal{Y}_x) = \hat{\mathcal{O}}_{\mathcal{Y}_x}(\mathcal{Y}_x)[\![q_{\text{dR}}^{1/q} - 1]\!],$$

and as before we write $F'_j = F'_j(q_{\text{dR}}^{1/q})$ *when we view* F'_j *in the power series ring* $\hat{\mathcal{O}}_{\mathcal{Y}_x}(\mathcal{Y}_x)[\![q_{\text{dR}}^{1/q} - 1]\!]$ *(the* $q_{\text{dR}}^{1/q}$*-expansion).*[1]

The next proposition says that around neighborhoods containing C^{good}, we have that the Maass-Shimura derivatives of suitable p-stabilizations coincide with those of $\flat$-stabilizations, and in particular applies the continuity of the function

$$j \mapsto F'_j(y'),$$

which will be crucial for the construction of the p-adic L-function.

Proposition 7.20. *Recall the map* $\theta : \mathcal{O}_\Delta \to \hat{\mathcal{O}}_{\mathcal{Y}}$. *We have for any* $s \in C_{m,n}^{\text{good}}(y)$, $j \geq 0$, *in the notation of Definition 6.18, the equality of values*

$$\theta(F'_j)(s) = q^k \begin{pmatrix} 1 & 0 \\ 0 & q \end{pmatrix}^* (q^j \theta_k^j F^{(p),\flat}(q_{\text{dR}}^{1/q}))(s). \tag{7.14}$$

In particular, this is true for $s = y'$. *Here* θ_k^j *is as in Definition 5.28.*

---

[1] *We note that in the above formula, we needed to use the equality*

$$\partial_k^j w^{(p)} = \delta_k^j F^{(p)} \omega_{\text{can}}^{\otimes k} \otimes (dz_{\text{dR}})^j,$$

instead of trivializing in terms of the basis $\omega_{\text{can}}^{\otimes k+2j}$, *in order to correctly compute the Hecke action of* $\begin{pmatrix} 1 & 0 \\ 0 & q \end{pmatrix}^*$; *this is because the Kodaira-Spencer isomorphism* $\omega^{\otimes 2} \cong \Omega^1$ *is not Hecke equivariant, but is after twisting the action on* $\omega^{\otimes 2}$ *by* $\det^{-1}$ *(cf. [13, Remark 2.21]).*

Proof. We will show equality of the elements in (7.14) via a descent argument. First, we can see that

$$\begin{pmatrix} 1 & 0 \\ 0 & q \end{pmatrix}^* (q^j \theta_k^j F_s^{(p),\flat}) \quad \text{and} \quad \begin{pmatrix} 1 & 0 \\ 0 & q \end{pmatrix}^* (q^j \theta_k^j F_s^{(p)})$$

are equal on a suitable $\mathcal{U}$ from Proposition 7.15, and are also equal on $\mathcal{Y}^{\mathrm{Ig}}$ via a computation of Serre-Tate coordinates. Using descent and the weight k property, this translates to an equality of rigid sections of $\mathcal{O}_{Y(\Gamma)}^{\mathrm{an}}$ on some affinoid subdomain of $Y(\Gamma)$ (containing the ordinary locus). Hence by rigidity these descended sections are equal, which gives (7.14).

By Proposition 7.15, we have that (6.31) is satisfied for all $s \in C_\Gamma^{\mathrm{good}}(y)$ with $b = 1$ if $p > 2$ and $b = 3$ if $p = 2$, and so shrinking $\mathcal{U}$ if necessary, recalling Definition 6.19 and Proposition 7.12, we see that

$$q^j \theta_k^j F^{(p),\flat} \in \hat{\mathcal{O}}_{\mathcal{Y}}(\mathcal{U}), \qquad q^j \theta_k^j F^{(p)} \in \hat{\mathcal{O}}_{\mathcal{Y}}(\mathcal{U})$$

have weight $k + 2j$ on $\mathcal{U}$. Hence by Proposition 7.18 applied to $\mathfrak{w} = q^j \partial_k^j w^{(p),\flat}$ $(q_{\mathrm{dR}}^{1/q})$,

$$\begin{pmatrix} 1 & 0 \\ 0 & q \end{pmatrix}^* (q^j \theta_k^j F^{(p),\flat}) \cdot \theta(\omega_{\mathrm{can}})^{\otimes k + 2j} \in (\omega \otimes_{\mathcal{O}_{Y(\Gamma)}} \hat{\mathcal{O}}_{\mathcal{Y}})^{\otimes k + 2j}(\mathcal{U}),$$

is a Γ-invariant section on $\mathcal{U}$. By the same proposition applied to $\mathfrak{w} = q^j \partial_k^j w^{(p)}$ $(q_{\mathrm{dR}}^{1/q})$,

$$\begin{pmatrix} 1 & 0 \\ 0 & q \end{pmatrix}^* (q^j \theta_k^j F^{(p)}) \cdot \theta(\omega_{\mathrm{can}})^{\otimes k + 2j} \in (\omega \otimes_{\mathcal{O}_{Y(\Gamma)}} \hat{\mathcal{O}}_{\mathcal{Y}})^{\otimes k + 2j}(\mathcal{U})$$

is also a Γ-invariant section on $\mathcal{U}$. Applying descent (Lemma 4.27) along the proétale map of adic spaces
$$\mathcal{U} \subset \hat{\mathcal{Y}} \to Y(\Gamma),$$

these sections descend to rigid analytic sections of $\mathcal{O}_{Y(\Gamma)}(U)$ for some admissible open $U \subset Y(\Gamma)^{\mathrm{ss}}$, the target denoting the supersingular locus of $Y(\Gamma)$.

By Theorem 5.9 and 6.9, we have that

$$\begin{pmatrix} 1 & 0 \\ 0 & q \end{pmatrix}^* (q^j \theta_k^j F^{(p)})|_{\mathcal{Y}^{\mathrm{Ig}}} = \begin{pmatrix} 1 & 0 \\ 0 & q \end{pmatrix}^* (q^j \theta_k^j F^{(p),\flat})|_{\mathcal{Y}^{\mathrm{Ig}}},$$

where each of these sections has weight k for the subgroup $B \subset GL_2(\mathbb{Z}_p)$ of upper triangular matrices (recall $\mathcal{Y}^{\mathrm{Ig}} = \{\hat{\mathbf{z}} = \infty\}$ and $\mathcal{Y}^{\mathrm{Ig}} \cdot B = \mathcal{Y}^{\mathrm{Ig}}$). Now multiplying by

$$\omega_{\mathrm{can}}^{\otimes k+2j}|_{\mathcal{Y}^{\mathrm{Ig}}} = \omega_{\mathrm{can}}^{\mathrm{Katz},\otimes k+2j}$$

to get B-invariant sections, and applying descent along the proétale cover of adic spaces

$$\mathcal{Y}^{\mathrm{Ig}} \twoheadrightarrow Y(\Gamma)^{\mathrm{ord}},$$

we see that the above equality descends to an equality of rigid analytic sections of

$$\mathcal{O}_{Y(\Gamma)}(Y(\Gamma)^{\mathrm{ord}}),$$

$Y(\Gamma)^{\mathrm{ord}}$ denoting the ordinary locus of $Y(\Gamma)$.

Hence we have two sections of $\mathcal{O}_{Y(\Gamma)}^{\mathrm{an}}$ on the admissible open $U \sqcup Y(\Gamma)^{\mathrm{ord}} \subset Y(\Gamma)$, which are equal on the open subdomain $Y(\Gamma)^{\mathrm{ord}}$, and hence are equal on all of $U \sqcup Y(\Gamma)^{\mathrm{ord}}$ by rigidity. Pulling back to $\mathcal{Y}$, we obtain (7.14). $\qquad\square$

Corollary 7.21. *Let*

$$\Gamma = \begin{cases} \Gamma(1,2) & p > 2 \\ \Gamma(3,2) & p = 2 \end{cases}$$

as above. Suppose w is a modular form on $Y(\Gamma_1(N))$, and view $w \in \omega^{\otimes k}(\Gamma)$. Let $(A, t, t_p) \in Y(\Gamma)(\overline{\mathbb{Q}}_p, \overline{\mathbb{Z}}_p)$ be a point corresponding to an elliptic curve with CM by $\mathcal{O} = \mathbb{Z} + c\mathcal{O}_K$, $(c, p) = 1$, and let $y = (A, \alpha) \in \mathcal{Y}(\overline{\mathbb{Q}}_p, \overline{\mathbb{Z}}_p)$ as in Propositions 7.3 and 7.7 lie above (A, t, t_p). Then for the corresponding point

$$y' = (A', \alpha') = (A, \alpha) \cdot \begin{pmatrix} q & 0 \\ 0 & 1 \end{pmatrix},$$

we have for any $j \geq 0$ (recalling $\theta_k^j = \theta \circ \delta_k^j$),

$$j \mapsto \theta(F_j')(y') \tag{7.15}$$

extends to a continuous function in $j \in \mathbb{Z}/(p-1) \times \mathbb{Z}_p$.

Proof. By Theorem 6.21 with $b = 1$ if $p > 2$ and $b = 3$ if $p = 2$, we have

$$j \mapsto q^j \theta_k^j F^{(p),b}(q_{\mathrm{dR}}^{1/q})(y')$$

extends to a continuous function in $j \in \mathbb{Z}/(p-1) \times \mathbb{Z}_p$, and hence by the continuity of the $GL_2(\mathbb{Q}_p)$-action, so does

$$j \mapsto \begin{pmatrix} 1 & 0 \\ 0 & q \end{pmatrix}^* \left(q^j \theta_k^j F^{(p),b} \left(q_{\mathrm{dR}}^{1/q} \right) \right)(y').$$

Applying (7.14), we then get the conclusion.   □

This key corollary will allow us to prove the interpolation of our p-adic L-function: we will use the auxiliary point y' and $\theta_k^j F'(y')$ to define our continuous p-adic L-function, and (7.15) will allow us to compare with $q^{k+2j}\theta_k^j F^{(p)}(y)$, and hence algebraic L-values using Theorem 5.48. We end this chapter with a final calculation.

Proposition 7.22. *We have*

$$\theta(F_j')(y') = q^{k+2j}\theta_k^j F^{(p)}(y). \tag{7.16}$$

Proof. By definition, we have

$$\theta(F_j')\omega_{\text{can}}^{\otimes k+2j}(y') = \begin{pmatrix} 1 & 0 \\ 0 & q \end{pmatrix}^* \theta(\partial_k^j \omega^{(p)})(y') = \theta_k^j F^{(p)}(y)\omega_{\text{can}}^{\otimes k+2j}(y)$$

By (7.13), the right-hand side when mapped to the fiber at y' (via the identification of fibers induced by $\begin{pmatrix} 1 & 0 \\ 0 & q \end{pmatrix}$) is

$$q^{k+2j}\theta_k^j F^{(p)}(y)\omega_{\text{can}}(y')^{\otimes k+2j}.$$

This gives (7.16).   □

Chapter Eight

Supersingular Rankin-Selberg p-adic L-functions

In this chapter, we construct our 1-variable anticyclotomic p-adic L-function attached to a given normalized eigenform (i.e., newform or normalized Eisenstein series). Again let

$$w \in \omega^{\otimes k}(Y)$$

correspond to a normalized new eigenform of level $\Gamma_1(N)$ and weight k (where $k \geq 0$ is an integer) and of nebentype ϵ_w, as in the statement of Theorem 5.48. Note then that letting N_{ϵ_w} denote the conductor of ϵ_w, we have

$$N_{\epsilon_w} | N.$$

8.1 PRELIMINARIES FOR THE CONSTRUCTION

Retaining the notation of the previous section, let $K/\mathbb{Q}$ denote an imaginary quadratic field with $K = \mathbb{Q}(\sqrt{-d})$ and $d > 0$ squarefree. Henceforth let d_K denote the fundamental discriminant of K. We will make the following *Heegner hypothesis* for N, namely

for each prime $\ell | N$, ℓ is split or ramified in K, and if $\ell^2 | N$ then ℓ is split in K.

We make the Heegner hypothesis primarily for convenience in this document, and plan to remove or weaken this assumption in forthcoming work. The Heegner hypothesis guarantees the existence of an integral ideal $\mathfrak{N} \subset \mathcal{O}_K$ such that

$$\mathcal{O}_K/\mathfrak{N} = \mathbb{Z}/N.$$

We henceforth fix a choice of such an $\mathfrak{N}$. Since $N_{\epsilon_w} | N$, this determines an ideal $\mathfrak{N}_{\epsilon_w} | \mathfrak{N}$ such that

$$\mathcal{O}_K/\mathfrak{N}_{\epsilon_w} = \mathbb{Z}/N_{\epsilon_w}.$$

Recall the fixed embedding $i_p : \overline{\mathbb{Q}} \hookrightarrow \overline{\mathbb{Q}}_p$ (from (3.1)), which determines a p-adic completion $K_p \hookrightarrow \overline{\mathbb{Q}}_p$ of K. Let q denote the order of the residue field of $\mathcal{O}_{K_p}$.

Definition 8.1. *Henceforth, given a complex-valued Hecke character*

$$\chi : K^\times \backslash \mathbb{A}_K^\times \to \mathbb{C}^\times$$

(which we henceforth assume arises from an algebraic Hecke character of "type A_0" in Weil's sense), we let $\check{\chi}$ denote its p-adic avatar. We let $\mathfrak{f}(\chi)$ denote the conductor of χ.

We call a Hecke character χ over K central with respect to w if it is of infinity type $(k+j, -j)$ for $j \in \mathbb{Z}$ and if it satisfies the following compatibility of central characters:

$$\chi|_{\mathbb{A}_\mathbb{Q}^\times} \mathrm{N}_k^{-k} = \epsilon_w,$$

where $\mathrm{N}_K : \mathbb{A}_K^\times \to \mathbb{C}^\times$ denotes the norm character normalized to have infinity type $(1, 1)$.

Let π_w denote the unitary automorphic representation of $GL_2(\mathbb{A}_\mathbb{Q})$ generated by w, and let $(\pi_w)_K$ denote its base change to $GL_2(\mathbb{A}_K)$. Central criticality forces the automorphic representation

$$(\pi_w)_K \times \pi_{\chi^{-1}}$$

to be self-dual (i.e., isomorphic to its contragredient), and the center of $L((\pi_w)_K \times \pi_{\chi^{-1}}, s)$ the functional equation to be $s = 1/2$. The Heegner hypothesis implies that the local root number

$$\epsilon_\ell(\pi_w \times \pi_{\chi^{-1}}, 1/2) = \epsilon_\ell((\pi_w)_K \otimes \chi^{-1}, 1/2) = \prod_{v|\ell} \epsilon_v((\pi_w)_K \otimes \chi^{-1}, 1/2) = +1$$

for all rational primes ℓ and primes v of K, where here

$$(\pi_w)_K$$

denotes the base change of π_w to K. Hence we satisfy the Saito-Tunnell local conditions so that the period integrals, whose squares we eventually relate to L-values via a calculation of Waldspurger, which we interpolate are not a priori 0. More precisely, we can view these period integrals as functionals on the (unitary) automorphic representation $(\pi_w)_K \times \chi^{-1}$ over $\mathbb{A}_K$ attached to the pair (w, χ^{-1}), or in other words as elements of the restricted tensor product

$$\bigotimes_v{}' \mathrm{Hom}_{K_v^\times}((\pi_w)_K \otimes \chi^{-1}, \mathbb{C}).$$

By the well-known theorem of Saito and Tunnell (see, for example, [69, Chapter 1, Theorem 1.3]), this space is of dimension less than or equal to 1, and is of dimension 1 if and only if the local root numbers $\epsilon_v(w, \chi^{-1}, 1/2) = +1$ for all finite places v of K, and $\epsilon_\infty(w, \chi^{-1}, 1/2) = -1$. In particular, the global root number satisfies

$$\epsilon(w, \chi^{-1}, 1/2) = -1.$$

If for χ central with respect to w of infinity type $(k + j, -j)$ we have $j \geq 0$ or $-1 \leq j \leq 1 - k$, then χ is *central critical with respect to* w, which means that in addition to the above properties the central value $L((\pi_w)_K \times \pi_{\chi^{-1}}, 1/2)$ is *critical* in the sense of Deligne: the Γ-factors appearing on either side of the functional equation for $L((\pi_w)_K \times \pi_{\chi^{-1}}, 1/2)$ have no poles.

Example 8.2. *As an example a family of central critical characters $\{\chi_j\}$ with $\epsilon_w = 1$, we can take the character defined on ideals by*

$$\chi_{-k/2+h_c wt}(\mathfrak{a}) = \mathbb{N}_K(\mathfrak{a})^{k/2} \left(\frac{a}{\bar{a}}\right)^{wt}$$

where $a \in K^\times$ is a generator of $\mathfrak{a}^{h_c}$, where

$$h_c = \#\mathcal{C}\ell(\mathcal{O}_c) \quad and \quad w = \#\mathcal{O}_K^\times / 2.$$

One can check that the value $\chi_j(\mathfrak{a})$ does not depend on the choice of generator a the principal ideal $\mathfrak{a}^{h_c}$. Then letting $j = -k/2 + h_K wt$, χ_j gives rise to a Hecke character which is unramified at all finite places, and which has infinity type $(k + j, -j)$.

Let

$$\mathcal{O}_c := \mathbb{Z} + c\mathcal{O}_K \subset \mathcal{O}_K, \qquad \hat{\mathcal{O}}_c^\times = \prod_{v \nmid \infty} (\mathcal{O}_c \otimes_{\mathcal{O}_K} \mathcal{O}_{K_v})^\times, \qquad \hat{\mathcal{O}}_K^\times = \prod_{v \nmid \infty} \mathcal{O}_{K_v}^\times,$$

$$\mathfrak{N}_c = \mathfrak{N} \cap \mathcal{O}_c, \qquad \mathfrak{N}_{\epsilon_w, c} = \mathfrak{N}_{\epsilon_w} \cap \mathcal{O}_c.$$

Henceforth, as in [5, Section 4] and [8, Section 8.2], we say a Hecke character $\phi: K^\times \backslash \mathbb{A}_K^\times \to \mathbb{C}^\times$ has *finite type* $(c, \mathfrak{N}, \epsilon_w)$ if we have $c | f(\chi) | c\mathfrak{N}_{\epsilon_w}$ and

$$\chi|_{\hat{\mathcal{O}}_c^\times} = \psi_{\epsilon_w}$$

where here ψ_{ϵ_w} is the composition

$$\hat{\mathcal{O}}_c^\times \twoheadrightarrow (\hat{\mathcal{O}}_c/(\mathfrak{N}_{\epsilon_w, c}\hat{\mathcal{O}}_c))^\times = (\mathcal{O}_K/\mathfrak{N}_{\epsilon_w})^\times = (\mathbb{Z}/N_{\epsilon_w}\mathbb{Z})^\times \xrightarrow{\epsilon_w^{-1}} \mathbb{C}^\times.$$

One can check that each member of the family of examples in Example 8.2 has finite type $(c, \mathfrak{N}, 1)$.

8.2 CONSTRUCTION OF THE p-ADIC L-FUNCTION

Let Σ denote the set of central critical characters χ of infinity type $(k+j, -j)$, where

$$j \in \mathbb{Z}, \quad \text{and} \quad j \geq 0 \quad or \quad -1 \geq j \geq 1-k,$$

and of finite type $(c, \mathfrak{N}, \epsilon_w)$, and which satisfy the following conditions on the local signs of the functional equation:

$$\epsilon_\ell(w, \chi^{-1}) = +1 \qquad \text{for all finite primes } \ell.$$

Under our assumptions, this holds automatically except for ℓ in the set

$$S_w := \{\ell : \ell | (N, d_K), \ell \nmid N_{\epsilon_w}\}.$$

Let

$$\Sigma_+ \subset \Sigma$$

denote the subset of such χ with $j \geq 0$, and let

$$\Sigma_- = \Sigma \setminus \Sigma_+.$$

Then given $\chi \in \Sigma_+$, χ is central critical with respect to w. Using Waldspurger's calculation involving the Rankin-Selberg formula, we will interpolate (square roots of) central values

$$L(\pi_w \times \pi_{(\chi\phi)^{-1}}, 1/2), \quad \chi \in \Sigma_+.$$

Let

$$\check{\Sigma} \subset \operatorname{Hom}_{\mathrm{cts}}(\Gamma^-, \mathcal{O}_{\mathbb{C}_p})$$

denote the set of p-adic avatars of elements of Σ, and similarly with $\check{\Sigma}_+$ and $\check{\Sigma}_-$. We consider $\check{\Sigma}$ as a subspace of the space

$$\operatorname{Fun}(\mathbb{A}_K^{\times, (p\infty)}, \mathcal{O}_{\mathbb{C}_p})$$

of functions $\mathbb{A}_K^{\times, (p\infty)} \to \mathcal{O}_{\mathbb{C}_p}$ equipped with the uniform convergence topology. We let $\overline{\check{\Sigma}}_+$ denote the closure of $\check{\Sigma}_+$ in $\operatorname{Fun}(\mathbb{A}_K^{\times, (p\infty)}, \mathcal{O}_{\mathbb{C}_p})$. One can in fact show, using Example 8.2, that

$$\check{\Sigma}_- \subset \overline{\check{\Sigma}}_+.$$

Henceforth, let $h_c = \#\mathcal{C}\ell(\mathcal{O}_c)$ and let $\{\mathfrak{a}\}$ be a fixed full set of representatives of $\mathcal{C}\ell(\mathcal{O}_c)$ which are prime to $\mathfrak{N}cp$. Recall the Shimura action of the ideals $\mathbb{I}^{p\mathfrak{N}}$

of $\mathcal{O}_c$ which are prime to $p\mathfrak{N}$ on the set of elliptic curves A with CM by $\mathcal{O}_c$:

$$\mathfrak{a} \star A = A/A[\mathfrak{a}], \qquad \pi_{\mathfrak{a}} : A \twoheadrightarrow \mathfrak{a} \star A$$

as in Definition 5.46. Then $\pi_{\mathfrak{a}}$ sends $\Gamma_1(N)$-level structures to $\Gamma_1(N)$-level structures and $\Gamma(p^\infty)$-level structures to $\Gamma(p^\infty)$-level structures, and so gives an action of $\mathbb{I}^{p\mathfrak{N}}$ on pairs (A, t) and triples (A, t, α)

$$\mathfrak{a} \star (A, t) = (\mathfrak{a} \star A, \pi_{\mathfrak{a}}(t)), \qquad \mathfrak{a} \star (A, t, \alpha) = (\mathfrak{a} \star A, \pi_{\mathfrak{a}}(t), \pi_{\mathfrak{a}}(\alpha)).$$

Given a differential $\omega \in \Omega^1_{A/\mathbb{C}_p}$, let $\omega_{\mathfrak{a}} \in \Omega^1_{(\mathfrak{a} \star A)/\mathbb{C}_p}$ be the unique differential such that

$$\pi_{\mathfrak{a}}^* \omega_{\mathfrak{a}} = \omega.$$

Now fix a CM elliptic curve A/H, a $\Gamma_1(N)$-level structure $t : \mathbb{Z}/N \xrightarrow{\sim} A[\mathfrak{N}] \subset A[N]$, and a $\Gamma(p^\infty)$-level structure $\alpha : \mathbb{Z}_p^{\oplus 2} \xrightarrow{\sim} T_p A$ which gives a point

$$(A, \alpha) = (A, t, \alpha) \in \mathcal{Y}(\mathbb{C}_p, \mathcal{O}_{\mathbb{C}_p})$$

above the point $(A, t) \in Y(\mathbb{C}_p, \mathcal{O}_{\mathbb{C}_p})$. Now write the pullback of the normalized eigenform w to $\mathcal{Y}_x$ as

$$w|_{\mathcal{Y}_x} = y_{\mathrm{dR}}^k f \cdot \omega_{\mathrm{can}}^{\otimes k},$$

where $f \in \mathcal{O}(\mathcal{Y}_x)$. Let $(y_{\mathrm{dR}}^k f)^\flat (q_{\mathrm{dR}})$ be as in (6.56). (We will retain the notation $y_{\mathrm{dR}}^k f$ instead of using F as in previous sections, as we will use F to denote the complex function on $\mathcal{H}^+$ corresponding to F below.)

Recall the complex analytic universal cover

$$\mathcal{H}^+ \to Y,$$

where $\mathcal{H}^+$ denotes the complex upper half-plane, and write

$$w|_{\mathcal{H}^+} = F \cdot (2\pi i dz)^{\otimes k}$$

where dz is induced by the standard holomorphic differential on $\mathbb{C}$ via the uniformization of the universal elliptic curve $\mathbb{C}/(\mathbb{Z} + \mathbb{Z}\tau)$.

Let $\chi \in \Sigma_+$ be a central critical character for w of infinity type $(k+j, -j)$ where $j \geq 0$ and of finite type $(c, \mathfrak{N}, \epsilon_w)$. We recall the classical result of Waldspurger relating the algebraic part of the central critical value

$$L(F, \chi^{-1}, 0) := L((\pi_w)_K \times \chi^{-1}, 1/2)$$

to the toric period, as explicitly calculated in our situation when w is a newform by Bertolini-Darmon-Prasanna.

First, for the case when w is an Eisenstein series, we recall the following definition.

Definition 8.3. *Let u, t be (not necessarily coprime) integers with $ut = N'|N$, and let $\psi_1 : (\mathbb{Z}/u)^\times \to \mathbb{C}^\times$, $\psi_2 : (\mathbb{Z}/t)^\times \to \mathbb{C}^\times$ be Dirichlet characters (extended by 0 to $\mathbb{Z}/u, \mathbb{Z}/t$). Recall the Eisenstein series $E_k^{\psi_1, \psi_2}$ attached to ψ_1, ψ_2, i.e., with q-expansion at ∞ given by*

$$E_k^{\psi_1, \psi_2}(q) = \delta_{\psi_1 = 1} \frac{1}{2} L(\psi_1, 1-k) + \sum_{n=1}^\infty \sigma_{k-1}^{\psi_1, \psi_2}(n) q^n,$$

$$\sigma_{k-1}^{\psi_1, \psi_2}(n) = \sum_{0 < d | n} \psi_1(n/d) \psi_2(d) d^{k-1},$$

where

$$\delta_{\psi = 1} = \begin{cases} 1 & \psi = 1 \\ 0 & else. \end{cases}$$

Then $E_k^{\psi_1, \psi_2}$ has level N' and nebentype $\psi_1 \psi_2$.

Theorem 8.4 ([5] Theorem 5.4, Proof of Theorem 4.1 of [28], Proposition 36 of [42], see also [44]). *Let $\chi \in \Sigma_+$ be of infinity type $(k+j, -j)$ as above, and suppose that c and d_K are odd, and let $w_K = \#\mathcal{O}_K^\times$. Then if F is a newform, we have*

$$C(w, \chi, c) \cdot L(F, \chi^{-1}, 0) = \sigma(w, \chi) \cdot \left(\sum_{[\mathfrak{a}] \in \mathcal{C}\ell(\mathcal{O}_c)} (\check{\mathbb{N}}_K^j \check{\chi})^{-1}(\mathfrak{a}) \partial_k^j F(\mathfrak{a} \star (A, t)) \right)^2,$$

$$(8.1)$$

where the representatives $\mathfrak{a}$ of classes of $\mathcal{C}\ell(\mathcal{O}_c)$ are chosen to be prime to $\mathfrak{N}_c$, ϵ_K is the quadratic character associated with the imaginary quadratic extension $K/\mathbb{Q}$, and

$$C(w, \chi, c) = \pi^{k+2j-1} \Gamma(j+1) \Gamma(k+j) w_k |d_K|^{1/2} \cdot c \cdot \mathrm{vol}(\mathcal{O}_c)^{-(k+2j)}$$
$$\cdot 2^{\#S_w - 2} \cdot \prod_{\ell | c} \frac{\ell - \epsilon_K(\ell)}{\ell - 1} \in \mathbb{C}^\times$$

and

$$\sigma(w, \chi) \in \mathbb{C}^\times, \qquad |\sigma(w, \chi)| = 1,$$

are constants depending on w, χ and c. (For a precise definition of $\sigma(w, \chi)$, see (5.1.11) of loc. cit. and the discussion preceding it.) Here, we note that the sum (8.1) is well-defined (i.e., independent of the choice of representatives $\mathfrak{a}$ of elements of $\mathcal{C}\ell(\mathcal{O}_c)$) by the central criticality and $(c, \mathfrak{N}, \epsilon_w)$-finite typeness of χ.

Now suppose $F = E_k^{\psi_1, \psi_2}$. Recall that we assumed the Heegner hypothesis for N, which specifies an integral ideal $\mathfrak{N}$ with $\mathcal{O}_K / \mathfrak{N} = \mathbb{Z}/N$. Let $\mathfrak{N}' = (N', \mathfrak{N}), \mathfrak{u} = (u, \mathfrak{N}), \mathfrak{t} = (t, \mathfrak{N})$. In this way, we can consider ψ_1, ψ_2 as Dirichlet characters $\psi_1 : (\mathcal{O}_K/\mathfrak{u})^\times = (\mathbb{Z}/u)^\times \to \mathbb{C}^\times, \psi_2 : (\mathcal{O}/\mathfrak{t})^\times = (\mathbb{Z}/t)^\times \to \mathbb{C}^\times$.

We have the following identity when $F = E_k^{\psi_1, \psi_2}$:

$$\frac{t^k \Gamma(k+j) \psi_1^{-1}(-\sqrt{d_K}) \chi^{-1}(\mathfrak{t})}{(2\pi i)^{k+j} \mathfrak{g}(\psi_1^{-1}) \sqrt{d_K}^j} L((\psi_1 \circ \mathrm{Nm}_{K/\mathbb{Q}})\chi^{-1}, 0)$$

$$= \sum_{[\mathfrak{a}] \in \mathcal{C}\ell(\mathcal{O}_c)} (\check{\mathbb{N}}_K^j \check{\chi})^{-1}(\mathfrak{a}) \eth_k^j F(\mathfrak{a} \star (A, t)) \tag{8.2}$$

where $\mathfrak{g}(\psi_2)$ is the usual Gauss sum attached to ψ_2, as defined in [42, Section 1], $\mathrm{Nm} : \mathbb{A}_K^\times \to \mathbb{A}_\mathbb{Q}^\times$ is the idélic norm, and $L(\chi, s)$ is the usual Hecke L-function.

Definition 8.5. *Following the notation of Bertolini-Darmon-Prasanna ([5, Theorem 5.5]), for $\chi \in \Sigma_+$ of infinity type $(k+j, -j)$, when w is an eigenform of weight k on $Y_1(N)$ with $w|_{\mathcal{H}^+} = F \cdot (2\pi i dz)^{\otimes k}$, we let*

$$L^{\mathrm{alg}}(F, \chi^{-1}, 0) := i_\infty^{-1} \left(\frac{1}{\Omega_\infty(A, t)^{k+2j}} \sum_{[\mathfrak{a}] \in \mathcal{C}\ell(\mathcal{O}_c)} (\check{\mathbb{N}}_K^j \check{\chi})^{-1}(\mathfrak{a}) \eth_k^j F(\mathfrak{a} \star (A, t)) \right).$$

Hence, when w is a newform,

$$L^{\mathrm{alg}}(F, \chi^{-1}, 0)^2 = i_\infty^{-1} \left(\Omega_\infty(A, t)^{-2(k+2j)} \cdot \sigma(w, \chi)^{-1} \cdot C(w, \chi, c) \cdot L(F, \chi^{-1}, 0) \right),$$

and when w is an Eisenstein series,

$$L^{\mathrm{alg}}(E_k^{\psi_1, \psi_2}, \chi^{-1}, 0)$$

$$= i_\infty^{-1} \left(\Omega_\infty(A, t)^{-(k+2j)} \cdot \frac{t^k \Gamma(k+j) \psi_1^{-1}(-\sqrt{d_K}) \chi^{-1}(\mathfrak{t})}{(2\pi i)^{k+j} \mathfrak{g}(\psi_1^{-1}) \sqrt{d_K}^j} \right.$$

$$\left. L((\psi_1 \circ \mathrm{Nm}_{K/\mathbb{Q}})\chi^{-1}, 0) \right).$$

Choice 8.6. *Fix $y = (A, t, \alpha) = (A, \alpha) \in \mathcal{Y}(\mathbb{C}_p, \mathcal{O}_{\mathbb{C}_p})$ with α as in Proposition 7.3 when p is ramified in K and as in Proposition 7.7 when p is inert in K; let*

$$y' = (A', \alpha') = (A, \alpha) \cdot \begin{pmatrix} q & 0 \\ 0 & 1 \end{pmatrix}$$

be as in those propositions. Recall that we often suppress the (tame) $\Gamma_1(N)$-level structures t in our notation.

Then Theorem 5.48 along with Propositions 5.47, 7.3 and 7.7 establish the equality of algebraic numbers when w is a newform:

$$L^{\mathrm{alg}}(F, \chi^{-1}, 0)$$

$$= i_\infty^{-1}\left(\frac{1}{\Omega_\infty(A,t)^{k+2j}} \cdot \left(\sum_{[\mathfrak{a}] \in \mathcal{C}\ell(\mathcal{O}_c)} (\check{\mathbb{N}}_K^j \check{\chi})^{-1}(\mathfrak{a}) \eth_k^j F(\mathfrak{a} \star (A,t))\right)\right)$$

$$= i_p^{-1}\left(\left(\frac{1}{\theta(\Omega_p)(A,t,\alpha)}\right)^{k+2j}\right.$$

$$\left.\left(\sum_{[\mathfrak{a}] \in \mathcal{C}\ell(\mathcal{O}_c)} (\check{\mathbb{N}}_K^j \check{\chi})^{-1}(\mathfrak{a}) \theta_k^j((y_{\mathrm{dR}}^k f)(q_{\mathrm{dR}}))(\mathfrak{a} \star (A,t,\alpha))\right)\right)$$

$$= i_p^{-1}\left(\left(\frac{1}{\theta(\Omega_p)(A,t,\alpha)}\right)^{k+2j}\right.$$

$$\left.\left(\sum_{[\mathfrak{a}] \in \mathcal{C}\ell(\mathcal{O}_c)} (\check{\mathbb{N}}_K^j \check{\chi})^{-1}(\mathfrak{a}) \theta \circ \partial_{k,\Delta}^{\diamond,j}((y_{\mathrm{dR}}^k f)(q_{\mathrm{dR}}))(\mathfrak{a} \star (A,t,\alpha))\right)\right)$$

$$= i_p^{-1}\left(\left(\frac{1}{\Omega_p(A,t,\alpha)}\right)^{k+2j}\right.$$

$$\left.\left(\sum_{[\mathfrak{a}] \in \mathcal{C}\ell(\mathcal{O}_c)} (\check{\mathbb{N}}_K^j \check{\chi})^{-1}(\mathfrak{a}) \partial_{k,\Delta}^{\diamond,j}((y_{\mathrm{dR}}^k f)(q_{\mathrm{dR}}))(\mathfrak{a} \star (A,t,\alpha))\right)\right),$$

$$(8.3)$$

and the following equality of algebraic numbers when w is an Eisenstein series:

$$L^{\mathrm{alg}}(E_k^{\psi_1,\psi_2}, \chi^{-1}, 0)$$

$$:= i_\infty^{-1}\left(\Omega_\infty(A,t)^{-(k+2j)} \cdot \sum_{[\mathfrak{a}] \in \mathcal{C}\ell(\mathcal{O}_c)} (\check{\mathbb{N}}_K^j \check{\chi})^{-1}(\mathfrak{a}) \eth_k^j F(\mathfrak{a} \star (A,t))\right)$$

$$= i_\infty^{-1}\left(\frac{1}{\Omega_\infty(A,t)^{k+2j}} \cdot \sum_{[\mathfrak{a}] \in \mathcal{C}\ell(\mathcal{O}_c)} (\check{\mathbb{N}}_K^j \check{\chi})^{-1}(\mathfrak{a}) \eth_k^j F(\mathfrak{a} \star (A,t))\right)$$

$$= i_p^{-1}\left(\left(\frac{1}{\theta(\Omega_p)(A,t,\alpha)}\right)^{k+2j} \sum_{[\mathfrak{a}] \in \mathcal{C}\ell(\mathcal{O}_c)} (\check{\mathbb{N}}_K^j \check{\chi})^{-1}(\mathfrak{a}) \theta_k^j((y_{\mathrm{dR}}^k f)(q_{\mathrm{dR}}))(\mathfrak{a} \star (A,t,\alpha))\right)$$

$$= i_p^{-1} \left(\left(\frac{1}{\theta(\Omega_p)(A,t,\alpha)} \right)^{k+2j} \right.$$

$$\left. \sum_{[\mathfrak{a}] \in \mathcal{C}\ell(\mathcal{O}_c)} (\check{\mathbb{N}}_K^j \check{\chi})^{-1}(\mathfrak{a}) \theta \circ \partial_{k,\Delta}^{\diamond,j}((y_{\mathrm{dR}}^k f)(q_{\mathrm{dR}}))(\mathfrak{a} \star (A,t,\alpha)) \right)$$

$$= i_p^{-1} \left(\left(\frac{1}{\Omega_p(A,t,\alpha)} \right)^{k+2j} \sum_{[\mathfrak{a}] \in \mathcal{C}\ell(\mathcal{O}_c)} (\check{\mathbb{N}}_K^j \check{\chi})^{-1}(\mathfrak{a}) \partial_{k,\Delta}^{\diamond,j}(y_{\mathrm{dR}}^k f)(q_{\mathrm{dR}})(\mathfrak{a} \star (A,t,\alpha)) \right).$$

$$(8.4)$$

Again the sum in the last lines of (8.3) and (8.4) are well-defined, which is seen as follows. For any $a \in K^\times$ which is prime to $pc\mathfrak{N}$, if we change all representatives

$$\mathfrak{a} \mapsto \mathfrak{a}a,$$

then using our previously fixed embeddings

$$K^\times \hookrightarrow GL_2(\mathbb{Q}) \xrightarrow{i_p} GL_2(\mathbb{Q}_p)$$

and invoking Proposition 5.17, we have, since $\theta_k^j(y_{\mathrm{dR}}^k f)$ is of weight $k+2j$ by Proposition 7.11,

$$\theta_k^j(y_{\mathrm{dR}}^k f)(\mathfrak{a}a \star (A,t,\alpha)) = \theta_k^j(y_{\mathrm{dR}}^k f)(\mathfrak{a} \star (A,t,\alpha \cdot a))$$
$$= a^{k+2j} \epsilon_w(a) \theta_k^j(y_{\mathrm{dR}}^k f)(\mathfrak{a} \star (A,t,\alpha)).$$

$$(8.5)$$

We also have

$$(\mathbb{N}_K^j \chi)^{-1}(\mathfrak{a}a) = a^{-(k+2j)} \epsilon_w^{-1}(a)(\mathbb{N}_K^j \chi)^{-1}(\mathfrak{a}),$$

since χ has infinity type $(k+j, -j)$ and finite type $(c, \mathfrak{N}, \epsilon_w)$. This with (8.5) gives the desired invariance.

Recall we let (Definition 7.17)

$$\Gamma = \begin{cases} \Gamma(1,2) & p > 2 \\ \Gamma(3,2) & p = 2. \end{cases}$$

Now recall the weight k modular form

$$w_j' \in \omega^{\otimes k}(Y(\Gamma)), \qquad F_j' \in \mathcal{O}_\Delta(\mathcal{Y}^{\mathrm{ss}})$$

from (7.12), associated to our above w. Pulling back along the projection

$$Y(\Gamma_1(qN) \cap \Gamma_0(p^2)) = Y(\Gamma) \to Y(0,2) = Y(\Gamma_1(N) \cap \Gamma_0(p^2)),$$

we also view $w^{(p)} \in \omega^{\otimes k}(Y(\Gamma))$.

Definition 8.7. *Recall that A/F is our fixed elliptic curve with CM by $\mathcal{O}_c \subset \mathcal{O}_K$, and $(A, \alpha) = (A, t, \alpha)$ (writing the former if we wish to suppress the tame $\Gamma_1(N)$-level structure $t : \mathbb{Z}/N \subset A[N]$) is chosen to be as in Choice 8.6, with corresponding (A', α'). For this choice, we define the* anticyclotomic p-adic L-*function attached to w as a function*

$$L_{p,\alpha}(w, \cdot) : \overline{\hat{\Sigma}}_+ \to \mathbb{C}_p$$

given by

$$L_{p,\alpha}(w, \check{\chi}) := \sum_{[\mathfrak{a}] \in \mathcal{C}\ell(\mathcal{O}_c)} (\check{\mathbb{N}}_K^j \check{\chi})^{-1}(\mathfrak{a}) \theta(F_j')(\mathfrak{a} \star (A', \alpha')) \tag{8.6}$$

where $\check{\chi}|_{\mathcal{O}_{K_p}^{\times}}(x_p) = x_p^{k+j}\overline{x_p}^{-j}$. Note that $L_{p,\alpha}(w, \cdot)$ is a continuous function by Proposition 7.15 and Theorem 6.21.

For the purposes of recovering classical algebraic L-values, we define the following averaged version of $L_{p,\alpha}$

Definition 8.8. *We define the* stabilized anticyclotomic p-adic L *function attached to w as*

$$\mathcal{L}_{p,\alpha}(w, \check{\chi}) := \frac{\#(\mathbb{Z}/N\mathfrak{p}\mathbb{Z})^{\times}}{\#(\mathcal{O}_K/N\mathfrak{p}\mathcal{O}_K)^{\times}}$$

$$\cdot \sum_{a \in (\mathcal{O}_K/N\mathfrak{p}\mathcal{O}_K)^{\times}/(\mathbb{Z}/p)^{\times}} \sum_{[\mathfrak{a}] \in \mathcal{C}\ell(\mathcal{O}_c)} (\check{\mathbb{N}}_K^j \check{\chi})^{-1}(\mathfrak{a}a) \theta(F_j')(\mathfrak{a}a \star (A', \alpha'))$$

$$= \frac{\#(\mathbb{Z}/N\mathfrak{p}\mathbb{Z})^{\times}}{\#(\mathcal{O}_K/N\mathfrak{p}\mathcal{O}_K)^{\times}} \sum_{[\mathfrak{a}] \in \mathcal{C}\ell(\mathcal{O}_{cN\mathfrak{p}})} (\check{\mathbb{N}}_K^j \check{\chi})^{-1}(\mathfrak{a}a) \theta(F_j')(\mathfrak{a}a \star (A', \alpha'))$$

$$\tag{8.7}$$

where $\check{\chi}|_{\mathcal{O}_{K_p}^{\times}}(x_p) = x_p^{k+j}\overline{x_p}^{-j}$,

$$N\mathfrak{p} := |\mathbb{N}_K(\mathfrak{p})| = \begin{cases} p^2 & p \text{ inert in } K \\ p & p \text{ ramified in } K, \end{cases}$$

the summation "$\sum_{a \in (\mathcal{O}_K/N\mathfrak{p}\mathcal{O}_K)^{\times}/(\mathbb{Z}/N\mathfrak{p})^{\times}}$" means summing over a set of representatives $a \in \mathcal{O}_K$ of the group $(\mathcal{O}_K/N\mathfrak{p}\mathcal{O}_K)^{\times}/(\mathbb{Z}/N\mathfrak{p})^{\times}$ where each a is coprime with $pc\mathfrak{N}$, and for such $a \in \mathcal{O}_K$,

$$a \star (A, t, \alpha) = (A, at, a\alpha)$$

is defined as in Definition 5.46 (using the identification $\mathrm{End}_F(A) = \mathcal{O}_c$). Since $\{\mathfrak{a}_i\}$ is a full set of representatives of $\mathcal{C}\ell(\mathcal{O}_c)$, then $\{\mathfrak{a}_i\mathfrak{a}\}$ is a full set of representatives of $\mathcal{C}\ell(\mathcal{O}_{c\mathrm{N}\mathfrak{p}})$, which gives the second equality in (8.7). We also note that the last equality in (8.7) comes from Theorem 5.48 for $j \geq 0$ and again extending by continuity.

We note that

$$\#(\mathcal{O}_K/\mathrm{N}\mathfrak{p}\mathcal{O}_K)^\times = \begin{cases} p^2(p^2-1) & p \text{ inert in } K \\ p-1 & p \text{ ramified in } K, \end{cases}$$

and

$$\#(\mathbb{Z}/\mathrm{N}\mathfrak{p}\mathbb{Z})^\times = \begin{cases} p(p-1) & p \text{ inert in } K \\ p-1 & p \text{ ramified in } K, \end{cases}$$

and so

$$\#(\mathcal{O}_K/\mathrm{N}\mathfrak{p}\mathcal{O}_K)^\times / \#(\mathbb{Z}/\mathrm{N}\mathfrak{p}\mathbb{Z})^\times = \begin{cases} p(p+1) & p \text{ inert in } K \\ 1 & p \text{ ramified in } K. \end{cases}$$

We remark that $\mathcal{L}_{p,\alpha}(w, \cdot)$ is well-defined, and in particular each $\mathfrak{a}\alpha$ satisfies Choice 8.6, which is seen as follows. Since $a \in \mathcal{O}_c = \mathrm{End}_F(A)$ was prime to pc, it preserves the $H^{1,0}$ and $H^{0,1}$ Hodge pieces, and hence the $\mathbf{z}_{\mathrm{dR}}(A,t,\alpha)$ and $\hat{\mathbf{z}}(A,t,\alpha)$, and hence also $y_{\mathrm{dR}}(A,t,\alpha)$ by (4.17). From Proposition 5.47, we have

$$\Omega_p(a(A,t,\alpha)) = \Omega_p(A,t,\alpha)$$

for all such a.

Recall that for a prime p,

$$q := \begin{cases} p & p > 2 \\ 8 & p = 2. \end{cases}$$

By (7.16), recalling that we wrote for our original modular form $w \in \omega^{\otimes k}$ $(Y_1(N))$,

$$w|_{\mathcal{Y}_x} = y_{\mathrm{dR}}^k f \cdot \omega_{\mathrm{can}}^{\otimes k}, \qquad y_{\mathrm{dR}}^k f \in \mathcal{O}_\Delta(\mathcal{Y}_x),$$

we have for $j \geq 0$

$$L_{p,\alpha}(w, \check{\chi}) = q^{k+2j} \sum_{[\mathfrak{a}] \in \mathcal{C}\ell(\mathcal{O}_c)} (\check{\mathbb{N}}_K^j \check{\chi})^{-1}(\mathfrak{a}) \theta_k^j((y_{\mathrm{dR}}^k f)^{(p)}(q_{\mathrm{dR}}))(\mathfrak{a} \star (A,t,\alpha))$$

$$= q^{k+2j} \sum_{[\mathfrak{a}]\in\mathcal{C}\ell(\mathcal{O}_c)} (\check{\mathbb{N}}_K^j\check{\chi})^{-1}(\mathfrak{a})\theta_k^j((y_{\mathrm{dR}}^k f)^{(p)}(q_{\mathrm{dR}}))(\mathfrak{a}\star(A,t,\alpha))$$

$$= q^{k+2j} \sum_{[\mathfrak{a}]\in\mathcal{C}\ell(\mathcal{O}_c)} (\check{\mathbb{N}}_K^j\check{\chi})^{-1}(\mathfrak{a})\theta\circ\partial_{k,\Delta}^{\diamond,j}((y_{\mathrm{dR}}^k f)^{(p)}(q_{\mathrm{dR}}))(\mathfrak{a}\star(A,t,\alpha))$$

$$(8.8)$$

(where the last equality above already follows from Theorem 5.48), and similarly

$$\mathcal{L}_{p,\alpha}(w,\check{\chi}) = \frac{\#(\mathbb{Z}/N\mathfrak{p}\mathbb{Z})^\times}{\#(\mathcal{O}_K/N\mathfrak{p}\mathcal{O}_K)^\times} q^{k+2j}$$

$$\cdot \sum_{a\in(\mathcal{O}_K/N\mathfrak{p}\mathcal{O}_K)^\times/(\mathbb{Z}/p)^\times}\sum_{[\mathfrak{a}]\in\mathcal{C}\ell(\mathcal{O}_c)} (\check{\mathbb{N}}_K^j\check{\chi})^{-1}(\mathfrak{a}a)\theta_k^j((y_{\mathrm{dR}}^k f)^{(p)}(q_{\mathrm{dR}}))(\mathfrak{a}a\star(A,t,\alpha))$$

$$= \frac{\#(\mathbb{Z}/N\mathfrak{p}\mathbb{Z})^\times}{\#(\mathcal{O}_K/N\mathfrak{p}\mathcal{O}_K)^\times} q^{k+2j}$$

$$\sum_{a\in(\mathcal{O}_K/N\mathfrak{p}\mathcal{O}_K)^\times/(\mathbb{Z}/p)^\times}\sum_{[\mathfrak{a}]\in\mathcal{C}\ell(\mathcal{O}_c)} (\check{\mathbb{N}}_K^j\check{\chi})^{-1}(\mathfrak{a}a)\theta_k^j((y_{\mathrm{dR}}^k f)^{(p)})(\mathfrak{a}a\star(A,t,\alpha))$$

$$= \frac{\#(\mathbb{Z}/N\mathfrak{p}\mathbb{Z})^\times}{\#(\mathcal{O}_K/N\mathfrak{p}\mathcal{O}_K)^\times} q^{k+2j} \sum_{[\mathfrak{a}]\in\mathcal{C}\ell(\mathcal{O}_{cN\mathfrak{p}})} (\check{\mathbb{N}}_K^j\check{\chi})^{-1}(\mathfrak{a})\theta_k^j((y_{\mathrm{dR}}^k f)^{(p)})(\mathfrak{a}\star(A,t,\alpha))$$

$$= \frac{\#(\mathbb{Z}/N\mathfrak{p}\mathbb{Z})^\times}{\#(\mathcal{O}_K/N\mathfrak{p}\mathcal{O}_K)^\times} q^{k+2j} \sum_{[\mathfrak{a}]\in\mathcal{C}\ell(\mathcal{O}_{cN\mathfrak{p}})} (\check{\mathbb{N}}_K^j\check{\chi})^{-1}(\mathfrak{a})\theta\circ\partial_{k,\Delta}^{\diamond,j}((y_{\mathrm{dR}}^k f)^{(p)})(\mathfrak{a}\star(A,t,\alpha))$$

$$(8.9)$$

(where the last equality again follows from Theorem 5.48), where $\check{\chi}|_{\mathcal{O}_{K_p}^\times}(x_p) = x_p^{k+j}\overline{x_p}^{-j}$.

The expression above can be related to certain constant multiples of the central values $L(F,\chi^{-1},0)$, using standard properties of Hecke operators (see Proposition 8.11), which will be used in proving the interpolation property.

Theorem 8.9. *The functions*

$$L_{p,\alpha}(w,\cdot):\overline{\overline{\Sigma}}_+\to\mathbb{C}_p, \qquad \mathcal{L}_{p,\alpha}(w,\cdot):\overline{\overline{\Sigma}}_+\to\mathbb{C}_p$$

are continuous.

Proof. Let $\check{\chi}_1,\check{\chi}_2\in\overline{\overline{\Sigma}}_+$ with

$$\check{\chi}_i|_{\mathcal{O}_{K_p}^\times}(x_p) = x_p^{k+j_i}\overline{x_p}^{-j_i}$$

for $i=1,2$ and which satisfy

$$\check{\chi}_1(\mathfrak{a})\equiv\check{\chi}_2(\mathfrak{a}) \pmod{p^m\mathcal{O}_{\mathbb{C}_p}} \tag{8.10}$$

for $\mathfrak{a} \in \mathbb{A}_K^{\times,(p\infty)}$. Then evaluating on idèles congruent with 1 (mod $\mathfrak{N}$), we see from (8.10) that

$$j_1 \equiv j_2 \quad (\mathrm{mod}\ (p-1)p^{m-1}). \tag{8.11}$$

Now the continuity of $L_{p,\alpha}(w,\cdot)$ and $\mathcal{L}_{p,\alpha}(w,\cdot)$ follows from the continuity statements in Theorem 6.21, Proposition 7.15, and the continuity of p-adic weights (8.10) established above. $\square$

8.3 INTERPOLATION

8.3.1 Interpolation formula

We now establish the interpolation properties of our p-adic L-function, which involves several intermediate calculations addressing the effect of p-stabilization on CM period sums.

Definition 8.10. *Let*

$$\Omega(A,t) := \Omega_\Gamma(A,t) \in \overline{\mathbb{Q}}^{\times}$$

denote the constant (5.75) defined with respect to Γ as in Definition 7.17 and with respect to

$$(A,t,t_p) = (A,t,\alpha_1 \quad (\mathrm{mod}\ q), \langle \alpha_2 \quad (\mathrm{mod}\ p^2)\rangle) \in Y(\Gamma)$$

and (A,t), where $y = (A,t,\alpha)$ is fixed as in Choice 8.6.

We start with the following proposition relating a p-adic period sum to $L(F,\chi^{-1},0)$ for $\chi \in \Sigma_+$.

Proposition 8.11. *For $\chi \in \Sigma_+$, we have when w is a newform*

$$\frac{\#(\mathbb{Z}/N\mathfrak{p}\mathbb{Z})^{\times}}{\#(\mathcal{O}_K/N\mathfrak{p}\mathcal{O}_K)^{\times}} \sum_{[\mathfrak{a}] \in \mathcal{Cl}(\mathcal{O}_{cN\mathfrak{p}})} (\check{\mathbb{N}}_K^j \check{\chi})^{-1}(\mathfrak{a}) \theta_k^j (y_{\mathrm{dR}}^k f)^{(p)}(\mathfrak{a} \star (A,t,\alpha))$$

$$= \frac{\theta(\Omega_p)(A,t,\alpha)^{k+2j}}{\Omega(A,t)^j} \Xi_p(w,\chi) \cdot i_p(L^{\mathrm{alg}}(F,\chi^{-1},0)), \tag{8.12}$$

and when w is an Eisenstein series

$$\frac{\#(\mathbb{Z}/N\mathfrak{p}\mathbb{Z})^{\times}}{\#(\mathcal{O}_K/N\mathfrak{p}\mathcal{O}_K)^{\times}} \sum_{[\mathfrak{a}] \in \mathcal{Cl}(\mathcal{O}_{cN\mathfrak{p}})} (\check{\mathbb{N}}_K^j \check{\chi})^{-1}(\mathfrak{a}) \theta_k^j (y_{\mathrm{dR}}^k f)^{(p)}(\mathfrak{a} \star (A,t,\alpha))$$

$$= \frac{\theta(\Omega_p)(A,t,\alpha)^{k+2j}}{\Omega(A,t)^j} \Xi_p(w,\chi) \cdot i_p(L^{\mathrm{alg}}(E_k^{\psi_1,\psi_2},\chi^{-1},0)), \tag{8.13}$$

where

$$\Xi_p(w, \chi) = \begin{cases} 1 - a_p(w)^2 \chi^{-1}(p)\frac{p-1}{p+1} - \frac{1}{p^2} & p \ inert \ in \ K \\ 1 - a_p(w)\chi^{-1}(\mathfrak{p})\frac{p-1}{p} - \frac{1}{p^2} & p \ ramified \ in \ K. \end{cases}$$

Proof. Recall that we have

$$w^{(p)} = w|_{V_p U_p - U_p V_p} = (1 - T_p V_p + [p]V_p^2)^* w.$$

Now write

$$w^{(p)}|_{\mathcal{H}^+} = F^{(p)} \cdot (2\pi i dz)^{\otimes k}.$$

Denote

$$Y_r = Y(\Gamma_1(N) \cap \Gamma_0(p^r)), \quad C_r = \lambda_r(\mathcal{C})$$

where

$$\lambda_r : \mathcal{Y} \twoheadrightarrow Y_r$$

is the natural projection. Recall that we denote the prime of $\mathcal{O}_K$ above p by $\mathfrak{p}$.

Suppose the base change of our fixed CM elliptic curve A is isomorphic to $\mathbb{C}/\mathfrak{a}_0$, where $\mathfrak{a}_0 \subset \mathcal{O}_K$ is an ideal prime to $cp\mathfrak{N}$. As a convention, given an integral ideal $\mathfrak{a}$ coprime with $p c \mathfrak{N}$, we let the triple

$$\mathfrak{a} \star (A, t, \alpha_1 \pmod p) \in C_1(\overline{K}_p, \mathcal{O}_{\overline{K}_p})$$

denote the CM point on Y_1 whose base change to $\mathbb{C}$ is

$$(\mathbb{C}/\mathfrak{a}^{-1}\mathfrak{a}_0\mathcal{O}_{cp}, t, \alpha_1 \pmod p).$$

Similarly, we let

$$\mathfrak{a} \star (A, t, \alpha_1 \pmod {p^2}) \in C_2(\overline{K}_p, \mathcal{O}_{\overline{K}_p})$$

denote the CM point on Y_2 whose base change to $\mathbb{C}$ is

$$(\mathbb{C}/\mathfrak{a}^{-1}\mathfrak{a}_0\mathcal{O}_{cp^2}, t, \alpha_1 \pmod {p^2}).$$

Given $(A, t, \alpha_1 \pmod {p^r})$ for $r \in \mathbb{Z}_{\geq 0}$ and $r' \in \mathbb{Z}_{\geq 0}$ with $r' \leq r$, we will let

$$(A, t, \alpha_1 \pmod {p^r}) = (A, t, \alpha_1 \pmod {p^{r'}}) \in Y_{r'}(\overline{K}_p, \mathcal{O}_{\overline{K}_p})$$

denote the same point.

We denote, for $(A, t, \alpha_1 \pmod {p^2}) \in Y_1(\overline{K}_p, \mathcal{O}_{\overline{K}_p})$,

$$\mathfrak{p}^{-1} \star (A, t, \alpha_1 \pmod {p^2}) = (A/A[\mathfrak{p}], t \pmod{A[\mathfrak{p}]}, (\alpha_1 \pmod{p^2}) \pmod{A[\mathfrak{p}]})$$

$$\tag{8.14}$$

where t (mod $A[\mathfrak{p}]$) denotes the image of t under the canonical projection $A \to A/A[\mathfrak{p}]$, and

$$\alpha_1 \quad (\text{mod } p^2) \quad (\text{mod } A[\mathfrak{p}]) : \mathbb{Z}/p \hookrightarrow (A/A[\mathfrak{p}])[p]$$

denotes the image of α_1 (mod p^2) under the canonical projection $A \to A/A[\mathfrak{p}]$, which induces a $\Gamma_0(p)$-level structure on $A/A[\mathfrak{p}]$. Note that we can evaluate a function $F|_{Y_1}$ for a function F on Y on the triple (8.14).

Similarly, $(A, t, \alpha_1 \ (\text{mod } p^2)) \in Y_2(\overline{K}_p, \mathcal{O}_{\overline{K}_p})$, we let

$$p^{-2} \star (A, t, \alpha_1 \quad (\text{mod } p^2)) = (A/A[p^2], p^2 t, \alpha_1 \quad (\text{mod } 1)). \tag{8.15}$$

Note that we can evaluate a function $F|_{Y_2}$ for a function F on Y on the triple (8.15).

We record the following computation as a lemma for future reference.

Lemma 8.12. *We have*

$$\sum_{[\mathfrak{a}] \in \mathcal{C}\ell(\mathcal{O}_{cN\mathfrak{p}})} (\mathbb{N}_K^j \chi)^{-1}(\mathfrak{a}) \partial_k^j F^{(p)}(\mathfrak{a} \star (A, t, \alpha_1 \quad (\text{mod } p^2)))$$

$$= \frac{\#(\mathcal{O}_K/N\mathfrak{p}\mathcal{O}_K)^\times}{\#(\mathbb{Z}/N\mathfrak{p}\mathbb{Z})^\times} \frac{\Xi_p(w, \chi)}{\Omega(A, t)^j} \sum_{[\mathfrak{a}] \in \mathcal{C}\ell(\mathcal{O}_c)} (\mathbb{N}_K^j \chi)^{-1}(\mathfrak{a}) \partial_k^j F(\mathfrak{a} \star (A, t)). \tag{8.16}$$

Proof. We compute

$$\sum_{[\mathfrak{a}] \in \mathcal{C}\ell(\mathcal{O}_{cN\mathfrak{p}})} (\mathbb{N}_K^j \chi)^{-1}(\mathfrak{a}) \partial_k^j F^{(p)}(\mathfrak{a} \star (A, t, \alpha_1 \quad (\text{mod } p^2)))$$

$$= \sum_{[\mathfrak{a}] \in \mathcal{C}\ell(\mathcal{O}_{cN\mathfrak{p}})} (\mathbb{N}_K^j \chi)^{-1}(\mathfrak{a})(1 - T_p V_p + \frac{1}{p}[p]V_p^2)^* \partial_k^j F(\mathfrak{a} \star (A, t, \alpha_1 \quad (\text{mod } p^2)))$$

$$= \sum_{[\mathfrak{a}] \in \mathcal{C}\ell(\mathcal{O}_{cN\mathfrak{p}})} (\mathbb{N}_K^j \chi)^{-1}(\mathfrak{a}) \partial_k^j F(\mathfrak{a} \star (A, t, \alpha_1 \quad (\text{mod } p^2)))$$

$$- a_p(w)p^j \sum_{[\mathfrak{a}] \in \mathcal{C}\ell(\mathcal{O}_{cN\mathfrak{p}})} (\mathbb{N}_K^j \chi)^{-1}(\mathfrak{a}) \partial_k^j F|_{Y_1}(\mathfrak{a}\mathfrak{p}^{-1} \star (A, t, \alpha_1 \quad (\text{mod } p^2)))$$

$$+ \epsilon_w(p)p^{k-1+2j} \sum_{[\mathfrak{a}] \in \mathcal{C}\ell(\mathcal{O}_{cN\mathfrak{p}})} (\mathbb{N}_K^j \chi)^{-1}(\mathfrak{a}) \partial_k^j F|_{Y_2}(\mathfrak{a}\mathfrak{p}^{-2} \star (A, t, \alpha_1 \quad (\text{mod } p^2)))$$

$$= \frac{\#(\mathcal{O}_K/N\mathfrak{p}\mathcal{O}_K)^\times}{\#(\mathbb{Z}/N\mathfrak{p}\mathbb{Z})^\times} \sum_{[\mathfrak{a}] \in \mathcal{C}\ell(\mathcal{O}_c)} (\mathbb{N}_K^j \chi)^{-1}(\mathfrak{a}) \partial_k^j F(\mathfrak{a} \star (A, t))$$

$$- a_p(w)p^j (N\mathfrak{p})^{-j} \chi^{-1}(\mathfrak{p}) \sum_{[\mathfrak{a}] \in \mathcal{C}\ell(\mathcal{O}_{cp})} (\mathbb{N}_K^j \chi)^{-1}(\mathfrak{a}) \partial_k^j F|_{Y_1}(\mathfrak{a} \star (A, t, \alpha_1 \quad (\text{mod } p)))$$

$$+ \epsilon_w(p)p^{k-1+2j} p^{-4j} \chi^{-1}(p^2) \sum_{[\mathfrak{a}] \in \mathcal{C}\ell(\mathcal{O}_{cN\mathfrak{p}})} (\mathbb{N}_K^j \chi)^{-1}(\mathfrak{a}) \partial_k^j F|_{Y_2}(\mathfrak{a} \star (A, t, \alpha_1 \quad (\text{mod } p^2)))$$

where the last equality follows because $\eth_k^j F$ is of level $\Gamma_1(N)$ and so does not depend on $\mathfrak{a} \star \alpha_1 \pmod{p^2}$, and $\eth_k^j F|_{Y_1}$ is of level $\Gamma_1(N) \cap \Gamma_0(p)$ and so only depends on $\mathfrak{a} \star \alpha_1 \pmod{p}$.

Note that if p is ramified in K, there is a canonical lifting $\mathcal{C}(\overline{K}_p, \mathcal{O}_{\overline{K}_p})$ given by

$$(A, t) \mapsto (A, t, A[\mathfrak{p}]).$$

Since $\alpha_1 \pmod{p} = A[\mathfrak{p}]$ in this case by our choice of α as in Choice 8.6, then we have

$$\sum_{[\mathfrak{a}] \in \mathcal{Cl}(\mathcal{O}_{cp})} (\mathbb{N}_K^j \chi)^{-1}(\mathfrak{a}) \eth_k^j F|_{Y_1}(\mathfrak{a} \star (A, t, \alpha_1 \pmod{p}))$$

$$= \sum_{[\mathfrak{a}] \in \mathcal{Cl}(\mathcal{O}_c)} (\mathbb{N}_K^j \chi)^{-1}(\mathfrak{a}) \eth_k^j F|_{Y_1}(\mathfrak{a} \star (A, t)).$$

Suppose p is inert in K. By the definition of T_p as a correspondence defined on a triple by summing over triples which are the images under degree p isogenies from the triple and dividing by p, we have

$$\sum_{[\mathfrak{a}] \in \mathcal{Cl}(\mathcal{O}_{cp})} (\mathbb{N}_K^j \chi)^{-1}(\mathfrak{a}) \eth_k^j F|_{Y_1}(\mathfrak{a} \star (A, t, \alpha_1 \pmod{p}))$$

$$= \sum_{[\mathfrak{a}] \in \mathcal{Cl}(\mathcal{O}_{cp})} (\mathbb{N}_K^j \chi)^{-1}(\mathfrak{a}) \eth_k^j (p T_p)^* F(\mathfrak{a} \star (A, t, \alpha_1 \pmod{p}))$$

$$= a_p(w) p^{j+1} \sum_{[\mathfrak{a}] \in \mathcal{Cl}(\mathcal{O}_c)} (\mathbb{N}_K^j \chi)^{-1}(\mathfrak{a}) \eth_k^j F(\mathfrak{a} \star (A, t, \alpha_1 \pmod{p})).$$

Now suppose p is ramified in K. Again by the definition of T_p, we have

$$\sum_{[\mathfrak{a}] \in \mathcal{Cl}(\mathcal{O}_{cp})} (\mathbb{N}_K^j \chi)^{-1}(\mathfrak{a}) \eth_k^j F|_{Y_2}(\mathfrak{a} \star (A, t, \alpha_1 \pmod{p}))$$

$$= \sum_{[\mathfrak{a}] \in \mathcal{Cl}(\mathcal{O}_{cp})} (\mathbb{N}_K^j \chi)^{-1}(\mathfrak{a}) \eth_k^j (p(T_p - [\mathfrak{p}]))^* F(\mathfrak{a}\mathfrak{p} \star (A, t, \alpha_1 \pmod{p}))$$

$$= p(a_p(w) p^j (\mathbb{N}\mathfrak{p})^j \chi(\mathfrak{p}) - \epsilon_w(p) p^{k-1+2j})$$

$$\sum_{[\mathfrak{a}] \in \mathcal{Cl}(\mathcal{O}_c)} (\mathbb{N}_K^j \chi)^{-1}(\mathfrak{a}\mathfrak{p}) \eth_k^j F(\mathfrak{a}\mathfrak{p} \star (A, t, \alpha_1 \pmod{p}))$$

$$= p(a_p(w) p^{2j} \chi(\mathfrak{p}) - \epsilon_w(p) p^{k-1+2j})$$

$$\sum_{[\mathfrak{a}] \in \mathcal{Cl}(\mathcal{O}_c)} (\mathbb{N}_K^j \chi)^{-1}(\mathfrak{a}) \eth_k^j F(\mathfrak{a} \star (A, t, \alpha_1 \pmod{p})).$$

Similarly, by the definition of T_p^2 as a correspondence defined on a triple by summing over triples which are the images under degree p^2 isogenies from the triple and dividing by p^2, we have

$$\sum_{[\mathfrak{a}]\in\mathcal{C}\ell(\mathcal{O}_{cp^2})} (\mathbb{N}_K^j \chi)^{-1}(\mathfrak{a}) \eth_k^j F|_{Y_2}(\mathfrak{a} \star (A, t, \alpha_1 \pmod{p^2}))$$

$$= \sum_{[\mathfrak{a}]\in\mathcal{C}\ell(\mathcal{O}_{cp})} (\mathbb{N}_K^j \chi)^{-1}(\mathfrak{a}) (p^2(T_p^2 - (p+1)[p]))^* \eth_k^j F(\mathfrak{a} \star (A, t, \alpha_1 \pmod{p^2}))$$

$$= p^2 (a_p(w)^2 p^{2j} - (p+1)\epsilon_w(p) p^{k-1+2j})$$

$$\sum_{[\mathfrak{a}]\in\mathcal{C}\ell(\mathcal{O}_c)} (\mathbb{N}_K^j \chi)^{-1}(\mathfrak{a}) \eth_k^j F(\mathfrak{a} \star (A, t, \alpha_1 \pmod{p^2})).$$

Finally, recall that since $\chi \in \Sigma_+$, we have

$$\chi(p) = \epsilon_w(p) p^k.$$

Hence, in the p inert in K case we have

$$\sum_{[\mathfrak{a}]\in\mathcal{C}\ell(\mathcal{O}_{cp^2})} (\mathbb{N}_K^j \chi)^{-1}(\mathfrak{a}) \eth_k^j F^{(p)}(\mathfrak{a} \star (A, t, \alpha_1 \pmod{p^2}))$$

$$= \frac{p^2(p^2-1)}{p(p-1)} \left(1 - \frac{p(p-1)}{p^2(p^2-1)} a_p(w) p^j p^{-2j} \chi^{-1}(p) p^{j+1} a_p(w) \right.$$

$$\left. + \frac{p(p-1)}{p^2(p^2-1)} \epsilon_w(p) p^{k-1+2j} p^{-4j} \chi^{-1}(p^2) p^2 \left(a_p(w)^2 p^{2j} - (p+1)\epsilon_w(p) p^{k-1+2j} \right) \right)$$

$$\cdot \sum_{[\mathfrak{a}]\in\mathcal{C}\ell(\mathcal{O}_c)} (\mathbb{N}_K^j \chi)^{-1}(\mathfrak{a}) \eth_k^j F(\mathfrak{a} \star (A, t))$$

$$= \frac{p^2(p^2-1)}{p(p-1)}$$

$$\left(1 - \frac{p}{p+1} a_p(w)^2 \chi^{-1}(p) + \frac{p^{-2j}}{p+1} \chi^{-1}(p) \left(a_p(w)^2 p^{2j} - (p+1)\chi(p) p^{-1+2j} \right) \right)$$

$$\cdot \sum_{[\mathfrak{a}]\in\mathcal{C}\ell(\mathcal{O}_c)} (\mathbb{N}_K^j \chi)^{-1}(\mathfrak{a}) \eth_k^j F(\mathfrak{a} \star (A, t))$$

$$= \frac{p^2(p^2-1)}{p(p-1)} \left(1 - a_p(w)^2 \chi^{-1}(p) \frac{p-1}{p+1} - \frac{1}{p^2} \right)$$

$$\sum_{[\mathfrak{a}]\in\mathcal{C}\ell(\mathcal{O}_c)} (\mathbb{N}_K^j \chi)^{-1}(\mathfrak{a}) \eth_k^j F(\mathfrak{a} \star (A, t, \alpha_1 \pmod{p^2}))$$

$$\overset{(5.75)}{=} \frac{\#(\mathcal{O}_K/N\mathfrak{p}\mathcal{O}_K)^\times}{\#(\mathbb{Z}/N\mathfrak{p}\mathbb{Z})^\times} \frac{\Xi_p(w,\chi)}{\Omega(A,t)^j} \sum_{[\mathfrak{a}]\in\mathcal{C}\ell(\mathcal{O}_c)} (\mathbb{N}_K^j \chi)^{-1}(\mathfrak{a}) \eth_k^j F(\mathfrak{a} \star (A, t)).$$

In the p ramified in K case, we have

$$\sum_{[\mathfrak{a}]\in\mathcal{Cl}(\mathcal{O}_{cp})}(\mathbb{N}_K^j\chi)^{-1}(\mathfrak{a})\mathfrak{d}_k^j F^{(p)}(\mathfrak{a}\star(A,t,\alpha_1 \pmod{p^2})))$$

$$=\frac{p(p-1)}{p-1}\left(1-\frac{p-1}{p(p-1)}a_p(w)p^{j+1}p^{-j}\chi^{-1}(\mathfrak{p})\right.$$

$$\left.+\frac{p-1}{p(p-1)}\epsilon_w(p)p^{k-1+2j}p^{-4j}\chi^{-1}(p^2)p\left(a_p(w)p^{2j}\chi(\mathfrak{p})-\epsilon_w(p)p^{k-1+2j}\right)\right)$$

$$\cdot\sum_{[\mathfrak{a}]\in\mathcal{Cl}(\mathcal{O}_c)}(\mathbb{N}_K^j\chi)^{-1}(\mathfrak{a})\mathfrak{d}_k^j F(\mathfrak{a}\star(A,t,\alpha_1 \pmod{p^2})))$$

$$=\frac{p(p-1)}{p-1}\left(1-a_p(w)\chi^{-1}(\mathfrak{p})+\epsilon_w(p)p^{k-2-2j}\chi^{-1}(p^2)p\right.$$

$$\left.\left(a_p(w)p^{2j}\chi(\mathfrak{p})-\epsilon_w(p)p^{k-1+2j}\right)\right)\cdot\sum_{[\mathfrak{a}]\in\mathcal{Cl}(\mathcal{O}_c)}(\mathbb{N}_K^j\chi)^{-1}(\mathfrak{a})\mathfrak{d}_k^j F(\mathfrak{a}\star(A,t))$$

$$=\frac{p(p-1)}{p-1}\left(1-a_p(w)\chi^{-1}(\mathfrak{p})\frac{p-1}{p}-\frac{1}{p^2}\right)$$

$$\sum_{[\mathfrak{a}]\in\mathcal{Cl}(\mathcal{O}_c)}(\mathbb{N}_K^j\chi)^{-1}(\mathfrak{a})\mathfrak{d}_k^j F(\mathfrak{a}\star(A,t,\alpha_1 \pmod{p^2})))$$

$$\overset{(5.75)}{=}\frac{\#(\mathcal{O}_K/Np\mathcal{O}_K)^\times}{\#(\mathbb{Z}/Np\mathbb{Z})^\times}\frac{\Xi_p(w,\chi)}{\Omega(A,t)^j}\sum_{[\mathfrak{a}]\in\mathcal{Cl}(\mathcal{O}_c)}(\mathbb{N}_K^j\chi)^{-1}(\mathfrak{a})\mathfrak{d}_k^j F(\mathfrak{a}\star(A,t)). \qquad \square$$

Now we finish the proof of Proposition 8.11. Suppose w is a newform. We have

$$i_p(L^{\mathrm{alg}}(F,\chi^{-1},0))=i_p\left(\sum_{[\mathfrak{a}]\in\mathcal{Cl}(\mathcal{O}_c)}(\mathbb{N}_K^j\chi)^{-1}(\mathfrak{a})\mathfrak{d}_k^j F(\mathfrak{a}\star(A,t))\right)$$

$$\overset{(8.16)}{=}i_p\left(\frac{\#(\mathbb{Z}/Np\mathbb{Z})^\times}{\#(\mathcal{O}_K/Np\mathcal{O}_K)^\times}\Xi_p(w,\chi)^{-1}\Omega(A,t)^j\right.$$

$$\left.\sum_{[\mathfrak{a}]\in\mathcal{Cl}(\mathcal{O}_{cNp})}(\mathbb{N}_K^j\chi)^{-1}(\mathfrak{a})\mathfrak{d}_k^j F^{(p)}(\mathfrak{a}\star(A,t,\alpha_2 \pmod{p^2})))\right)$$

$$\overset{(8.3)}{=}\left(\frac{\#(\mathbb{Z}/Np\mathbb{Z})^\times}{\#(\mathcal{O}_K/Np\mathcal{O}_K)^\times}(\theta(\Omega_p)(A,t,\alpha))^{-(k+2j)}\Xi_p(w,\chi)^{-1}\Omega(A,t)^j\right)$$

$$\cdot\sum_{[\mathfrak{a}]\in\mathcal{Cl}(\mathcal{O}_{cNp})}(\check{\mathbb{N}}_K^j\check{\chi})^{-1}(\mathfrak{a})\theta_k^j(y_{\mathrm{dR}}^k f)^{(p)}(\mathfrak{a}\star(A,t,\alpha))$$

$$(8.17)$$

which is what we wanted. The statement for w an Eisenstein series is entirely analogous, invoking (8.4). $\qquad \square$

Proposition 8.13. *The function*

$$\check{\Sigma} \to \mathbb{C}_p: \quad \check{\chi} \mapsto \Xi_p(w, \check{\chi})$$

is continuous.

Proof. Note that since $p \nmid N$ that $\chi \in \Sigma_+$ is ramified only possibly at primes dividing $\mathfrak{N}_{\epsilon_w}$. Since $\chi \in \Sigma$, we have

$$\chi^{-1}(p) = p^{-k}\epsilon_w^{-1}(p) \tag{8.18}$$

which is constant, and so the proposition follows in the p inert case from the definition of $\Xi_p(w, \check{\chi})$.

When p is ramified in K, we have

$$\chi^{-1}(\mathfrak{p}) = \chi^{-1}(\dots, 1, \sqrt{-d}, 1, \dots) = (\sqrt{-d})^{-k}\epsilon_w^{-1}(\sqrt{-d}) \prod_{v|d, v \nmid p\mathfrak{N}_{\epsilon_w}} \chi_v^{-1}(\sqrt{-d}) \tag{8.19}$$

if p is ramified in $K = \mathbb{Q}(\sqrt{-d})$ with d squarefree, where $(\dots, 1, \sqrt{-d}, 1, \dots)$ denotes the idèle which is $\sqrt{-d}$ at the $\mathfrak{p}^{\text{th}}$ place and 1 everywhere else; here recall that we have made the identification

$$\epsilon_w: (\mathcal{O}_K/\mathfrak{N})^\times = (\mathbb{Z}/N)^\times \to \overline{\mathbb{Q}}^\times.$$

We remark that since

$$\check{\Sigma} \subset \text{Fun}(\mathbb{A}_K^{\times,(p\infty)}, \mathcal{O}_{\mathbb{C}_p})$$
$$= \{\text{functions } \mathbb{A}_K^{\times,(p\infty)} \to \mathcal{O}_{\mathbb{C}_p} \text{ with the uniform convergence topology}\}$$

is given the uniform convergence topology, then

$$\check{\Sigma} \to \mathbb{C}_p: \quad \check{\chi} \mapsto \prod_{v|d, v \nmid p\mathfrak{N}_{\epsilon_w}} \chi_v^{-1}(\sqrt{-d})$$

is a continuous function, and hence by (8.19) we have that

$$\check{\Sigma} \to \mathbb{C}_p: \quad \check{\chi} \mapsto \check{\chi}^{-1}(\mathfrak{p}) \tag{8.20}$$

is continuous. Now the proposition follows from the definition of $\Xi_p(w, \check{\chi})$. $\square$

8.3.2   Interpolation formula

We finally establish the precise interpolation property of our p-adic L-functions.

Theorem 8.14. *Suppose that the fundamental discriminant d_K is odd. Then for all $j \geq 0$, and for any $\chi \in \Sigma_+$ of infinity type $(k+j, -j)$, when $w \in \omega^{\otimes k}(Y_1(N))$ is a newform $(w|_{\mathcal{H}^+} = F \cdot (2\pi i dz)^{\otimes k})$ we have*

$$\mathcal{L}_{p,\alpha}(w, \check{\chi}) = \frac{(q\theta(\Omega_p)(A, t, \alpha))^{k+2j}}{\Omega(A, t)^j} \Xi_p(w, \chi) \cdot i_p \left(L^{\mathrm{alg}}(F, \chi^{-1}, 0) \right), \qquad (8.21)$$

and when $w \in \omega^{\otimes k}(Y_1(N))$ is an Eisenstein series $(w|_{\mathcal{H}^+} = E_k^{\psi_1, \psi_2} \cdot (2\pi i dz)^{\otimes k})$ we have

$$\mathcal{L}_{p,\alpha}(w, \check{\chi}) = \frac{(q\theta(\Omega_p)(A, t, \alpha))^{k+2j}}{\Omega(A, t)^j} \Xi_p(w, \chi) \cdot i_p \left(L^{\mathrm{alg}}(E_k^{\psi_1, \psi_2}, \chi^{-1}, 0) \right), \qquad (8.22)$$

where

$$\Xi_p(w, \chi) = \begin{cases} 1 - a_p(w)^2 \chi^{-1}(p) \frac{p-1}{p+1} - \frac{1}{p^2} & p \text{ inert in } K \\ 1 - a_p(w) \chi^{-1}(\mathfrak{p}) \frac{p-1}{p} - \frac{1}{p^2} & p \text{ ramified in } K. \end{cases}$$

Proof. (8.21) and (8.22) follow from (8.3), (8.4), (8.9), Definitions 8.7 and 8.8. $\qquad\square$

Chapter Nine

The p-adic Waldspurger formula

Recall that w is a newform of weight $k \geq 2$ on $Y = Y_1(N)$, with p-stabilization

$$w^{(p)} = (U_p V_p - V_p U_p)^* w = w|_{U_p V_p - V_p U_p} = w|_{1 - T_p V_p + [p] V_p^2}$$

as in Definition (6.28), which is an eigenform of weight k on $Y(\Gamma_1(N) \cap \Gamma_0(p^2))$. In this section, we derive p-adic special value formulas for our anticyclotomic p-adic L-function defined in Definition 8.7. These will give a first arithmetic application of our p-adic L-function. The main ingredient for this is Coleman's theory of p-adic integration of 1-forms on p-adic analytic spaces, along with studying the p-adic variation of p-adic Maass-Shimura derivatives

$$\{\theta_k^j (y_{\mathrm{dR}}^k f)^{(p)} (q_{\mathrm{dR}})\}$$

in a neighborhood containing *both* $\mathcal{Y}^{\mathrm{Ig}}$ and the supersingular CM torus $\mathcal{C}$. We then perform p-adic integration by taking a p-adic limit for $j = -r + p^m (p - 1)$ as $m \to \infty$ for $1 \leq r \leq \frac{k}{2}$ in the weight space $\mathbb{Z}/(p-1) \times \mathbb{Z}_p$. Using the fact that the q_{dR}-expansion encodes the Serre-Tate expansion on $\mathcal{Y}^{\mathrm{Ig}}$ (see Theorem 5.9), we then can compute the restriction to $\mathcal{Y}^{\mathrm{Ig}}$ of the resulting limits

$$\theta_k^{-r} (y_{\mathrm{dR}}^k f)^{(p)} \tag{9.1}$$

(using computations already done in [5], [46] and [8]) as *Coleman primitives* of $w^{(p)}$; explicitly, they are related to certain p-*adic Abel-Jacobi maps* attached to $w^{(p)}$, which in the weight $k = 2$ is simply the formal logarithm attached to $w^{(p)}$. By descent, we can then descend (9.1) to a *rigid section* on an admissible open U (containing a copy Y^{ord} and our supersingular CM points of interest) on a certain finite cover $Y' \to Y$.

By the above, on an admissible subopen U' of U, this coincides with a p-adic Abel-Jacobi map (or more simply, in the weight $k = 2$ case, a p-adic formal logarithm attached to $w^{(p)}$). By rigidity, this equality on U' extends to all of U, and hence we get the equality of (9.1) and these p-adic Abel-Jacobi maps on the supersingular CM points of our interest. The evaluation of the former on these CM points are simply special values of our p-adic L-function, and the evaluation of the latter can be computed as the image of a certain *generalized Heegner cycle* under the p-adic Abel-Jacobi map, which in the weight $k = 2$ is simply the formal logarithm $\log_{w^{(p)}}$ evaluated at a Heegner point. Using properties of Hecke operators, the latter value can be related in a simple way to $\log_w$ evaluated at a Heegner point.

9.1   COLEMAN INTEGRATION

We give an overview of Coleman's theory of locally analytic integration of rigid 1-forms ([14]), highlighting its compatibility with the logarithm.

Let $U \subset Y$ be an affinoid subdomain over a local field F with good reduction (that is, its image under the reduction map is smooth).

Definition 9.1. *We call a chain of affinoid subdomains $U \subset W_0' \subset W_n' \subset \ldots \subset W_{n-1}' \subset W$ of Y a degree n Frobenius proper neighborhood (U, W_i', W, ϕ) if there is a morphism $\phi : W_n' \to W$ mapping $\phi : W_{i-1}' \to W_i'$ whose restriction to U is a lifting of the relative Frobenius morphism (on special fibers).*

Definition 9.2. *Suppose $w \in \omega^{\otimes r} \otimes \Omega_Y^1(Y)$. A polynomial $P(X) \in F[X]$ of degree n is called a Coleman polynomial for w if no root of $P(X)$ is a root of unity, and if there is a degree n Frobenius proper neighborhood (U, W_i', W, ϕ) such that*

1. *$P(\phi^*)w$ is an exact $r + 2$-form, and*
2. *$P(\phi^*)$ is an isomorphism on the space $\operatorname{Sym}^r \mathcal{H}_{\mathrm{dR}}^1(\mathcal{A})(Y)^{\nabla=0}$.*

Theorem 9.3 (Lemma 3.14 of [5]). *Suppose $w \in \omega^{\otimes r} \otimes \Omega_Y^1(Y)$. Then there is a Coleman polynomial for w.*

Theorem 9.4 (Theorem 3.15 of [5]). *Suppose $w \in \omega^{\otimes r} \otimes \Omega_Y^1(Y)$. Choose a Coleman polynomial for w, and let n be its degree. Then there exists a locally analytic section F_w of $\operatorname{Sym}^r \mathcal{H}_{\mathrm{dR}}^1(\mathcal{A})$ such that*

1. *$\nabla F_w = w$, and*
2. *$P(\phi^*)F_w \in \operatorname{Sym}^r \mathcal{H}_{\mathrm{dR}}^1(\mathcal{A})(W_0')$, i.e., is rigid analytic.*

In fact, F_w is unique up to additive constant. It is called a Coleman primitive of w.

9.2   COLEMAN PRIMITIVES IN OUR SITUATION

Retain the notation and setting of the last section. In particular $p \nmid Nc$, $(c, N) = 1$, N satisfies the Heegner hypothesis with respect to K which determined an ideal $\mathfrak{N} | N$, and A was a fixed elliptic curve with CM by $\mathcal{O}_c$ and $\Gamma_1(N)$-level structure t determined by the cyclic ideal $\mathfrak{N}$. We will give special value formulas of our p-adic L-function. We view this as analogous to the p-adic Waldspurger formulas of [5] and [46] in the case where p is inert or ramified in K. For the Eisenstein case, we view this formula as an analogue of the "p-adic Kronecker limit formula" proven in [35, Chapter X]. In fact our approach can be used to recover these special value formulas, using the fact that the q_{dR}-expansion recovers the Serre-Tate expansion (Theorem 5.9) on the cover

$$\mathcal{Y}^{\mathrm{Ig}} \to Y^{\mathrm{ord}}$$

and Theorem 5.41.

As before, let w be a weight $k \geq 2$ normalized $\Gamma_1(N)$-new eigenform with nebentype ϵ_w. In this section, we will briefly consider compactified modular curves

$$X(\Gamma) = \overline{Y(\Gamma)},$$

where we recall that $Y(\Gamma)$ is the modular curve associated with a finite-index subgroup

$$\Gamma \subset SL_2(\mathbb{Z}).$$

Let

$$X_1(N) = X(\Gamma_1(N)), \quad J_1(N) = \mathrm{Jac}(X_1(N)),$$
$$J_1(N, p^r) = \mathrm{Jac}(X(\Gamma_1(N) \cap \Gamma_0(p^r))).$$

Then let

$$h : X_1(N) \to J_1(N)$$

denote the (analytification of the usual algebraic) Abel-Jacobi map over $\mathbb{Q}$, sending the cusp at infinity

$$(\infty) \mapsto 0.$$

Thus

$$w = h^* \omega$$

where

$$\omega = g(q)dq/q \in \Omega^1_{J_1(N)/\mathbb{Q}}$$

and

$$g(q) = \sum_{n=0}^{\infty} a_n q^n$$

is the q-expansion of w. Then the p-stabilization $w^{(p)}$ as in (6.28) satisfies

$$h_2^* w^{(p)} = w^{(p)}$$

where

$$w^{(p)} = g^{(p)}(q)dq/q \in \Omega^1_{J(N,p^2)/\mathbb{Q}}$$

and

$$g^{(p)}(q) = \sum_{n=0,p\nmid n}^{\infty} a_n q^n$$

is the q-expansion of $w^{(p)}$, viewed as a form of level $\Gamma_1(N) \cap \Gamma(p^2)$; it is an old eigenform. Let

$$\log_{w^{(p)}} : J(N,p^2)(\mathbb{C}_p, \mathcal{O}_{\mathbb{C}_p}) \to \mathbb{C}_p$$

be the formal logarithm attached to $w^{(p)}$, obtained by integrating the q-expansion

$$g^{(p)}(q)dq/q$$

in the residue disc containing the origin, specifying that it takes the value 0 at the origin, and then extending linearly. From its q-expansion, we see that $\log_{w^{(p)}}$ is rigid on the residue disc containing the origin and hence extends uniquely via linearity to a locally analytic function on all of $J(N,p^2)$. Pulling back by h_2^*, we have a locally analytic function

$$\log_{w^{(p)}} := h_2^* \log_{w^{(p)}} : X(N,p^2)(\mathbb{C}_p, \mathcal{O}_{\mathbb{C}_p}) \to \mathbb{C}_p.$$

Theorem 9.5 ([46] Proposition A.0.1). *We have that*

$$\log_{w^{(p)}} \quad \textit{is a Coleman primitive of } w^{(p)} \textit{ on } Y_2^{\mathrm{ord}}.$$

In fact, it is a rigid primitive. Moreover, we have that

$$\log_w \quad \textit{is a Coleman primitive of } w \textit{ on } Y^{\mathrm{ord}}.$$

Let H_c denote the ring class field attached to $\mathcal{O}_c$, and for $1 \le r \le \infty$ let $H_c(p^r)$ denote the compositum of H_c.

Now fix

$$(A, t, \alpha) \in \mathcal{C}(\overline{K}_p, \mathcal{O}_{\overline{K}_p})$$

as in Choice 8.6. Given a point

$$P \in X(N,p^n)(\mathbb{C}_p, \mathcal{O}_{\mathbb{C}_p}),$$

let $[P]$ denote its associated divisor. Let $\chi : \mathcal{C}\ell(\mathcal{O}_c) \to \overline{\mathbb{Q}}^\times$ be a character.

Definition 9.6. *In the notation of Definition 5.46, and a finite-order character* $\chi : \mathrm{Gal}(H_c/K) \cong C\ell(\mathcal{O}_c) \to \overline{\mathbb{Q}}^\times$, *define a Heegner point*

$$P_K(\chi) := \sum_{[\mathfrak{a}] \in C\ell(\mathcal{O}_c)} \chi^{-1}([\mathfrak{a}])([\mathfrak{a} \star (A, t)] - [(\infty)]) \in J_1(N)(H_c)^\chi$$

where

$$J_1(N)(H_c)^\chi$$

denotes the χ-isotypic component of $J_1(N)(H_c)$ under the action of the group $\mathrm{Gal}(H_c/K)$.

Note that when $\chi = 1$, we have

$$P_K(1) \in J_1(N)(K).$$

Now we consider generalizations of the p-adic logarithm and Heegner points in the case of higher weight.

Definition 9.7. *Let $r := k - 2$, and fix a field extension $F \supset H$. Henceforth, fix an elliptic curve A_0/F with CM by the* full *ring of integers $\mathcal{O}_K$ (and not just $\mathcal{O}_c$, as was assumed for our previous A).[1] Recall as in [5, Section 2] and the introduction of [42] that the Kuga-Sato variety is given by the r-fold product of universal elliptic curves*

$$W_r = \mathcal{A} \times_Y \cdots \times_Y \mathcal{A},$$

and the generalized Kuga-Sato variety *is the canonical desingularization X_r of*

$$W_r \times A_0^r$$

as defined in [5, Appendix]. Let $(A_0, t_0) \in Y_1(N)(\overline{\mathbb{Q}})$. Suppose we have an isogeny $\phi : A_0 \to A'$ of elliptic curves with CM by $\mathcal{O}_c \subset \mathcal{O}_K$. Then $(A', \phi(t_0)) \in Y(F)$ (supposing F is large enough to be an appropriate field of definition) determines a fiber of $X_r \to Y$ and hence a product of elliptic curves $(A')^r$ and a closed embedding

$$(A')^r \hookrightarrow W_r$$

defined over F. Let

$$\Gamma_\phi \hookrightarrow X_r$$

denote the r-fold product of the graph of $\phi : A_0 \to A'$.

In [5, (2.2.1)], a correspondence ε_X is defined on X_r. Given a number field L and a smooth variety X/L and a field $E/\mathbb{Q}$, an E-algebraic cycle over L is a

---

[1] *Note that [5] uses the notation to denote an elliptic curve with CM by $\mathcal{O}_c$, and A to denote an elliptic curve with full CM by $\mathcal{O}_K$, whereas here we have taken the reverse notation.*

$\mathrm{Gal}(\overline{F}/F)$-stable E-linear combination of codimension j subvarieties over L of X. Let

$$\mathrm{CH}^j(X)(L)_E$$

denote the Chow group of codimension j algebraic cycles defined over L on X, which is the quotient of the free abelian group on algebraic cycles over L by the relation of rational equivalence. Let

$$\mathrm{CH}^j_0(L)_E \subset \mathrm{CH}^j(L)_E$$

denote the subgroup of null-homologous algebraic cycles over L. We have a generalized Heegner cycle attached to ϕ, which is given by

$$\Delta_\phi = \varepsilon_X \Gamma_\phi \in \mathrm{CH}^{r+1}_0(X_r)(L).$$

Let

$$\phi_0 : A_0 \to A$$

denote the dual of the natural isogeny $A \to A_0$ (induced by $\mathbb{C}/\mathcal{O}_c \to \mathbb{C}/\mathcal{O}_K$); this has degree c. Given an ideal $\mathfrak{a} \subset \mathcal{O}_c$ which is prime to $\mathfrak{N}c$, we let

$$\phi_\mathfrak{a} : A_0 \xrightarrow{\phi_0} A \to A/A[\mathfrak{a}]$$

be the natural projection, note that $\phi_\mathfrak{a}(t_0) : \mathbb{Z}/N \xrightarrow{\sim} (A/A[\mathfrak{a}])[\mathfrak{N}]$, and so we get a corresponding $\Delta_{\phi_\mathfrak{a}}$. Suppose we are given a character $\chi \in \Sigma$. Let $\{\mathfrak{a}\}$ be a full set of representatives of elements of $\mathcal{C}\ell(\mathcal{O}_c)$ with each $\mathfrak{a} \subset \mathcal{O}_c$ prime to $\mathfrak{N}c$. If $\chi : \mathrm{Gal}(H_c/K) \to E^\times$, then have

$$\Delta_\chi := \sum_{[\mathfrak{a}] \in \mathcal{C}\ell(\mathcal{O}_c)} \chi^{-1}(\mathfrak{a}) \Delta_{\phi_\mathfrak{a}} \in \mathrm{CH}^{r+1}_0(X_r)(H_c)_E^\chi. \qquad (9.2)$$

Definition 9.8. *For any field extension $E/\mathbb{Q}$, let*

$$\mathrm{AJ}_F(\cdot) : \mathrm{CH}^{r+1}_0(X_r)(F)_E \to (S_2(\Gamma_1(N), F) \otimes_F \mathrm{Sym}^r H^1_{\mathrm{dR}}(A_0/F))_E^\vee$$

denote the E-linear p-adic Abel-Jacobi map, as defined in [5, Section 3.4]. Here, we view $(S_2(\Gamma_1(N), F) \otimes \mathrm{Sym}^r H^1_{\mathrm{dR}}(A_0/F))_E$ is the E-vector space generated by a free basis consisting of elements of $S_2(\Gamma_1(N), F)$, and the superscript "$\vee$" denotes the E-linear dual.

Now fix generators

$$\omega_{A_0} \in \Omega^1_{A_0/F}, \quad \eta_{A_0} \in H^{0,1}(A_0/F), \quad \langle \omega_{A_0}, \eta_{A_0} \rangle_{\mathrm{Poin}} = 1.$$

As described in the introduction of [42], for large enough F, there are correspondences ε_w on W_r and ε_χ on A_0^r such that

$$\varepsilon_w \varepsilon_\chi H_{\text{ét}}^{r+1}(X_r \times_F \overline{F}, \mathbb{Q}_\ell(r+1)) = V_w(r+1)|_{\text{Gal}(\overline{K}/K)} \times V_\chi$$

where V_w is the ℓ-adic $\text{Gal}(\overline{\mathbb{Q}}/\mathbb{Q})$-representation associated with w, and V_χ is the ℓ-adic $\text{Gal}(\overline{K}/K)$-representation associated with χ.

Theorem 9.9 ([5], Proposition 3.24). *The normalized $\Gamma_1(N)$-new eigenform w has a Coleman primitive F_w on $Y(\Gamma_1(N) \cap \Gamma_0(p^2))^{\text{ord}}$. Moreover, $w^{(p)}$ has a rigid analytic primitive $F_{w^{(p)}}$ on $Y(\Gamma_1(N) \cap \Gamma_0(p^2))^{\text{ord}}$, and*

$$F_w^{(p)} = F_{w^{(p)}}.$$

Write

$$w|_{Y^{\text{Ig}}} = f \cdot \omega_{\text{can}}^{\text{Katz}, \otimes k}, \quad w^{(p)}|_{Y^{\text{Ig}}} = f^{(p)} \cdot \omega_{\text{can}}^{\text{Katz}, \otimes k}.$$

We have for any $0 \le j \le r := k-2$, and the period Ω_p defined in loc. cit,

$$\langle \phi_0^* F_{w^{(p)}}, \omega_{A_0}^j \eta_{A_0}^{r-j} \rangle_{\text{Poin}} = j! \Omega_p^{k-2(j+1)} \theta_{\text{AS}}^{-(j+1)} f^{(p)2} \qquad (9.3)$$

and

$$(\mathbb{N}(\mathfrak{a})c)^{-j} \text{AJ}_F(\Delta_{\phi_\mathfrak{a}}) = \langle \phi_\mathfrak{a}^* F_{w^{(p)}}, \omega_{A_0}^j \eta_{A_0}^{r-j} \rangle_{A_0},^3 \qquad (9.4)$$

where

$$\langle \cdot, \cdot \rangle_{A_0} : \text{Sym}^r H_{\text{dR}}^1(A_0/F) \times \text{Sym}^r H_{\text{dR}}^1(A_0/F) \to F$$

is the natural pairing induced by restricting the Poincaré pairing

$$\langle \cdot, \cdot \rangle_{A_0} : H_{\text{dR}}^r(A_0^r/F) \times H_{\text{dR}}^r(A_0^r/F) \to H_{\text{ét}}^{2r}(A_0^r/F) = F$$

to $\text{Sym}^r H_{\text{dR}}^1(A_0/F) \subset H_{\text{dR}}^r(A_0^r/F)$.

---

[2] *Suppose that A, the elliptic curve with CM by $\mathcal{O}_c$, were ordinary at p. In this case, we note that the period $\theta(\Omega_p(A, t, \alpha))$, with $\alpha : \mathbb{Z}_p^{\oplus 2} \xrightarrow{\sim} T_p A$ such that $(A, t, \alpha) \in \mathcal{Y}^{\text{Ig}}$, so that α_1 induces an isomorphism of p-divisible groups $\mu_{p^\infty} \cong \hat{A}[p^\infty]$ as in (5.2.2) of loc. cit., coincides with the period Ω_p such that $\omega_{\text{can}}^{\text{Katz}} \Omega_p = \omega_0$ (here $\omega_0 \in \Omega_{A/\mathcal{O}_F}^1$ is the previously fixed generator) from loc. cit., since $\omega_{\text{can}}|_{\mathcal{Y}^{\text{Ig}}} = \omega_{\text{can}}^{\text{Katz}}$.*

[3] *The factor of c^{-j} comes from pulling back via the degree c isogeny $\mathbb{C}/\mathcal{O}_K \to \mathbb{C}/\mathcal{O}_c$ dual to the natural projection $\mathbb{C}/\mathcal{O}_c \to \mathbb{C}/\mathcal{O}_K$, see [5, Lemma 3.22].*

9.3   THE p-ADIC WALDSPURGER FORMULA

Theorem 9.10 ("p-adic Waldspurger formula," cf. Theorem 5.13 of [5], Theorem 1.5.3 of [46]). *Fix a CM triple* $(A, t, \alpha) \in \mathcal{C}(\overline{K}_p, \mathcal{O}_{\overline{K}_p}) = \mathcal{C}(K_p^{\mathrm{ab}}, \mathcal{O}_{K_p^{\mathrm{ab}}})$ *as in Choice 8.6. We have, for any* $\chi \in \check{\Sigma}_- \subset \check{\overline{\Sigma}}_+$ *of infinity type* $(k - (j+1), j+1)$ *where* $0 \leq j \leq r := k - 2$, *that*

$$\mathcal{L}_{p,\alpha}(w, \check{\chi}) = \frac{(q\theta(\Omega_p)(A, t, \alpha))^{k-2(j+1)}}{\Omega(A, t)^{-(j+1)}} \Xi_p(w, \check{\chi}) \cdot \frac{c^{-j}}{j!} \mathrm{AJ}_F(\Delta_{\chi \mathbb{N}_K^{-1}})(w \wedge \omega_{A_0}^j \eta_{A_0}^{r-j}),$$

where

$$\Xi_p(w, \check{\chi}) = \begin{cases} 1 - a_p(w)^2 \check{\chi}^{-1}(p) \frac{p-1}{p+1} - \frac{1}{p^2} & p \text{ inert in } K \\ 1 - a_p(w) \check{\chi}^{-1}(\mathfrak{p}) \frac{p-1}{p} - \frac{1}{p^2} & p \text{ ramified in } K. \end{cases}$$

Let χ *take values in* E. *In particular, if*

$$\mathcal{L}_{p,\alpha}(w, \check{\chi}) \neq 0, \tag{9.5}$$

then

$$\epsilon_w \epsilon_\chi \mathbb{N}_K^{-1} \Delta_{\chi} \mathbb{N}_K^{-1} \in \epsilon_w \varepsilon_{\chi \mathbb{N}_K^{-1}} \mathrm{CH}_0^{r+1}(X_r)(F)_E^{\chi \mathbb{N}_K^{-1}}$$

is nontrivial.

When $k = 2$ *or equivalently* $r = 0$, *so that*

$$\chi \mathbb{N}_K^{-1} : \mathrm{Gal}(H_c / K) \to \overline{\mathbb{Q}}^\times$$

is a finite-order character (viewed as a Galois character under Artin reciprocity), then

$$\Delta_{\chi \mathbb{N}_K^{-1}} = P_K(\chi \mathbb{N}_K^{-1})$$

is a Heegner point as described in Definition 9.6. Then if (9.5) holds, $P_K(\chi \mathbb{N}_K^{-1})$ *projects via a modular parametrization to a non-torsion point in* $A_w(H_c)^{\chi \mathbb{N}_K^{-1}}$, *where* $A_w/\mathbb{Q}$ *is the* GL_2-*type abelian variety associated uniquely up to isogeny with* w. *Then by the Gross-Zagier formula and Kolyvagin, we have*

$$\mathrm{rank}_{\mathbb{Z}} A_w(H_c)^{\chi \mathbb{N}_K^{-1}} = \dim_{\mathbb{Q}} A_w = \mathrm{ord}_{s=1} L(A_w, \chi, s).$$

Proof. We need to slightly refine the statement of Theorem 6.21. Recall the q_{dR}-expansion map from Definition 6.6.

$$\mathcal{O}_{\mathcal{Y}_x} \hookrightarrow \hat{\mathcal{O}}_{\mathcal{Y}_x} [\![q_{\mathrm{dR}} - 1]\!] \tag{9.6}$$

which is an injective map of proétale sheaves. Letting $y = (A, t, \alpha)$ as in Choice 8.6, with associated

$$y' = y \cdot \begin{pmatrix} q & 0 \\ 0 & 1 \end{pmatrix},$$

recalling that $b = 1$ if $p > 2$ and $b = 3$ if $p = 2$ for y', by Propositions 7.3 and 7.7, in the notation of Theorem 6.21, as well as our notation $q = p$ if $p > 2$ and $q = 8$ if $p = 2$. From Lemma 6.16, we have

$$(y_{\mathrm{dR}}^k f)_{y'}(q_{\mathrm{dR}}^{1/q}) \in \hat{\mathcal{O}}_{\mathcal{Y}, y'}^+ [\![q_{\mathrm{dR}}^{1/q} - 1]\!][1/p]. \tag{9.7}$$

Recall the affinoid $\mathcal{U} \subset \mathcal{Y}_x$ from Proposition 7.15, and which has $\mathcal{C}^{\mathrm{good}} \subset \mathcal{U}$. Then as in (7.14) (note that $F = y_{\mathrm{dR}}^k f$ in the notation there), we had

$$q^j \theta_k^j (y_{\mathrm{dR}}^k f)^{(p)} |_{\mathcal{U}} = q^j \theta_k^j (y_{\mathrm{dR}}^k f)^{(p)} |_{\mathcal{U}} (q_{\mathrm{dR}}) = q^j \theta_k^j (y_{\mathrm{dR}}^k f)^{(p), \flat} (q_{\mathrm{dR}})$$
$$\in \hat{\mathcal{O}}_{\mathcal{Y}}^+ (\mathcal{U}) [\![q_{\mathrm{dR}} - 1]\!][1/p]. \tag{9.8}$$

Now consider

$$\mathcal{U}^{\mathrm{ss}} := \mathcal{U} \cap \mathcal{Y}^{\mathrm{ss}},$$

which is affinoid since it is the intersection of two affinoids (recall that

$$\mathcal{Y}_x = \{ \hat{\mathbf{z}} \neq 0 \},$$

which is evidently affinoid, and $\mathcal{U}$ is a rational open and so is affinoid).

Now consider the subspace

$$\mathcal{Y}^{\mathrm{Ig}} \subset \mathcal{Y}_x,$$

and recall that

$$\mathcal{Y}^{\mathrm{Ig}} \to Y(\Gamma)^{\mathrm{ord}}$$

is a B-proétale covering, where $B \subset GL_2(\mathbb{Z}_p)$ is the subgroup of upper triangular matrices (here Γ is as in Definition 7.17, and note that $\Gamma \subset B$), and on this subspace the restriction

$$q^j \theta_k^j (y_{\mathrm{dR}}^k f)^{(p)} (q_{\mathrm{dR}}^{1/q}) |_{\mathcal{Y}^{\mathrm{Ig}}}$$

as defined in Chapter 7, Section 7.2. By Proposition 7.11, $q^j \theta_k^j (y_{\mathrm{dR}}^k f)^{(p)} (q_{\mathrm{dR}}^{1/q}) |_{\mathcal{Y}^{\mathrm{Ig}}}$ has weight $k + 2j$ for B on $\mathcal{Y}^{\mathrm{Ig}}$ for all $j \in \mathbb{Z}/(p-1) \times \mathbb{Z}_p$.

We have the locus

$$\mathcal{Y}^{\mathrm{Ig}} \sqcup \mathcal{U} \subset \mathcal{Y}_x.$$

Then for $j \geq 0$ we have the section

$$H_j := q^j \theta_k^j (y_{\mathrm{dR}}^k f)^{(p)} (q_{\mathrm{dR}}^{1/q})|_{\mathcal{Y}^{\mathrm{Ig}} \sqcup \mathcal{U}}$$

which can be viewed in either $\hat{\mathcal{O}}_{\mathcal{Y}}(\mathcal{Y}^{\mathrm{Ig}})$ or $\hat{\mathcal{O}}_{\mathcal{Y}}(\mathcal{U})$, and which is of weight $k + 2j$ for B on $\mathcal{Y}^{\mathrm{Ig}}$, and of weight $k + 2j$ for $GL_2(\mathbb{Z}_p)$ on $\mathcal{U}$. Recall we wrote $w^{(p)}|_{\mathcal{Y}_x} = f^{(p)} \mathfrak{s}^{\otimes k}$, so that since

$$\mathfrak{s}|_{\mathcal{Y}^{\mathrm{Ig}}} \overset{y_{\mathrm{dR}}|_{\mathcal{Y}^{\mathrm{Ig}}} = 1}{=} \omega_{\mathrm{can}}|_{\mathcal{Y}^{\mathrm{lt}}} = \omega_{\mathrm{can}}^{\mathrm{Katz}},$$

we have

$$w^{(p)}|_{\mathcal{Y}^{\mathrm{Ig}}} = f^{(p)}|_{\mathcal{Y}^{\mathrm{Ig}}} \omega_{\mathrm{can}}^{\mathrm{Katz}, \otimes k}.$$

By Proposition 5.9 and Theorem 5.41, we have for $j \geq 0$

$$H_j|_{\mathcal{Y}^{\mathrm{Ig}}} = q^j \theta_{\mathrm{AS}}^j f^{(p)} (T^{1/q}). \tag{9.9}$$

The left-hand side is continuous in j by Theorem 6.21. For the continuity of the right-hand side, note the formula

$$q^j \theta_{\mathrm{AS}}^j = q^j \left(\frac{T d}{dT} \right)^j = \left(\frac{T^{1/q} d}{dT^{1/q}} \right),$$

which acts continuously on the $T^{1/q}$-expansion of the right-hand side, by (6.30). Hence, (9.9) extends to all weights $j \in \mathbb{Z}/(p-1) \times \mathbb{Z}_p$. Recalling F_j' from Definition 7.19, we have

$$\theta(F_j') = q^k \begin{pmatrix} 1 & 0 \\ 0 & q \end{pmatrix}^* H_j \tag{9.10}$$

which, by (the proof of) Proposition 7.18, also has weight $k + 2j$ for B on $\mathcal{Y}^{\mathrm{Ig}}$ and weight $k + 2j$ for Γ on $\mathcal{U}$.

Recall the proétale projection map

$$\lambda : \mathcal{Y} \to Y(\Gamma),$$

and let

$$U = \lambda(\mathcal{U}),$$

and so

$$Y(\Gamma)^{\mathrm{ord}} = \lambda(\mathcal{Y}^{\mathrm{Ig}}).$$

Moreover, recall the Galois group of

$$\mathcal{Y}^{\mathrm{Ig}} \to Y(\Gamma)^{\mathrm{ord}}$$

is B. Now by descent (Lemma 4.27) along B on $\mathcal{Y}^{\mathrm{Ig}} \to Y(\Gamma)^{\mathrm{ord}}$ and by descent on the Γ-cover

$$\mathcal{U} \to U,$$

we have a (rigid analytic) section for $-k/2 \le j$,

$$G_j := H_j \cdot \theta(\omega_{\mathrm{can}})^{\otimes k+2j}|_{\mathcal{Y}^{\mathrm{Ig}} \sqcup \mathcal{U}} \in \omega^{\otimes k+2j} \otimes_{\mathcal{O}_{Y(\Gamma)}} \hat{\mathcal{O}}_{Y(\Gamma)}(Y(\Gamma)^{\mathrm{ord}} \sqcup U)$$

on the admissible open

$$Y(\Gamma)^{\mathrm{ord}} \sqcup U \subset Y(\Gamma).$$

Let $F_{w^{(p)}}$ be the Coleman primitive of $w^{(p)}$ as in Theorem 9.9. Now we note that for $-k/2 \le j$, G_j is a rigid analytic section of $\hat{\mathcal{O}}_{Y(\Gamma)}$ on $Y(\Gamma)^{\mathrm{ord}} \sqcup U$ which is equal to the rigid analytic section

$$q^j \theta_{\mathrm{AS}}^j f^{(p)}(T^{1/q}) \cdot \omega_{\mathrm{can}}^{\mathrm{Katz}, \otimes k+2j}$$

of $\hat{\mathcal{O}}_{Y(\Gamma)}$ on $Y(\Gamma)^{\mathrm{ord}}$ by (9.9) (extending by continuity to $-k/2 \le j \le -1$). Hence by rigid analytic continuation, for $j \ge -k/2$, we have that G_j is a rigid section on the affinoid subdomain $Y(\Gamma)^{\mathrm{ord}} \sqcup U \subset Y(\Gamma)$ such that

$$G_j|_{Y(\Gamma)^{\mathrm{ord}}} = q^j \theta_{\mathrm{AS}}^j f^{(p)}(T^{1/q}) \cdot \omega_{\mathrm{can}}^{\mathrm{Katz}, \otimes k+2j}, \tag{9.11}$$

and in particular, for $j = -1$,

$$G_{-1}|_{Y(\Gamma)^{\mathrm{ord}}} = q^{-1} \theta_{\mathrm{AS}}^{-1} f^{(p)}(T^{1/q}) \cdot \omega_{\mathrm{can}}^{\mathrm{Katz}, \otimes k-2} = F_{w^{(p)}}(T^{1/q}) \tag{9.12}$$

where the last equality follows since $F_{w^{(p)}}$ is characterized by $d_{\mathrm{AS}}(F_{w^{(p)}}) = w^{(p)}$. Hence by the uniqueness of Coleman primitives, we have that G_{-1} is a Coleman primitive of $w^{(p)}$ on $Y(\Gamma)^{\mathrm{ord}} \sqcup U$.

Hence, pulling back to $\mathcal{Y}^{\mathrm{Ig}} \sqcup \mathcal{U} \to Y(\Gamma)^{\mathrm{ord}} \sqcup U$, we have for $j \ge -k/2$,

$$H_j|_{\mathcal{Y}^{\mathrm{Ig}}} = q^j \theta_{\mathrm{AS}}^j f^{(p)}(T^{1/q}), \tag{9.13}$$

which extends by continuity to all $j \in \mathbb{Z}/(p-1) \times \mathbb{Z}_p$ (since

$$\{j \ge -k/2\} \subset \mathbb{Z}/(p-1) \times \mathbb{Z}_p$$

is dense).

Pulling back to $\mathcal{Y}^{\mathrm{Ig}} \sqcup \mathcal{U} \to Y(\Gamma)^{\mathrm{ord}} \sqcup U$, we have

$$H_{-1}|_{\mathcal{Y}^{\mathrm{Ig}}} = q^{-1} \theta_{\mathrm{AS}}^{-1} f^{(p)}(T^{1/q}) \overset{(9.3)}{=} \langle F_{w^{(p)}}(T^{1/q}), \eta_A^r \rangle_{\mathrm{Poin}}. \tag{9.14}$$

Hence by [5, Proposition 3.24], we have for $0 \le j \le r$,

$$H_{-(j+1)}|_{y^{\mathrm{Ig}}} = q^{-(j+1)}\theta_{\mathrm{AS}}^{-(j+1)}f^{(p)}(T^{1/q}) \overset{(9.3)}{=} \frac{1}{j!}\langle F_{w^{(p)}}(T^{1/q}), \omega_{A_0}^j \eta_{A_0}^{r-j}\rangle_{\mathrm{Poin}}.$$
(9.15)

This, combined with (9.12) and analytic continuation, implies

$$H_{-(j+1)} = \frac{1}{j!}\langle G_{-1}, \omega_{A_0}^j \eta_{A_0}^{r-j}\rangle_{\mathrm{Poin}}.$$
(9.16)

The identity (9.16), along with (9.10) extended by continuity to all $j \in \mathbb{Z}/(p-1) \times \mathbb{Z}_p$ implies that for $0 \le j \le r$,

$$\theta(F_j')(y') = q^k \begin{pmatrix} 1 & 0 \\ 0 & q \end{pmatrix}^* H_j(y') = q^k \begin{pmatrix} 1 & 0 \\ 0 & q \end{pmatrix}^* \frac{1}{j!}\langle G_{-1}(y'), \omega_{A_0}^j \eta_{A_0}^{r-j}\rangle_{\mathrm{Poin}}$$
$$= q^{k-2(j+1)}\frac{1}{j!}\langle G_{-1}(y), \omega_{A_0}^j \eta_{A_0}^{r-j}\rangle_{\mathrm{Poin}}, \quad (9.17)$$

where the last equality follows by the same calculation as in the proof of (7.16) (i.e., considering the action of pullback corresponding to $\begin{pmatrix} 1 & 0 \\ 0 & q \end{pmatrix}$ on ω_{can}).

Similarly, repeating the above argument with y' replaced by $\mathfrak{a} \star y' \in \mathcal{U}(\mathbb{C}_p, \mathcal{O}_{\mathbb{C}_p})$ for any integral ideal $\mathfrak{a} \subset \mathcal{O}_c$ prime to $p\mathfrak{N}c$, we have for $0 \le j \le r$,

$$\theta(F_j')(\mathfrak{a} \star y') = q^{k-2(j+1)}\frac{1}{j!}\langle G_{-1}(\mathfrak{a} \star y), \omega_{A_0}^j \eta_{A_0}^{r-j}\rangle_{\mathrm{Poin}}.$$
(9.18)

Now by (8.9), for χ with $\check{\chi}|_{\mathcal{O}_{K_p}^\times}(x_p) = x_p^{k-(j+1)}\bar{x}_p^{j+1}$, we have

$$\mathcal{L}_{p,\alpha}(w, \check{\chi}) = \frac{\#(\mathbb{Z}/N\mathfrak{p}\mathbb{Z})^\times}{\#(\mathcal{O}_K/N\mathfrak{p}\mathcal{O}_K)^\times} \sum_{[\mathfrak{a}] \in \mathcal{C}\ell(\mathcal{O}_{cN\mathfrak{p}})} (\check{\mathbb{N}}_K^{-(j+1)}\check{\chi})^{-1}(\mathfrak{a})\theta(F_j')(\mathfrak{a} \star y')$$

$$\overset{(9.18)}{=} \frac{\#(\mathbb{Z}/N\mathfrak{p}\mathbb{Z})^\times}{\#(\mathcal{O}_K/N\mathfrak{p}\mathcal{O}_K)^\times}q^{k-2(j+1)}$$
$$\sum_{[\mathfrak{a}] \in \mathcal{C}\ell(\mathcal{O}_{cN\mathfrak{p}})} (\check{\mathbb{N}}_K^{-(j+1)}\check{\chi})^{-1}(\mathfrak{a})\frac{1}{j!}\langle G_{-1}(\mathfrak{a} \star y), \omega_{A_0}^j \eta_{A_0}^{r-j}\rangle_{\mathrm{Poin}}$$

$$\overset{(9.3),(9.4),\omega_{\mathrm{can}}|_{y^{\mathrm{Ig}}}=\omega_{\mathrm{can}}^{\mathrm{Katz}}}{=} q^{k-2(j+1)}\theta(\Omega_p)(A,t,\alpha)^{k-2(j+1)}\frac{\#(\mathbb{Z}/N\mathfrak{p}\mathbb{Z})^\times}{\#(\mathcal{O}_K/N\mathfrak{p}\mathcal{O}_K)^\times}\frac{c^{-j}}{j!}$$
$$\mathrm{AJ}_F(\Delta_{\chi\mathbb{N}_K^{-1}})(w^{(p)} \wedge \omega_{A_0}^j \eta_{A_0}^{r-j})$$

$$= q^{k-2(j+1)}\frac{\theta(\Omega_p)(A,t,\alpha)^{k-2(j+1)}}{\Omega(A,t)^{-(j+1)}}\Xi_p(w, \check{\chi}) \cdot \frac{c^{-j}}{j!}$$
$$\mathrm{AJ}_F(\Delta_{\chi\mathbb{N}_K^{-1}})(w \wedge \omega_{A_0}^j \eta_{A_0}^{r-j})$$
(9.19)

where the last equality follows from the same calculation as in Proposition 8.11 (where, as in that proposition, $\Omega(A,t)$ accounts for the discrepancy between the stalk of differentials of $\Omega^1_{Y_1(N)}$ at (A,t) mapped into the stalk of $\Omega^1_{Y(\Gamma)}$ at (A,t,t_p) via pullback through $Y(\Gamma) \to Y$, for a point (A,t,t_p) above (A,t)), which is in turn essentially the same kind of calculation as in [5, Lemma 3.23].

Now for the final statement of the theorem in $k=2$ (and $r=k-2=0$) case, we note that by (9.19),

$$P_K(\chi \mathbb{N}_K^{-1}) \in J_1(N)(H_c)^{\chi \mathbb{N}_K^{-1}}$$

is nontrivial in the w-isotypic Hecke component Eichler-Shimura decomposition. This latter component is isogenous to

$$A_w(H_c)^{\chi \mathbb{N}_K^{-1}}$$

by Eichler-Shimura theory. Hence, given the nonvanishing of (9.19), we see that $P_K(\chi \mathbb{N}_K^{-1})$ projects to a nontrivial point in $A_w(H_c)^{\chi \mathbb{N}_K^{-1}}$. Now the arithmetic results follow from the work of Gross-Zagier and Kolyvagin.   □

9.4 p-ADIC KRONECKER LIMIT FORMULA

We conclude with another consequence of the arguments of the previous sections, in the case where w corresponds to a weight $k=2$ Eisenstein series $E_{2,1,\psi}$ ($\psi_1 = 1, \psi_2 = \psi$): namely, a supersingular analogue of the p-adic Kronecker limit formula of [21, Theorem II.5.2] and [35, Chapter X]. See also the discussion of [4, Section 1.2].

We briefly recall Siegel modular units, following Section 1.1 of loc. cit. Given an integer $N \geq 4$ and an integer a, $(a,N)=1$, fixing a primitive N^{th} root of unity ζ_N, the Siegel modular unit is an element

$$g_a \in \mathcal{O}^\times_{Y_1(N)}(Y_1(N))$$

with q-expansion

$$g_a(q) = q^{1/12}(1 - \zeta_N^a) \prod_{n=0}^{\infty} (1 - q^n \zeta_N^a)(1 - q^n \zeta_N^{-a}).$$

It has p-stabilization

$$g_a^{(p)} = g_a^{(p)}|_{V_p U_p - U_p V_p} \in \mathcal{O}^\times_{Y_1(N)}(Y_1(N) \cap Y_0(p^2)),$$

which is a locally analytic function on $Y(\Gamma_1(N) \cap \Gamma_0(p^2))$. As $g_a^{(p)}$ is a global section on $Y(\Gamma_1(N) \cap \Gamma_0(p^2))$, it in particular has a local power series expansion on every residue disc D of $Y(\Gamma_1(N) \cap \Gamma_0(p^2))$. Letting T_D be a parameter on D, we can write

$$g_a^{(p)} = \sum_{n=0}^{\infty} a_n T_D^n,$$

with $a_n \in W(\overline{\mathbb{F}}_p)[1/p]$ (since D is a rigid analytic space defined over $W(\overline{\mathbb{F}}_p)[1/p]$). Since

$$g_a^{(p)} \in \mathcal{O}_{Y(\Gamma_1(N) \cap \Gamma_0(p^2))}^{\times}(Y(\Gamma_1(N) \cap \Gamma_0(p^2)))$$

is a unit, this means there exists $c \in p^{\mathbb{Z}} W(\overline{\mathbb{F}}_p)^{\times}$ such that

$$c g_a^{(p)}(T_D) \in W(\overline{\mathbb{F}}_p)[\![T_D]\!].$$

Since

$$c g_a^{(p)}|_D \in \mathcal{O}_{Y(\Gamma_1(N) \cap \Gamma_0(p^2))}^{\times}(D) \cap W(\overline{\mathbb{F}}_p)[\![T_D]\!],$$

letting $u = c g_a^{(p)}|_D(0)$, we have $u \in W(\overline{\mathbb{F}}_p)^{\times}$, and so

$$u^{-1} c g_a^{(p)}|_D(T_D) \in 1 + T_D W(\overline{\mathbb{F}}_p)[\![T_D]\!].$$

Now let $\log_p : \mathbb{C}_p^{\times} \to \mathbb{C}_p$ be the usual Iwasawa branch of the p-adic logarithm homomorphism, i.e., defined by

$$\log_p(1+X) = \sum_{n=1}^{\infty} (-1)^{n+1} \frac{X^n}{n}$$

as a homomorphism $1 + \mathfrak{m}\mathcal{O}_{\mathbb{C}_p} \to \mathbb{C}_p$ ($\mathfrak{m} \subset \mathcal{O}_{\mathbb{C}_p}$ the maximal ideal), extended by linearity to $\log_p : \mathcal{O}_{\mathbb{C}_p}^{\times} \to \mathbb{C}_p$, and finally to $\log_p : \mathbb{C}_p^{\times} \to \mathbb{C}_p$ by declaring $\log_p(p) = 0$. Then we define

$$\log_p g_a^{(p)}|_D := \log_p(uc^{-1}) + \log_p(u^{-1} c g_a^{(p)}|_D(T_D)).$$

Since $\{D\}$ is an admissible covering of $Y(\Gamma_1(N) \cap \Gamma_0(p^2))$, this defines $\log_p g_a^{(p)}$ as a locally analytic function on $Y(\Gamma_1(N) \cap \Gamma_0(p^2))$. Letting

$$g(\psi) = \sum_{a=1}^{N-1} \psi(a) \zeta_N^a$$

denote the (unnormalized) Gauss sum, we have a locally analytic function

$$h_{\psi}^{(p)} := \frac{1}{g(\psi^{-1})} \sum_{a=1}^{N-1} \psi^{-1}(a) \log_p g_a^{(p)}$$

which is a locally analytic function on $Y(\Gamma_1(N) \cap \Gamma_0(p^2))$, and which satisfies (cf. proof of Theorem 1.3 of loc. cit.)

$$dh_\psi^{(p)} = E_{2,1,\psi}^{(p)}.$$

And so $h_\psi^{(p)}$ is a Coleman primitive of $(E_2^{1,\psi})^{(p)}$.

Theorem 9.11 ("p-adic Kronecker limit formula," cf. Chapter X of [35], Theorem II.5.2 of [21], Theorem 1.2 of [4]). *Suppose the new eigenform w is an Eisenstein series of weight $k = 2$ corresponding to a Dirichlet character ψ of conductor $N \geq 4$, $p \nmid N$, i.e., $w|_{\mathcal{H}^+} = E_2^{1,\psi} \cdot (2\pi i dz)^{\otimes 2}$, so that w has nebentypus ψ. Suppose χ has infinity type $(1,1)$ and finite type $(1, \mathfrak{n}, \psi)$. Then we have*

$$\mathcal{L}_{p,\alpha}(w, \mathbb{N}_K) = \Omega(A, t) \frac{\Xi p(w, \check{\mathbb{N}}_K)}{g(\psi^{-1})} \sum_{\mathfrak{a} \in \mathcal{Cl}(\mathcal{O}_K)} (\chi^{-1} \mathbb{N}_K)(\mathfrak{a}) \sum_{a=1}^{N-1} \psi^{-1}(a) \log_p g_a(\mathfrak{a} \star (A, t)).$$

Proof. This follows from the same argument as in the proof of Theorem 9.10, using the fact that $h_\psi^{(p)}$ is a Coleman primitive of $E_2^{1,\psi}$. $\qquad\qquad\square$

Bibliography

[1] A. Agboola, B. Howard, *Anticyclotomic Iwasawa theory of CM elliptic curves II*, Math. Res. Lett. 12 (2005), no. 5–6, 611–621.

[2] Y. Amice, J. Vélu, *Distributions p-adiques associées aux séries de Hecke*, Journées arithmétiques de Bordeaux (Bordeaux, 1974), 119–131. MR 51: 12709.

[3] M. Artin, A. Grothendieck, J.-L. Verdier, *Séminaire de Géométrie Algébrique du Bois Marie - 1963-64 - Théorie des topos et cohomologie étale des schémas - SGA 4*, vol. 1. Lecture Notes in Mathematics (in French). 269. Berlin; New York: Springer-Verlag. xix+525.

[4] M. Bertolini, F. Castella, S. Dasgupta, H. Darmon, K. Prasanna, V. Rotger, *p-adic L-functions and Euler systems: A tale in two trilogies*, Automorphic forms and Galois representations, London Mathematical Society Lecture Notes Series 414, 52–101 (2014).

[5] M. Bertolini, H. Darmon, K. Prasanna, *Generalized Heegner cycles and p-adic Rankin L-series*. Duke Math. J. 162 (2013), no. 6, 1033–1148.

[6] M. Bhargava, C. Skinner, W. Zhang, *A majority of elliptic curves over Q satisfy the Birch and Swinnerton-Dyer conjecture*, preprint, arXiv:1407. 1826 (2014).

[7] K. Büyükboduk, A. Lei, *Iwasawa theory of elliptic modular forms over imaginary quadratic fields at non-ordinary primes*, preprint, arXiv:1605. 05310 (2016).

[8] M. Brakočević, *Anticyclotomic p-adic L-function of central critical Rankin-Selberg L-value*. Int. Math. Res. Not. IMRN 2011, no. 21, 4967–5018.

[9] A. Caraiani, P. Scholze, *On the generic part of the cohomology of compact unitary Shimura varieties*. Ann. of Math. (2) 186 (2017), no. 3, 649–766.

[10] F. Castella, M. Çiperiani, C. Skinner, F. Sprung, *On the Iwasawa main conjectures for modular forms at non-ordinary primes*, preprint, arxiv.org: 1804.10993.

[11]   F. Castella, X. Wan, *Perrin-Riou's main conjecture for elliptic curves at supersingular primes*, preprint, arXiv:1607.02019.

[12]   P. Chojecki, *On nonabelian Lubin-Tate theory and analytic cohomology.* Proc. Amer. Math. Soc. 146 (2018), 459–471.

[13]   P. Chojecki, D. Hansen, C. Johansson, *Overconvergent modular forms and perfectoid Shimura curves.* Doc. Math. 22 (2017), 191–262.

[14]   R. Coleman, *Torsion points on curves and p-adic abelian integrals.* Ann. of Math. (2) 121 (1985), no. 1, 111–168.

[15]   P. Colmez, *Intégration sur les variétés p-adiques.* Astérisque No. 248 (1998), viii.+155pp.

[16]   B. Conrad, *Basic Generalities on Adic Spaces*, Perfectoid Seminar, http://math.stanford.edu/ conrad/Perfseminar/Notes/L14.pdf.

[17]   B. Conrad, *Higher-level canonical subgroups in abelian varieties*, http://math.stanford.edu/ conrad/papers/subgppaper.pdf.

[18]   B. Conrad, *Modular curves and rigid analytic spaces*, http://math.stanford.edu/ conrad/papers/genpaper.pdf.

[19]   A. J. de Jong, *Étale fundamental groups of non-archimedean analytic spaces*, Compositio Math. (1995).

[20]   P. Deligne, Critère d'existence de points, Appendix to Exp. VI in *Théroei de Topos et Cohomologie Étale des Schémas*, Séminaire de Géométrie Algébrique, dir. par M. Artin, A. Grothendieck, J. L. Verdier, Springer LNM 269 and 270, 1972.

[21]   E. de Shalit, *Iwasawa Theory of Elliptic Curves with Complex Multiplication.* Perspectives in Mathematics, 3. Academic Press, Inc., Boston, MA, 1987. x+154 pp. ISBN: 0-12-210255-X.

[22]   B. Dwork, *Norm residue Symbol in local number Fields.* Abh. Math. Sem. Univ. Hamburg 22 1958 180-90.

[23]   G. Faltings, *Coverings of p-adic period domains*, J. Reine Angew. Math. (2010).

[24]   G. Faltings, *p-adic Hodge theory.* J. Amer. Math. Soc. 1 (1988), no. 1, 255–299.

[25]   L. Fargues, A. Genestier, V. Lafforgue, *L'isomorphisme entre les tours de Lubin-Tate et de Drinfeld*, Progr. Math. 262, Birkhäuser Verlag, Basel (2008). xxii+406 pp. ISBN: 978-3-7643-8455-5.

[26] M.H.J. Hopkins, B. Gross, *Equivariant vector bundles on the Lubin-Tate moduli space*, Topology and representation theory (Evanston, IL, 1992), Contemp. Math., vol. 158, pp. 23–88, Amer. Math. Soc., Providence, RI (1994).

[27] B. H. Gross, D. B. Zagier, *Heegner points and derivatives of L-series*, Invent. Math., 84(2): 225–320, 1986.

[28] H. Hida, J. Tilouine, *Anticylcotomic Katz p-adic L-functions and congruence modules*, Ann. Scient. Ec. Norm. Sup. 26 (1993), 189–259.

[29] S. Howe, *Transcendence of the Hodge-Tate filtration*, https://arxiv.org /abs/1610.05242.

[30] R. Huber, *Etale cohomology of Rigid Analytic Varieties and Adic Spaces.* Aspects of Mathematics, E30. Friedr. Vieweg & Sohn, Braunschweig, 1996. x+450 pp. ISBN: 3-528-06794-2.

[31] R. Huber, *A generalization of formal schemes and rigid analytic varieties*, Math. Z., 217(4): 513–551, 1994.

[32] K. Kato, *p-adic Hodge theory and values of zeta functions of modular forms*, preprint series, Graduate School of Mathematical Sciences, University of Tokyo.

[33] N. Katz, *p-adic L-functions for CM fields.* Invent. Math. 49 (1978), no. 3, 199–297.

[34] N. Katz, *Serre-Tate local moduli.* Algebraic surfaces (Orsay, 1976-78), pp. 138–202, Lecture Notes in Math., 868, Springer, Berlin-New York, 1981.

[35] N. Katz, *p-adic interpolation of real analytic Eisenstein series.* Ann. of Math. (2) 104 (1976), no. 3, 459–571.

[36] N. M. Katz, *p-adic properties of modular schemes and modular forms*, Modular functions of one variable, III (Proc. Internat. Summer School, Univ. Antwerp, Antwerp, 1972), Lecture Notes in Mathematics, 350, Berlin, New York: Springer-Verlag, pp. 69–190, (1973).

[37] K. Kedlaya, Q. Liu, *Relative p-adic Hodge theory: Foundations.* Astérisque No. 371 (2015), 239 pp. ISBN: 978-2-85629-807-7.

[38] B. D. Kim, *Signed Selmer groups over the $\mathbb{Z}_p^2$-extension of an imaginary quadratic field*, Canad. J. Math. 66 (2014), no. 4, 826–843, MR 3224266.

[39] S. Kobayashi, *Iwasawa theory for elliptic curves at supersingular primes*, Invent. Math. 152 (2003), no. 1, 1–36. MR 1965358 (2004b:11153).

[40]   V. A. Kolyvagin, *Euler systems*, The Grothendieck Festschrift, Vol. II, 435–483, Progr. Math. 87, Birkhäuser Boston, Boston, MA, 1990.

[41]   U. Köpf, *Über eigenliche Familien algebraischer Varietäten über affinoiden Räumen*, Schr. Math. Inst. Univ. Münster (2), (heft 7):iv+72, 1974.

[42]   D. Kriz, *Generalized Heegner cycles at Eisenstein primes and the Katz p-adic L-function*. Algebra Number Theory 10 (2016), no. 2, 309–374.

[43]   D. Kriz, *A New p-adic Maass-Shimura Operator and Supersingular p-adic L-functions*, Ph.D. thesis, Princeton University, 2018, 197 pp.

[44]   D. Kriz, C. Li, *Goldfeld's conjecture and congruences between Heegner points*, Forum Math. Sigma, 7 (2019), e15, 80pp. doi:10.1017/fms.2019.9.

[45]   A. Lei, D. Loeffler, S. Zerbes, *Wach modules and Iwasawa theory for modular forms*, Asian J. Math. 14 (2010), no. 4, 475–528. MR 2774276.

[46]   Y. Liu, S. Zhang, W. Zhang, *A p-adic Waldspurger formula*. Duke Math. J. 167 (1018), no. 4, 743–833.

[47]   B. Mazur, J. Tate, J. Teitelbaum, *On p-adic analogues of the conjectures of Birch and Swinnerton-Dyer*, Invent. Math. 84 (1986), 1-48 MR 87e:11076.

[48]   W. Messing, *The crystals associated to Barsotti-Tate groups*, Lecture Notes in Math. 264, Springer (1972).

[49]   B. Perrin-Riou, *Fonctions L p-adiques, théroei d'Iwasawa et points de Heegner*, Bull. Soc. Math. France 115 (1987), no. 4, 399–456. MR 928018 (89d:11094).

[50]   R. Pollack, *On the p-adic L-function of a modular form at a supersingular prime*, Duke Math. J. 118 (2003), 525–558.

[51]   R. Pollack, K. Rubin, *The main conjecture for CM elliptic curves at supersingular primes*, Ann. of Math. 159 (2004), 447–464.

[52]   M. Rapoport, T. Zink, *Period spaces for p-divisible groups*, Annals of Mathematics Studies, no. 141, Princeton University Press, Princeton, NJ (1996).

[53]   K. Rubin, *Local units, elliptic units, Heegner points and elliptic curves*, Invent. Math. 88 (1987), no. 88, 407–422.

[54]   K. Rubin, *Congruences for special values of L-functions of elliptic curves with complex multiplication*. Invent. Math. 71 (1983), no. 2, 339–364.

[55]   K. Rubin, *The "main conjectures" for imaginary quadratic fields*, Invent. Math. 103 (1991), no. 1, 25–68. MR 1079839 (92f:11151).

[56] P. Scholze, *On torsion in the cohomology of locally symmetric spaces*. Ann. of Math. (2) 182 (2015), no. 3, 945–1066.

[57] P. Scholze, *p-adic Hodge theory for rigid analytic varieties*. Forum Math. Pi 1 (2013), e1, 77 pp.

[58] P. Scholze, *p-adic Hodge theory for rigid analytic varieties - corrigendum*. Forum Math. Pi 4 (2016), e6, 4 pp.

[59] P. Scholze, *Perfectoid spaces*, Publ. Math. de l'IHES, 116(1):245–313, (2012).

[60] P. Scholze, *Perfectoid spaces: a survey*, http://www.math.uni-bonn.de /people/scholze/CDM.pdf.

[61] P. Scholze, J. Weinstein, *Berkeley lectures on p-adic geometry*, Annals of Math. Studies (2020).

[62] P. Scholze, J. Weinstein, *Moduli of p-divisible groups*. Camb. J. Math 1 (2013), no. 2, 145–237.

[63] P. Schneider, J. Teitelbaum, *p-adic Fourier theory*, Doc. Math. 6 (2001), 447–481.

[64] C. Skinner, E. Urban, *The Iwasawa Main Conjectures for GL_2*, Invent. Math. 195 (2014), no. 1, 1–277. MR 3148103.

[65] F. Sprung, *Iwasawa theory for elliptic curves at supersingular primes: A pair of main conjectures*, Journal of Number Theory, Volume 132, Issue 7, July 2012, pages 1483–1506.

[66] M. M. Višik, *Nonarchimedean measures associated with Dirichlet series* (in Russian), Mat. Sb. (N.S.) 99 (141), no. 2 (1976), 248–260, 296. MR 54:23.

[67] X. Wan, *Iwasawa main conjecture for supersingular elliptic curves*, preprint, arXiv:1411.6352.

[68] J. Weinstein, *Nonabelian local class field theory and the geometry of Lubin-Tate spaces*, FRG lecture notes, http://math.bu.edu/people/jsweinst/ FRGLecture.pdf.

[69] X. Yuan, S. Zhang, W. Zhang, *The Gross-Zagier Formula on Shimura Curves*. Annals of Mathematics Studies, 184. Princeton University Press, Princeton, NJ, 2013. x+256 pp. ISBN: 978-0-691-15592-0.

Index

Lightning Source UK Ltd.
Milton Keynes UK
UKHW020726121121
393837UK00004B/16

9 780691 216461